W9-BKT-362

A User's Guide for Planet Earth

A User's Guide for Planet Earth

Fundamentals of Environmental Science

Revised Second Edition

Dork Sahagian

Bassim Hamadeh, CEO and Publisher
Carrie Montoya, Manager, Revisions and Author Care
Kaela Martin, Project Editor
Jeanine Rees, Production Editor
Jess Estrella, Senior Graphic Designer
Alexa Lucido, Licensing Manager
Natalie Piccotti, Director of Marketing
Kassie Graves, Vice President of Editorial
Jamie Giganti, Director of Academic Publishing

Copyright © 2021 by Dork Sahagian. All rights reserved. No part of this publication may be reprinted, reproduced, transmitted, or utilized in any form or by any electronic, mechanical, or other means, now known or hereafter invented, including photocopying, microfilming, and recording, or in any information retrieval system without the written permission of Cognella, Inc. For inquiries regarding permissions, translations, foreign rights, audio rights, and any other forms of reproduction, please contact the Cognella Licensing Department at rights@cognella.com.

Trademark Notice: Product or corporate names may be trademarks or registered trademarks, and are used only for identification and explanation without intent to infringe.

Cover images:
Copyright © 2011 iStockphoto LP/JohnnyLye.
Copyright © 2015 iStockphoto LP/LeoPatrizi.
Copyright © 2016 iStockphoto LP/Stock_Colors.
Copyright © 2016 iStockphoto LP/Yangna.
Copyright © 2017 iStockphoto LP/Mimadeo.
Copyright © 2017 iStockphoto LP/Primeimages.

Printed in the United States of America.

Brief Contents

Detailed Contents

Preface

THERE ARE MANY well-written and comprehensive books available from numerous authors and publishers on the subject of Environmental Science. However, there are two major issues outstanding for which this book was written. The first is that most introductory Environmental Science textbooks, while beautifully produced, include so many case studies, examples, horror stories, and human interest anecdotes, that the main scientific points can become lost or at least diluted in large volumes of text and pretty pictures. For this reason, this book focuses on the science involved, in an attempt to make the main points clearer to the student (and professor) in a one-semester introductory course. As such, this book is by no means comprehensive. This book was conceived in days of virtually unlimited free information access, when any topic can be searched on-line and facts and figures, along with a multitude of photographs, can be found for any environmental event, regulation, or technology (with the usual precautions about web searches, "alternative facts," etc.). As such, no attempt is made to repeat all that in this textbook. Rather, we focus on helping the student understand the processes that control the physical, chemical, biological, and human components of the environment, and how they interact to drive the behavior of the Earth System. "How does the greenhouse effect work?" "What do air conditioners have to do with skin cancer?" These are the kinds of questions the student will be able to answer after reading this book, rather than "How big was the last oil spill?"

The second reason for writing this somewhat unusual Environmental Science textbook is the more mundane question of money. Environmental Science textbooks have reached astronomical prices, a situation poorly suited for a field that is changing so rapidly that books become outdated by mid-semester. Without a used book market in which students can engage, these expensive books end up on dusty shelves, or worse yet are tossed (hopefully recycled) after the class is done. This is no way for Environmental Science to be conducted. With this less rotund and much less expensive volume, the intent is that course material can be presented clearly and concisely, and should it be rendered out of date in the coming months by analysis of another coastal storm, photovoltaic discovery, or satellite observation, for instance, that refines our understanding of the environment, it is less of a loss to the student, school library, or college professor. It will be readily updated.

Both these reasons bring to mind the landmark book and subsequent documentary film by former Vice President Al Gore, *An Inconvenient Truth*. Learning the facts and concepts involved in environmental science should not be inconvenient or expensive. As such, one might consider this book a more "convenient truth" that is accessible to students and the general public alike. With this in mind, we will explore the various topics that help us understand not only the specific details of environmental problems, but also the way these interact to control the behavior of the Earth System.

Like the first edition, this second edition of the book is organized in such a way that provides the student with an exploration of several key background issues that lay the foundation for understanding the processes and drivers that control the behavior of the environment. It then delves into the main environmental systems of ecosystems, biogeochemical cycles, water, agriculture, oceans, human health, energy, and climate, culminating in a discussion of the Earth System. Definitions of key bolded terms are found in the margins of the same page, as are a few worked problems as examples for the student. Journal articles from the primary literature are included at the end of each chapter for References and research.

This book was written with college freshmen in mind. It assumes no prior courses in environmental science, Earth science, or related disciplines.

CHAPTER 1

What Is the Environment, How Do We Study It, and Why Does Anyone Care?

What is the environment? Why should we care?
As conditions are changing we'd better beware.
Let's take a good look
As we read through this book
To learn how to save our food, water, and air!

SIMPLY PUT, THE environment is the physical and biological surroundings in which life on the planet is sustained. This includes climate, ecosystems, water, and all the interactions between these systems and energy from the sun that supports the functioning of the Earth System. The physical environment would persist in the absence of life, of course, but as living organisms ourselves, we are most interested in how life and the physical environment interact (Figure 1.1) and in what ways, as humans who can (and do) have a great impact on the environment, we can develop sustainable approaches to land use, emissions, and general resource utilization.

FIGURE 1.1 Birds

Environmental Goods and Services

Centuries ago, in Europe and other parts of the world, the environment was a frightening concept that brought storms, floods, predators, disease, and death to people and their families, and thus was something to fight against in order to sustain humanity. As knowledge and technology advanced, humans became very adept at altering their environment to suit their needs, such as personal security from predators, disease, and the elements; availability of food for a growing population; and resources for a growing industrial complex that became increasingly efficient at converting natural resources into useful or otherwise desired products. For centuries, the environment was the venue where such natural resources could be found, and found they were, underground in mines, at the surface in soils, in streams, groundwater and the ocean, in the myriad ecosystems globally, and in the atmosphere. The environment became a fountain of goods and services to provide people with whatever they wanted, with little thought given to replacing anything that was removed or destroyed in the process. There was plenty more where that came from. Humanity (in its increasingly vast numbers) became so efficient at exploiting natural resources that the rate at which resources were desired began to exceed the rate at which the global ecosystems could provide them, leading to concerns about sustainability (Figure 1.2).

The types of environmental goods and services most useful to the technologically developing societies in recent centuries include (see Table 1.1):

FIGURE 1.2 EPA research ship, *Lake Guardian*

GOODS:

- Fresh water (clean and plentiful)
- Fertile soil (for growing crops)
- Fish (from the ocean)
- Wood and fiber (for construction, clothing, etc.)
- Fuels (for heating, transportation, electricity generation, industry)
- Special or exotic materials (for medicines, specialized applications, etc.)
- Minerals (for construction, products, various industries, ornaments)

SERVICES:

- Net Primary Production (Conversion of CO_2, H_2O and sunlight into biomass and oxygen)
- Water cycling (water availability for drinking, agriculture, and transportation)
- Climate regulation (temperature, rainfall, storm tracks, winds, etc.)
- Atmospheric composition regulation (CO_2, O_2, contaminant and radiation absorption)
- Physical protection (from storms, tidal surges, floods)
- Water purification (removal of sediments, contaminants)
- Soil retention (prevention of wind and water erosion)
- Waste regulation (breaking down biological wastes)
- Nutrient cycling (N, K, P, etc.)

Equilibrium of Processes within the Environment

There is a multitude of processes constantly operating throughout the Earth System, and the interaction between these controls the nature of the environment at all scales. Examples of processes include radiative balance between incoming solar radiation and outgoing infrared from the earth; the cycling of carbon and nutrients between various **reservoirs** in the solid, liquid, gaseous, and living parts of the environment; predation between animals; and the movement of water in the hydrologic cycle. These and many other processes interact with each other, such that a steady state is maintained in the amount of material stored in various reservoirs (e.g., CO_2 in the atmosphere, nutrients in the soil, fish in the sea, water in aquifers, etc.) and in the fluxes of energy and materials between reservoirs.

Reservoirs: Places where a stock (amount of stuff) is stored, at least temporarily. Reservoirs are characterized by how much they contain, as well as the fluxes in and out.

Describe two environmental goods and two services YOU value most highly. Why are these most important to you?

TABLE 1.1 ECOSYSTEM FUNCTIONS AND SERVICES

Ecosystem functions and services	Examples
Nutrient cycling (storage, processing, and acquisition of nutrients within the biosphere)	Nitrogen cycle; phosphorus cycle
Net primary production (conversion of sunlight into biomass)	Plant growth
Pollination and seed dispersal (movement of plant genes)	Insect pollination; seed dispersal by animals
Habitat (the physical place where organisms reside)	Refugium for resident and migratory species; spawning and nursery grounds
Hydrological cycle (movement and storage of water through the biosphere)	Evapotranspirrtation; stream runoff; groundwater retention
Water supply (filtering, retention, and storage of fresh water)	Provision of fresh water for drinking; medium for transportation; irrigation
Food (provisioning of edible plants and animals for human consumption)	Hunting and gathering of fish, game, fruits, and other edible animals and plants; small-scale subsistence farming and aquaculture
Raw materials (building and manufacturing; fuel and energy; soil and fertilizer)	Lumber, skins, plant fibers, oils, dyes; fuelwood, organic matter (e.g., peat); topsoil, leaves, litter, excrement
Genetic resources (genetic resources)	Genes to improve crop resistance to pathogens and pests and other commercial applications
Medicinal resources (biological and chemical substances for use in drugs and pharmaceuticals)	Quinine; Pacific yew; Echinacea
Gas regulation (regulation of the chemical composition of the atmosphere and oceans)	Biotic sequestration of carbon dioxide and release of oxygen; vegetative absorption of volatile organic compounds
Climate regulation (regulation of local to global climate processes)	Direct influence of land cover on temperature, precipitation, wind, and humidity
Disturbance regulation (dampening of environmental fluctuations and disturbance)	Storm surge protection; flood protection
Biological regulation (species interactions)	Control of pests and diseases; reduction of herbivory (crop damage)
Water regulation (flow of water across the planet surface)	Modulation of the drought-flood cycle; purification of water
Nutrient regulation (maintenance of major nutrients within acceptable bounds)	Prevention of premature eutrophication in lakes; maintenance of soil fertility
Ornamental resources (resources for fashion, handicraft, jewelry, pets, worship, decoration, and souvenirs)	Feathers used in decorative costumes; shells used as jewelry

Modified after: Farber, S., R. Costanza, D. L. Childers, J. Erickson, K. Gross, M. Grove, C. S. Hopkinson, J. Kahn, S. Pincetl, A. Troy, P. Warren, and M. Wilson. 2006. "Linking Ecology and Economics for Ecosystem Management: A Services-Based Approach with Illustrations from LTER Sites." *BioScience* 56:117–129.

The ability of the Earth System to maintain equilibrium conditions depends on the presence of negative feedbacks within the various component subsystems, as we will explore in Chapter 3. With a suite of such stabilizing feedbacks, the Earth has maintained a surface temperature between the freezing and boiling points of water for billions of years. (However, there is evidence of several "snowball Earth" events that occurred between about 750 and 650 million years ago, each lasting perhaps millions of years, during which the planet was almost completely frozen over until feedbacks drove warming to melt the ice.)

Human Perturbations

The equilibrium of Earth System processes has been maintained for billions of years because with few exceptions, the fluxes (and changes in fluxes) of materials between reservoirs have been slow relative to the size of the reservoirs. Humans were the first species (that we know of) that learned to influence the environment at rates that drastically altered some key fluxes, and thus reservoirs, throughout the Earth System. This was accomplished in two basic ways: cutting and burning, with the invention of the saw (axe) and the discovery of fire. In the broader sense in the modern world, this involves land use and emissions, the two primary means by which people affect (usually adversely) the environment.

Emissions: Chemicals (usually liquid or gaseous) that are expelled into the environment (atmosphere or water bodies) as a result of human activities, such as fossil fuel burning. Of particular interest are emissions of greenhouse gases into the atmosphere.

A major human perturbation is **emissions** from burning of fuels such as coal, oil, gas, and wood (Figure 1.3). While wood burning has occurred since man discovered fire, it was originally done at a slow rate, such that the regrowth of trees could keep up with cutting. In many places today, it is cut much faster than it can regrow, and is thus non-renewable. (Renewable and non-renewable energy sources will be explored in Chapter 11.) Further, and much more significantly, fossil fuels were discovered. These could produce a great deal of energy, as they contained the equivalent of millions of years of the Sun's energy concentrated and stored in the chemistry of coal, oil, and natural gas that could be combined with atmospheric oxygen to produce heat, CO_2, and in the case of oil and gas, water vapor. While the CO_2 and water were discarded to the atmosphere (Figure 1.4), the heat was used to cook, heat buildings, drive turbines to generate electricity, and operate motors for vehicles of all kinds.

FIGURE 1.3 Oil Platform

Early industrial societies were focused on the value of the energy to be extracted from fossil fuels, but did not consider the role of the rest of the combustion process in altering the environment. While the greenhouse effect and the role of CO_2

has been understood since before the industrial revolution, it was assumed impossible for people to actually alter anything so vast as the atmosphere and the planetary energy balance. It has since been revealed that the burning of fossil fuels has released sufficient CO_2 (and other greenhouse gases) into the atmosphere to drastically alter the radiative balance between the earth and the sun. Since the industrial revolution (mid-19th century), we have increased atmospheric CO_2 concentration from about 280 ppm to a 2018 level of over 400 ppm. This rate of change of atmospheric composition is unprecedented in Earth history. In effect, we are conducting an experiment on the planet. Take a planet, drastically alter atmospheric composition to perturb the radiative balance with its star (also change the land cover so you can further alter the atmospheric chemistry), and drive the system until temperature, ice cover, land cover, biome distribution, water temperature, and air/water circulation are profoundly out of equilibrium with each other. Stand by and see what happens. (Scientists love this sort of thing.) Of course, it might be more convenient to conduct this experiment on a planet we didn't actually need to live on ...

Land use is another major human perturbation and stems from the evolution of societies from hunting/gathering to agriculture. Climate became relatively stable after the last glacial period at the start of the Holocene Epoch, about 12,000 years ago. This is said to have enabled the people of the time to stop chasing migrating resources north and south, and settle down to plant crops for a more stable and productive food source by 10,000 years ago. This, of course, led to increasing population, as more people could be fed from specialized crops than from the limited number of (parts of) plants that could be eaten in the natural ecosystems available to them. This led to further destruction of the ecosystems in favor of planting more crops to feed more people. At the time, people were unaware of the various goods and services being provided to them by the environment, and some cultures collapsed due to inadvertently "biting the hand that fed them" (e.g., Easter Island). At a much larger scale, this is beginning to occur globally now, which is the cause for concern by many scientists and others throughout the world.

Land use: The destruction of natural ecosystems by conversion to some other land cover, such as agriculture, industry, or residential (mostly agriculture).

Land use over the last few centuries has involved massive deforestation, first of northern regions (e.g., Europe, North America) and more recently the tropics, where highly productive rainforests are found. Elimination of northern forests in favor of farmland began to erode soil that had taken centuries to millennia to develop, and it is still being eroded today (with about half of America's fertile topsoil already gone). However, with modern industrial farming methods, large corporate farms have taken over the market, leaving subsistence or otherwise small farmers to find other employment. This has led to regrowth of many of the forests in eastern North America. Meanwhile, deforestation of the tropics has accelerated, but the soil in rainforests is very thin and fragile, with little carbon and few nutrients stored in the soil (most is stored in litterfall and decaying vegetation). Once the surface is cleared

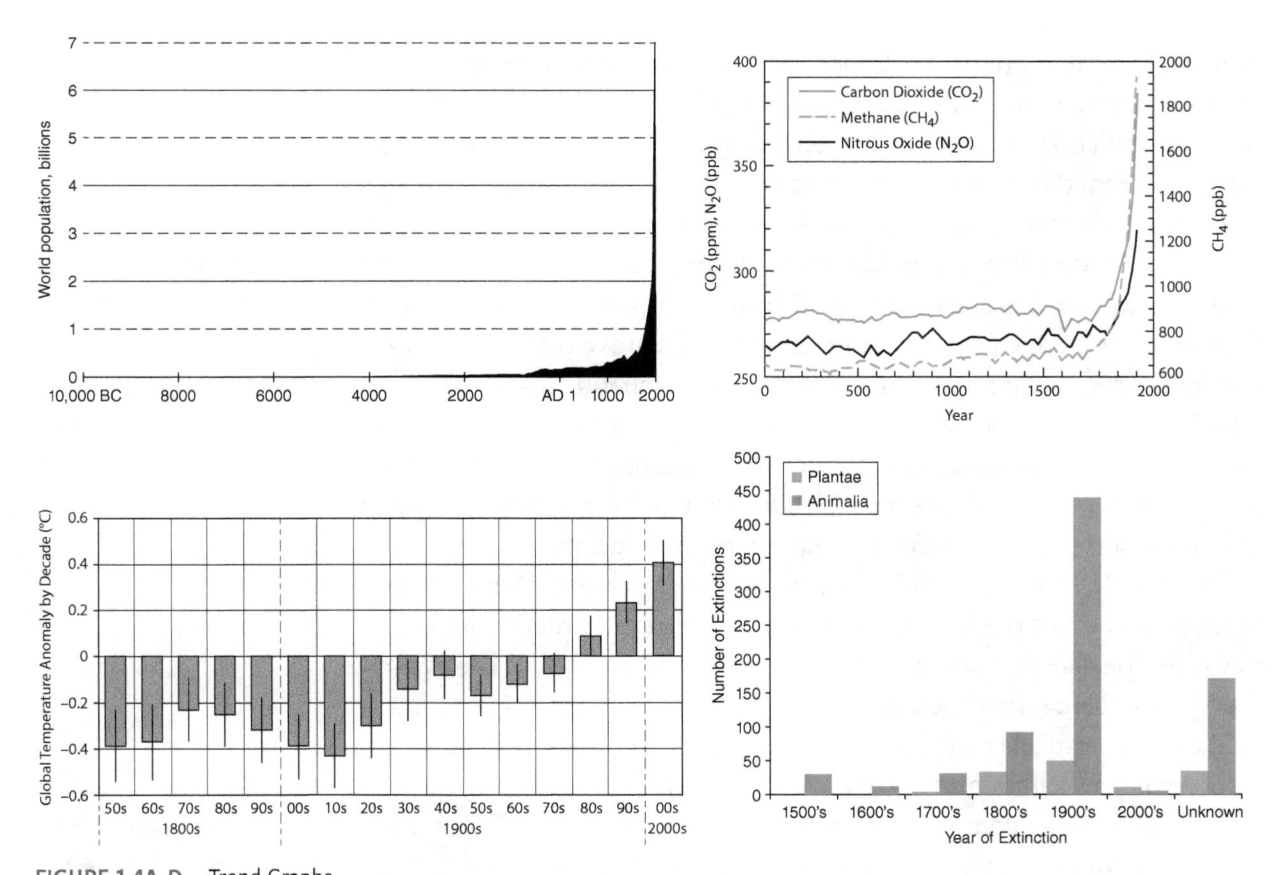

FIGURE 1.4A-D Trend Graphs

Critical observational trends show the importance of humanity in the Earth System. The recent human population explosion has led to altered atmospheric chemistry, while intensified land use for agriculture has caused a new "mass extinction" of species globally. Greenhouse gas emissions and land use have combined to lead to observed climate change, which is predicted to change further in the coming decades.

for agriculture, the soil fails very quickly, and in just a few years, the farms become unproductive, with loss of the ecosystem as well as all the services it had been providing. It is difficult for such an ecosystem to redevelop once it is destroyed over a large area.

The loss of forests and other ecosystems eliminates all the goods and services they provide, most notably at the global level, the sequestration of CO_2 out of the atmosphere. As such, the emissions from fossil fuel burning that would potentially be alleviated by ecosystem uptake remain in the atmosphere, leading to an exacerbated greenhouse effect and global warming. This will be explored further in Chapters 12 and 13.

The Commons

In an "open world" in which the impact of humans is negligible, and the amount of energy and resources available is effectively infinite, people would be able to take what they need from the environment, discard what they don't want, and the Earth System could easily regenerate, absorb, or reprocess in a way that would make people feel as if they could live as they like forever. This was the case for millennia, until recent centuries, when technology advanced and human population exploded. When anyone is authorized to remove natural resources from the environment but no one is responsible for maintaining or replacing those resources, the resources become depleted, contaminated, or destroyed, making them unavailable for the future. This is called the "**tragedy of the commons**" and has already occurred in many parts of the environment. The commons is any communal natural resource (such as a community grazing pasture, or the fish in the ocean, or even the atmosphere). To date, there has been little regulation of the commons, and these resources are being depleted to the point of uselessness, in addition to obliterating the parts of the global ecosystem of which they are a part, thus losing many essential ecosystem goods and services upon which we rely as a modern society.

Tragedy of the commons: Destruction of publicly available environmental goods or services caused by over-utilization due to free availability to all with no one responsible for maintenance and sustainability.

The tragedy of the commons was a term coined by Garrett Hardin in 1968, when he presented the idea that any property held in common, but that no one is responsible to maintain, will deteriorate (Hardin, 1968). His example was of cattle herders with access to a community grazing ground (Figure 1.5). While it was available to all to use, there was no limit, nor was there anyone responsible for maintaining (or even monitoring) the usability of the common resource. From the perspective of an individual herder, it was to his advantage to add more cattle to graze. After all, he did not have to pay for the grazing rights, did not have to replace the grass that was eaten, and was able to profit personally from the sale of the milk and meat from the cattle.

FIGURE 1.5 Cattle herd—grass fed beef

The cost of overgrazing the common land was not borne by the individual who added more cattle, but to the community as a whole. Thus each individual kept the profit, yet transferred the cost to the community. Naturally, since everyone was doing this, there was no net gain for anyone, and the commons was destroyed—the tragedy of the commons. This should have been no surprise, as even Aristotle of ancient Greece knew that "What is common to the greatest number has the least care bestowed upon it." The tragedy of the commons emerges from the practice of externalizing costs of an activity while internalizing the profits. This is what the herders were doing, and what ocean fishermen do, as well as oil companies, farmers, and many others, who exploit natural resources at a profit while transferring the environmental and other costs to others, either in time or in space. Especially important commons being degraded at present are the oceans (fish, ability to absorb carbon from the atmosphere and contaminants from the land), atmosphere (ability to hold

CO_2 and pollutants that harm human and ecosystem health), forests (carbon storage, water storage and mediation, biogeochemical cycling, and much more), soils (carbon storage, nutrient storage, ability for future generations to grow crops), and many others.

Externalization of Environmental Costs

It is well known in the business world that one maximizes profit by internalizing that profit while externalizing costs of operation. If the costs of doing business can be borne by someone else who does not demand a share of the profit, it is the ideal scenario for a business person who looks no farther than the immediate bottom line. This, in part, leads to the tragedy of the commons if the cost is borne by the community. There are numerous environmental costs associated with modern society, and while individuals profit from their own activities, the costs are externalized to others, either alive today, or who have not yet been born, but will have to pay in loss of environmental services, restoration of degraded ecosystems, or simply in more expensive ways of doing things that would otherwise be cheaper if the appropriate environmental goods and services were still available (Constanza et al., 1998; Constanza et al., 1997). In a sense, this is a market failure, as the costs of goods and services that externalize environmental costs do not fully reflect the costs of doing business.

For example, the cost of fossil fuels such as gasoline does not reflect the cost of air, water, and soil contamination caused by producing and ultimately burning the fuels. These costs are borne by others, either through current taxes or by the degradation of the services that would otherwise have been enjoyed by the provision of clean air, water, and soil. There is a more serious externalization involved in fossil fuels, however, as will be explored in Chapter 11. Companies that sell fossil fuels charge the cost merely of delivering those fuels to the customer (and a bit of processing). The customer does not pay for the production of these fuels, as they were produced by concentration of solar energy over tens to hundreds of millions of years of geologic processes. If the costs of concentration of solar energy in this manner were internalized into the cost of electricity generated by burning coal, for example, such electricity would be far more expensive than that generated from wind or photovoltaic cells. As such, externalizing to the geologic past the production costs of fossil fuels artificially depresses the cost of those fuels to merely the cost of delivery—a small fraction of the actual cost.

Another example is the water mined from aquifers that do not recharge at a significant rate, such as the Ogallala Aquifer in the U.S. Great Plains. If farmers had to pay for (and charge for) the cost of replacing that water, the food they produced from it would be vastly more expensive than it is now. How much more would a fish cost if, rather than simply catching one from the ocean, a fisherman had to raise it from an egg, ensure that its marine environment was preserved and healthy, refrain from catching it until it had reproduced in sufficient numbers, and then replace it upon removal? While aquaculture and mariculture take a few steps in that direction, they have their own environmental impacts, as discussed in Chapter 9.

Sustainability

There has been a great deal of discussion of late regarding the desirability of being sustainable. The exact meaning of that has been somewhat unclear. According to the United Nations (at the Rio Summit in 1987), "sustainable development is development that meets the needs of the present without compromising the ability of future generations to meet their own needs." How many future generations? What will be "their own needs" relative to the current generation's "needs"? Shall we declare that "they" can get by with much less than "we" need? Is "sustainable development" an oxymoron? These questions remain a matter of debate that becomes moot when there are no effective mechanisms to ensure that current populations will enable future generations to consume as much energy and resources as is done at present. Native Americans considered activities sustainable if they could continue for seven generations. While this is only about 150 years, it is a far longer-term view than any current policy found in "modern" societies.

Various aspects of sustainability have arisen to the point that while it is becoming a household word, it has so many meanings and interpretations that its meaning is becoming bent, if not broken. Some companies sell products that they market as sustainable without providing any indications of what that means. All products require energy and resources to produce. As such, the "greenest" product is the one you leave on the market shelf so that no more need to be made. This is certainly not what the selling company has in mind. Consequently, sustainability and economic growth sometimes appear to collide head-on. This need not be the case, however. Use and reuse of materials is sustainable indefinitely, given careful mass management (i.e., recycling) and sufficient energy to reform old or dysfunctional products into new and marketable products (Figure 1.6). While management can be arranged by human diligence, energy

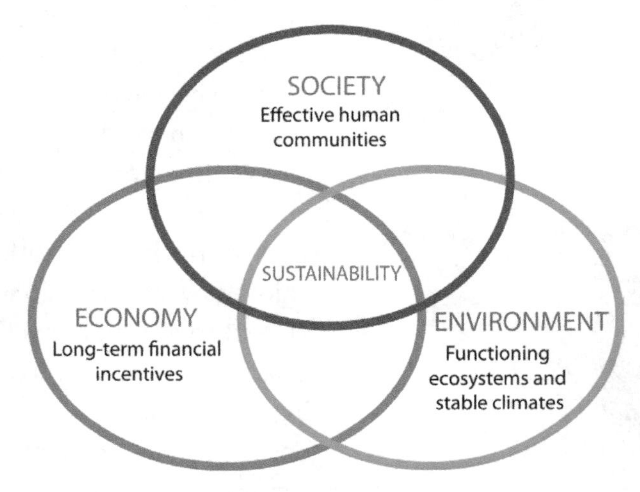

FIGURE 1.6 Sustainability Venn Diagram
The ability to devise a sustainable society depends on three basic factors. While the environmental factor is the focus of this book, social and economic factors also play a role. In the wrong context, "Sustainable Development" can be an oxymoron (e.g., increasing population and per capita consumption), but when considered as the development of knowledge, intellect, and efficient harmonization of human and natural systems, it is a laudable (and necessary) goal.

is more difficult to come by, except, at present, from fossil fuels, the burning of which is far from sustainable for a host of reasons. Until sustainable energy sources of sufficient magnitude are developed, some means will need to be found to reduce consumption of not only energy, but also the products that are made from it. This will be a very difficult political and economic nut to crack, and many are pessimistic about modern society's ability to do it or even willingness to try.

Tools for Studying the Environment

How do scientists monitor the environment to observe changes and perturbations? What means are there for understanding the critical processes so that reliable model-based predictions can be made of future environmental conditions and processes? As in any science, a set of basic tools is necessary for scientists to use to study the system at hand. Some of these are conceptual, while others are technical. Conceptually, a somewhat trivialized hierarchy of science can be viewed as follows:

For math, you just need math.

For physics, you need math.

For chemistry, you need physics and math.

For biology, you need chemistry and physics and math.

For Earth science, you need biology and chemistry and physics and math.

For environmental science, you need Earth science and biology and chemistry and physics and math.

As such, the environmental scientists must draw on a broad body of knowledge to even begin to approach the complexity of the problems found in the environment. This makes environmental science highly interdisciplinary (not just multidisciplinary) and the environmental scientist must take a problem-based approach that may call upon any of a wide spectrum of tools to solve. Some of these tools are as follows:

Fieldwork: Most observations have historically been accomplished by direct observation and monitoring in the field. The field may be a forest, wetland, or any other biome, or it may be the deep sea, watershed, individual stream, or other physiographic entity. In some cases, it may be a water treatment plant or cement manufacturing facility. In the field, scientists may use a variety of tools such as rock hammers, drills, coring tubes, gas collection enclosures, automated water collection devices, in situ chemical analysis instruments, light meters, and a plethora of other instrumentation, some of it very specialized, that enables scientists to gather the data needed to answer pressing environmental questions in their research. In all cases,

FIGURE 1.7 Satellite in orbit

FIGURE 1.8 NASA satellite missions

careful observation, monitoring, and recording of data are needed to solve any well-posed problem.

Remote Sensing: Some of the most exciting new data sets now come from a long-distance extension of fieldwork with a bird's-eye view of the earth from airplanes, balloons, and orbiting satellites in space (Figures 1.7 and 1.8). Remote sensing provides an entire new dimension for environmental data collection because the global data sets they can provide would be impractical at best, and in most cases impossible to collect by instrumentation on the ground.

Yet most satellites can do nothing more than look down at the earth. However, they can "look" in virtually any wavelength. This includes the visible, of course, and one obvious observation is the tracking of storms and weather systems from the cloud formations visible from space, leading to a quantum leap forward in weather forecasting ability. Land cover can be determined simply by looking down at the ground during cloudless times. Another type of observation in the visible band, for example, is the greenness of vegetation in an ecosystem, as characterized by the **Fraction of Photosynthetically Active Radiation** (**FPAR**) and Leaf Area Index (LAI) that can be used to calculate the **Net Primary Productivity** (**NPP**) of an ecosystem. Ocean color reflects photosynthetic activity of phytoplankton. Microwaves (radar) are used to examine surface-water heights, as another example. Sensors in the infrared can measure temperature. These and many other examples use the spectrum of electromagnetic radiation to observe a whole host of environmental parameters of interest to scientists because of the information they provide about the conditions and processes operating in the Earth System. Other satellite sensors measure the Earth's magnetic field. Still others are used to measure the details of Earth's gravity field. These do not measure gravity per se. Rather, they orbit in pairs and record their exact height above the surface and the exact distance between them. When passing over an area with a large mass at or near the Earth's surface—such as a glacier, excess

Fraction of Photosynthetically Active Radiation (FPAR): The portion of the electromagnetic spectrum arriving from the sun (with wavelengths between 0.4- and 0.7-microns) that is actually used (absorbed) by plants.

Net Primary Productivity (NPP): The rate or flux of carbon (in the form of CO_2) drawn out of the atmosphere and converted to biomass (e.g., simple sugars) by photosynthesis in plants after subtraction of the nighttime respiration that "burns" these sugars during plant metabolic processes and growth.

groundwater, or dense part of the crust—the leading satellite moves faster and pulls away from the following satellite. The rear catches up as it approaches the mass and they regain their normal spacing. The distance between them can be measured with microwaves, and new technologies are being developed to use laser ranging to even more accurately determine the inter-satellite distance and thus infer the details of the Earth's gravity field. All of these remote-sensing techniques are being applied and refined for ever-better observations of the Earth's environment.

Experimentation: Frequently, experiments on parts of the environment are needed to better understand structure or processes. If ecologists need to determine the effect of ozone on leaf photosynthesis and respiration, for instance, they can place plants in an ozone-rich (or -poor) enclosure and observe photosynthetic activity, leaf growth, and various other measurements of interest. In other cases, it may be necessary to determine the role of a contaminant such as lead or arsenic in disrupting aquatic organisms, in which case the organisms can be placed in specially designed contaminant-rich (or -poor) facilities for observation. There are many other examples, and each scientist designs experiments specifically for the problem at hand.

Numerical Modeling: Computer models are numerical representations of some part of the environment and are used for various purposes to better understand the functions and processes operating in the Earth's environment. There are different types of models used for different purposes. Diagnostic models focus on some number of processes and delve into the details that control the various parts of the system being explored. They diagnose the system.

Models range in complexity from simple conceptual models to full-blown numerical models of the Earth System that purport to reproduce all of the important systems and processes in sufficient detail to be used as specific tools for global and even regional projections and the actions that may be based on them. Conceptual models can be as simple as "Daisyworld," a fictitious planet occupied by black and by white daisies. Such conceptual models help the student (and researcher) understand the connections and interactions within the system without getting lost in the geographic or process details. They also can be constructed and run very quickly on the most basic computers. At the other extreme are Global Climate Models (GCMs) that include a great many processes and interactions at as high spatial and temporal resolution as modern computers will allow. Between these two extremes are Earth System Models of Intermediate Complexity (EMICs) that seek to capture only the key driving functions within the system, and keep spatial and temporal resolution to the point that they can be run for long model times, thus enabling paleo records to be reconstructed, and future scenarios to be analyzed with multiply varying parameters. All of these types of models are being used by the scientific community, and each sheds its own light on our understanding of the Earth System and the environment that it provides for all life.

Finding Solutions

In the 19th and early 20th centuries, scientists were employed to find minerals, petroleum, and generally provide the intellectual means to extract natural resources for industrialization and economic growth. As such, they were like **industrialists**, creating new wealth by exploiting existing resources. By the end of the century, the by-products of industrial and other human activity led to a new and perhaps more pressing need for scientists to understand the environmental consequences of this activity. Rather than being industrialists, they began to be treated more like **physicians**, in the hopes that as a community, they could fix the environmental problems that had begun to be recognized at all space and timescales. This new type of role for scientists has brought them closer to the public eye than ever before, and many tensions have developed between the pure scientific endeavors of researchers and the policy community that is expected to act on the basis of scientific findings for the greater public good. It has also led to the realization that the general citizenry needs to understand the basics of environmental science in order to support appropriate policy and mold societal activities for the rest of the twenty-first century and beyond.

Environmental problems stemming from human activities cannot, however, be solved by scientists who can merely inform the public about the environmental consequences of specific human activities. The implied need for alterations of behavior can be addressed only by policy makers and the broader public itself. Consequently, it is more important than ever for scientists and policy makers to work together, each informing the other regarding the realities of their own realm of expertise. Scientists cannot influence the political process any more than politicians can influence scientific investigation. Historically, scientists have kept as far as possible from the policy process in an attempt to maintain an image of objectivity. Indeed, for decades, those few scientists who attempted to put the results of their research into the context of political action opened themselves up to criticism and accusations of bias, political sell-out, and loss of scientific credibility. This is beginning to change, as scientists and policy makers begin to work more closely together to address the very pressing environmental problems facing society today. Although much more progress will need to be made before basing policy on scientific results becomes a leading consideration of politicians, inroads are already being made. Some key environmental policies have met with great success. A famous example is the Montreal Protocol, which banned the production and use of Chlorofluorocarbons (CFCs) because it was discovered by scientists studying the Antarctic atmosphere that they were destroying the stratospheric ozone layer that protects us from harmful ultraviolet radiation. More locally, the American clean air act led to much healthy air to breathe, especially in cities, and also led to unleaded gasoline, dramatically reducing the amount of lead in the environment (see Chapter 10 regarding human health).

What can WE do as individuals? Most people have behaved under that assumption that their own individual activities are too insignificant to influence the local, regional or global environment. The tragedy of the commons is merely one aspect of that assumption. However, environmental impact stems from the sum of ALL human activity, including each of us individually. It is our own personal consumption and waste that controls the environment that supports us. As such, we can each consume less energy, food and materials. With regard to energy, we can promote and use public transportation, tolerate slightly warmer temperatures in summer and cooler temperatures in winter (just wear a sweater!), and ensure that our living and working spaces are most efficiently managed. Americans are notorious for the vast quantities of food that is wasted, and this is an easy issue to address as individuals. If you don't really need it, don't take it. Further, we can eat lower on the food chain (see chapter 5) thus requiring less agricultural lands to produce what we need. Regarding overconsumption of materials, Americans (and others) have become so accustomed to planned obsolescence that shopping has become a recreational activity. This is unsustainable, as we burn through resources and the energy required to convert them into the products that we imagine that we need in vast quantities, but for only short periods of time. As such, rather than characterizing ourselves as "consumers," we can begin to play the role of "preservers," "sustainers," or "restorers." This would involve a major shift in the mindset of many individuals, but one that will be necessary to enable future generations to have access to the goods and services that the global environment can provide. As a general rule, it could be said that we should all **"Do what would be best if everyone did it."**

References

Costanza, R., d'Arge, R., de Groot, R., Farber, S., Grasso, M., Hannon, B., . . . & van den Belt, M. (1997). The value of the world's ecosystem services and natural capital. *Nature*, 387, 253–260.

Farber, S., Costanza, R., Childers, D. L., Erickson, J., Gross, K., Grove, M., . . . & Wilson, M. (2006). Linking ecology and economics for ecosystem management. *BioScience*, 56(2), 117–129.

Hardin, G. (1968). Tragedy of commons. *Science*, 162(3859), 1243-1248.

Figure Credits

Fig. 1.1: Copyright © Ken Billington (CC BY-SA 3.0) at https://commons.wikimedia.org/wiki/File:Sanderling_(Calidris_alba)_(6).JPG.

Fig. 1.2: https://commons.wikimedia.org/wiki/File:Lake_Guardian_-h.jpg

Fig. 1.3: Copyright © Divulgação Petrobras / Abr (CC BY 3.0 BR) at https://commons.wikimedia.org/wiki/File:Oil_platform_P-51_(Brazil).jpg.

Fig. 1.4A: http://commons.wikimedia.org/wiki/File:Population_curve.svg

Fig. 1.4B: http://www.epa.gov/climatechange/images/science/GHGConc2000-large.jpg

Fig. 1.4C: https://www2.ucar.edu/climate/faq/how-much-has-global-temperature-risen-last-100-years

Fig. 1.4D: https://commons.wikimedia.org/wiki/File:Average_global_temperature_anomalies_by_decade_between_1850s-2000s,_estimated_by_the_Hadley_Centre.gif

Fig. 1.5: Copyright © Wald1siedel (CC BY-SA 4.0) at https://commons.wikimedia.org/wiki/File:Pinzgauer_Rinder_im_Wolfbachtal_3.JPG.

Fig. 1.7: Copyright © AIRBUS Defence & Space (CC BY-SA 4.0) at https://commons.wikimedia.org/wiki/File:6._Azersky.jpg.

Fig. 1.8: https://www.nasa.gov/images/content/665717main1_a-train-image-29jun2012-430.jpg

CHAPTER 2

Science and a Critical Approach to the Environment

The nature of science is simply expressed.
Make some hypotheses, put them to test.
Some notions may fail,
While others prevail,
To construct a theory from those that are best.

Observations: The collection of information regarding the environment by various means, such as visible light (just looking by eye or with the help of telescopes/microscopes), other portions of the electromagnetic spectrum with special instruments, sound (by listening by ear or with special instruments to record earthquakes, for example), or any other means of gathering information regarding environmental conditions or processes.

Hypothesis: A proposed explanation for an observation or set of observations that remains to be tested either by further observations or predictions, that can be verified by various means.

SCIENCE IS MUCH more than simply a body of technical knowledge. It is a way of thinking that follows a specific process to glean meaning from observations of the natural world. The first step in the process is to make **observations**, the records of which become data to be used in the subsequent steps. Observations can be made of natural systems in the field, such as forests or rocks, or they can be of laboratory experiments, in which variables are carefully controlled and adjusted. Some observations can be verified by further observations by others. For example, experiments must be repeatable by others with the same results to be valid. Some observations are not repeatable. The light spectrum emitted by a supernova, for example, cannot be repeated for the convenience of future scientists. However, the records of the one-time observation must be made available so that others can draw the same conclusions, or refine initial interpretations based on a deepening of understanding of the processes involved. As observations are repeated and re-interpreted, this first step in the scientific process is refined in preparation for the next step.

Based on such observations, the next step is to formulate a **hypothesis** or set of hypotheses that seek to explain the observations in terms of the processes that led to the conditions being observed. Hypotheses are based on questions that arise from the initial observations. For example, it is observed that some robins are not flying south for the winter from New England, but are wintering in the north. A scientist may wonder why. A hypothesis from this simple observation could be that "Some robins can remain through the winter because winters are getting warmer." Like any hypothesis, this must be put to the test. If the hypothesis is true, then there should be a relation between winter weather and the number of robins that remain. At the very least, there should be an observable trend for warming in the climate record over the period of time that robins were observed to begin wintering in the north. Other hypotheses could also be formulated from the same observation of robins, however, and these can also be tested. If there is no correlation between climate and robin behavior, perhaps there is another explanation. A very different hypothesis could be generated: "Robins are able to remain in the north during winter because they can find food in cities and towns from trash, litter, and other anthropogenic sources, and keep warm on windowsills and other parts of buildings." This can also be tested by further observations. In order to be a proper scientific hypothesis, a hypothesis must be disprovable. If there is no way to test it and for it to fail, then it is not science.

Based on hypothesis testing and a set of successful hypotheses, conclusions can be drawn. These conclusions may be subsequently disproved, or more commonly, refined, by more accurate observations, more sophisticated tests, or better theoretical constructs. When a large set of hypotheses regarding a major system within the natural environment is successfully tested, it may be possible to formulate a **theory**. A theory not only explains existing observations that may not have appeared related, but also can be used to predict future observations that are not directly related to the initial observations that led to the hypotheses that tested successfully. As such, a theory is the highest form of scientific knowledge. It is the end of the line of the scientific process. The theory of plate tectonics, the theory of relativity, and the theory of evolution are among the most significant theories of all time. It is important to note that a theory is not perfect. It can be refined and improved as better observations and interpretations emerge over time. As such, science is not static, but is a dynamic process, always changing and improving.

Theory: A body of knowledge and understanding about a process, based on rigorous testing and confirmation of numerous related hypotheses. It is the highest form of scientific understanding.

Scientific Reasoning

There are two scientific approaches to drawing a conclusion. These are inductive and deductive reasoning. **Inductive reasoning** is the generalization of a limited set of observations in order to predict subsequent observations. If you observe the sun rise each day (for your whole life), and generalize this to predict that the sun will rise tomorrow, you are using inductive reasoning. If you notice that oak, and pine, and birch wood all float in water, you may generalize this to say that "All wood floats." However, in this case you would be incorrect, as there are some kinds of wood that do not float, such as some kinds of ebony; the most you could conclude is that "many types of wood float in water." **Deductive reasoning**, on the other hand, draws a conclusion from a set of defined premises and observations (Figure 2.1). For example:

Inductive reasoning: The generalization of a limited set of observations in order to predict subsequent observations.

Deductive reasoning: A conclusion drawn from a set of defined premises.

> Premise: All mammals have hair.
>
> Observation: Camels are mammals.
>
> Conclusion: Camels have hair.

FIGURE 2.1 Camel

However, conclusions drawn are only as valid as the premises upon which they are based, and false premises are very common (Figure 2.2). For example:

Premise: Only humans make and use tools.

Observation: Chimpanzees make tools from sticks to draw ants from anthills.

Conclusion: Chimpanzees are human.

FIGURE 2.2 Chimpanzee using tools

In this case, the premise was false and led to an invalid conclusion. Always beware of false premises.

Disprovability

In order for a hypothesis to be scientifically valid, it must be testable and it must be possible for it to either pass or fail. In fact, any statement must be at least potentially disprovable in order to be considered scientific. For example, the long-standing interpretation that the sun revolved around the earth was a perfectly valid scientific interpretation of the observations available to people at the time. With the invention of the telescope, better observation made it possible to disprove that interpretation and formulate the heliocentric model of the solar system. Some statements are promoted as scientific, but not open to disproof, and thus are not science at all. For example, to claim that Zen Buddhism is the only true way to spiritual enlightenment is not something that can be disproved by any sort of experiment, repeatable observation, or by any means at all. This is not science.

To state that "Bach wrote better music than Handel" is also not open to disproof. It is a matter of opinion, personal preference, and perhaps the parts of their writings that you have heard. On the other hand, to state that "Most people think Beethoven wrote beautiful symphonies" can be tested through broad public surveys and analysis of the responses, and thus, can be considered a scientific statement.

It is critical not to confuse science with belief. They have nothing in common. One can believe that Lady Gaga sang snappier songs than Katy Perry, or that Monet created more beautiful paintings than Degas. One can believe that chartreuse is the prettiest color there is. These are personal opinions, however well-founded or not they may be. You can argue with them if you have a different opinion. This is not science.

Premise: Greenhouse gases such as CO_2 trap heat in the atmosphere. Observations: Atmospheric CO_2 has increased since the mid-20th century. Conclusion:

Scientists do not "believe" anything—they observe, assess, hypothesize, test, analyze, concatenate, reassess, and perhaps finally conclude based on a clear and reproducible line of reasoning. As such, one cannot state that "I believe in global warming"—it is a non sequitur. So is asking world-renowned scientists for their "scientific opinions." There is no such thing. A scientist would interpret observations to draw a conclusion, but this is not belief.

Science is not subject to opinion or belief, but it is certainly subject to testing and disproof. In practice, scientists conduct research, write papers describing their results, and send them to journals for consideration. This consideration involves "peer review" a critical process in which other scientists carefully check the data, methods, results, and conclusions presented in the paper to ensure that the most valid conclusions are drawn, and that the paper can serve as the basis for further investigation and refinement by future scientists.

There was a time when the leading scholars of the day thought that the earth was the center of the universe, and that the sun, moon, and stars all revolved around it. Elaborate constructs were imagined to explain how this was accomplished (some involving chariots of the gods, and the like), and entrained into a belief system that was not based on science. Along came Galileo, an actual scientist who invented a better observation tool (the telescope), and found that the earth actually revolved around the sun, as did the other planets, but not the stars (not knowing anything about other galaxies at the time). He ran into trouble, however, because the Catholic Church was deeply entrenched in a belief system that had incorporated the earth as the center of the universe. They had confused science with belief. While there are few who would assert that they did this on purpose in order to obfuscate the truth and thus gain a stronger stranglehold on the population, it serves as a prime example of how science and personal or organized belief systems must not be confused. In the modern day, we see some signs of such confusion in the denial of climate change, and while there are few geocentrists any more, there are still some who do not accept scientific principles (or observations) and prefer to base their thinking on a belief that humanity does not or cannot impact global climate. This will pass, as it did in Galileo's day; the difference with this example being that only Galileo suffered (directly) as a result of confusing science with belief, yet the entire global ecosystem and all of human society will feel the impact of climate change in the coming centuries.

Yet, the scientific conclusion that anthropogenic greenhouse gas emissions have caused and will continue to cause global warming is subject to disproof as was the geocentric view of the universe. Someone could come along and discover that the observed warming was actually caused by a previously unobserved and unexplained variation in solar output, or an increase in cosmic ray energy incidence on the atmosphere, or some other completely unexpected discovery. We may even discover that everything we knew about electronic energy levels and absorption of infrared photons by CO_2 molecules is wrong. While this is exceedingly unlikely, even the remote possibility of disproof (and/or refinement) is what makes it science (Figure 2.3).

FIGURE 2.3 Basic flowchart of the scientific method

Uncertainty

Uncertainty is a concept that has caused a great deal of confusion and even problems between scientists and the lay public because the two groups assign it very different meanings. To many people, uncertainty means ignorance, lack of knowledge, or perhaps even inability to decide due to lack of moral fiber. This is completely different from how a scientist uses the term. Uncertainty is a quantification of the precision of a measurement. It is also used in assessing the performance of numerical models in regard to the probability distribution of outcomes around a mean.

In general, there are three types of uncertainty (Pielke et al., 2003):

Remediative uncertainty can be reduced through better or further research, providing more or better data or observations. This is the least problematic type of uncertainty, as it can be reduced (but never completely eliminated). Examples are measurements of global average sea level, the paleo record of past CO_2 concentrations, or the amount of non-point source pollution entering a river. In a very simple example, consider measuring the length of a football field. You can do it by figuring that your stride is about 3 feet, and walking across it while counting your steps. Your uncertainty in the measurement will be large. The precision of your measurement could be improved with better tools, such as a long tape measure, thus reducing uncertainty. Laser-ranging equipment could reduce the uncertainty further.

Irremediative uncertainty is caused by non-linear feedbacks that cannot be dispelled through better measurement. These uncertainties arise from the amplification of small and essentially random fluctuations within a system that then drive the subsequent direction of system. An example is the track of a future storm. While a hurricane track is carefully analyzed once a storm is already moving, its track cannot be exactly predicted due to the many factors that emerge and expand to determine its motion after it has already started moving. Another is the track of the Gulf Stream, which, while generally moving from southern Florida to somewhere near the British Isles, cannot be exactly predicted, nor can its meanders and spin-off ring currents, as they depend on tiny variations in waves, winds, and perhaps even fish that cause perturbations in the flow that can amplify to determine the subsequent track of the stream.

Cognitive/social uncertainty arises from future actions of intelligent agents (such as people or governments). In predictive models that depend on the actions of humans, it is necessary to include what people will do, yet as difficult as it is to predict the action of a molecule, plant, or animal, it is all the more difficult to predict human behavior. As such, incorporating into climate models the amount of CO_2 that will be emitted in the future is difficult. It depends on energy use and emerging sources, which also depend on human decisions yet to come. A major difficulty arises in the feedback between model results and human decision-making, as the model results may affect decisions, thereby invalidating the model input parameters and thus results. (Recall Isaac Asimov's "Foundation" trilogy.) In a more day-to-day example, consider walking across a crowded mall, in which people are moving in all directions. You cannot plan your exact route in advance (consider programming a remote control car to go the distance) because you don't know in advance how people will change their paths when they notice or suddenly remember something. You must alter your course as you go, taking into account the ever-changing environment of the crowd around you. This is the essence of "**fuzzy logic**" in which decisions are made on the fly as you go along.

> You are invited to a concert but you are not sure what time it starts. What kind of uncertainty is this?

Each of these types of uncertainty is involved in environmental science in various ways, and only the first can be reduced by further observation and data collection. Scientists live and work with these uncertainties, knowing that they won't go away. Sometimes the non-scientific public does not understand the distinction between them, or that decisions can (and must) be made in the presence of uncertainty. In military maneuvers, it is well known that if a general waits until all information has been gathered so that there is no remaining uncertainty before making a decision, the battle will already have been lost. An example of this concept will emerge prominently in the discussion of climate change.

Accuracy and Precision

Accuracy is a measure of how correct we are in a measurement, prediction, or conclusion, relative to a previously and generally accepted value, or one that will be accepted in the future. Precision is a measure of how exactly we think we know that measurement, prediction, or conclusion. Consider the student who set his alarm clock to exactly 11:59:00 am in order to wake up and rush to class by noon. This left very little room for error. In fact, upon arriving in class, he discovered that he neglected to set his clock for daylight savings time and he missed class altogether. The time setting was precise, but very inaccurate.

In a measurement, a numerical result will be obtained with a specified amount of uncertainty. On a graph, this would appear as error bars. These signify that you have measured something to within a certain range of its actual value. This is usually expressed as, for instance, "38.3 ±0.1," indicating that the actual value could be anywhere from 38.2 to 38.4. When the error bounds approach the magnitude of the value itself, reliability is considered to be low, indeed.

In a sense, the way measurements are expressed reflects the difference between accuracy and precision. Precision is indicated in the number of decimal points that the measurement provides. In the above case, 38.31068392 would be much more precise, but if the error bounds are still ±0.1, the measurement would be no more accurate. All of those extra decimal places included in the expression of the measurement are considered **spurious precision** and should never be included. Precision is expressed in the error bounds, and if we were very sure about a particularly careful measurement, we might express it as 38.31068392 ± 0.00000001, for example.

Spurious precision: The pretense of knowing very exactly a quantity or measurement by expressing its value to more decimal places than can be discerned on the basis of the measurement or observation involved.

Assumptions

In daily life we assume a great number of things rather than checking and rechecking every event, occurrence, and measurement. We assume that when a traffic light turns red in front of us, it will turn green again within a few moments. (We do not assume that everyone will stop for the red light in the other direction once it turns green for us, however. We check that!) We assume that a staircase has evenly spaced steps. We used to assume that a TV show we plan to watch or record will start on the hour or half hour.

In science, we make a great many assumptions as well, and under normal circumstances, they are valid. For example, we assume that the ruler we just bought is accurate, having been calibrated against the best available standard. If we have exceedingly expensive equipment that must maintain a strong vacuum to preserve a sensitive sample, we do NOT assume that a blackout is impossible, and we install an uninterruptable power supply as a backup.

A businessman, a philosopher, and a scientist were traveling by train across the Scottish countryside when the businessman spotted a black sheep grazing on a hillside. He exclaimed to the other two "I see that the sheep in Scotland are black!" The philosopher responded, "Well, all you can really say is that ONE Scottish sheep is black." Having a scientific awareness of assumptions, the scientist finally corrected them, suggesting "Actually, all you really know is that one side of one sheep in Scotland is black."

As an experimental test of your own assumptions, consider a simple math problem:

$$2 + 2 = x$$

Solve for x. If you said that x = 4, what assumption are you making? (Hint—consider your fingers and toes.)

Theory in Science

The concept of a theory is one of the most misunderstood in attempts to communicate between scientists and the public. A theory is the highest form of scientific understanding. It can be formulated only through testing, demonstrating, prediction of previously unobserved phenomena, and is as sure as anything gets in science (Figure 2.3). There was once a time when "laws" were postulated, such as Newton's "laws of gravity," but no scientist makes laws any more (Congress takes care of all that ...). While Newton made laws of gravity, a more refined approach was taken by Einstein, producing the more modern theory of relativity, which greatly extended our understanding of gravity.

Many non-scientists think of a theory as a personal belief. So a non-scientist might say "That's just YOUR theory. I have a different opinion." They would be making two mistakes in this. First, a single person does not formulate a theory. It takes entire scientific communities generations to accomplish this (even Einstein had help from years of hypothesis testing by others). Second, there is no such thing as a scientific opinion. It is an oxymoron. Scientists measure, observe, calculate, test hypotheses, and draw conclusions, but they do not offer personal opinions.

In order to develop a theory, the highest level of understanding in science, an entire process of scientific investigation is necessary first. This begins with observation of the natural world (Figure 2.3). On the basis of some initial observations, a hypothesis can be developed. A set of related hypotheses can be tested by various observational means, including experiments, and if the hypotheses are demonstrated to be correct, a theory can be finally developed. This theory is not the last word, however, as additional observations, insights, or experiments can necessitate refinement or even abandonment of a theory. Newton had gravity pretty well figured out in the seventeenth century. Then Einstein came along in the twentieth century and showed that the simple rules by which Newton (and the people of the world) thought gravity worked were not quite right, and through his theory of general relativity, took the next step. Modern physicists continue to refine this. So even though a theory is the highest level of scientific understanding, it can always be refined and improved as we learn more.

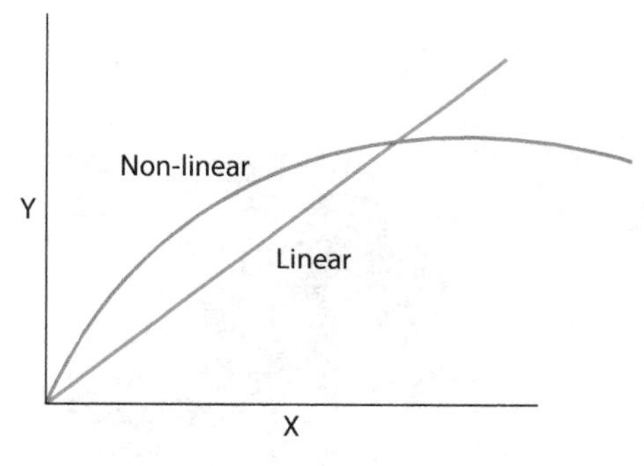

FIGURE 2.4 Linear and non-linear relationships

Nonlinearities in the Earth System

The Earth System includes interactions between forcing factors and responses that can be linear, or non-linear (Figure 2.4). Linear interactions are those in which the response to a change in a forcing factor is proportional to the change in forcing. A simple example of this is a spring, in which the extension of the spring is directly proportional to the force applied to it. In simple terms, $F = kx$, where F is the force applied, k is the spring constant related to the strength of the spring, and x is the distance the spring is stretched. A change in the distance a spring is stretched is proportional to the change in the force applied. This is true up to a point. If such a great force is applied that the spring stretched beyond a critical point (essentially straightened out), the simple linear relation no longer applies, and the spring is pushed beyond its working limit.

A non-linear system, the response of the system is NOT proportional to the change in the forcing. This is the realm of positive or negative feedbacks. In such cases, a small change in driving force can have either an amplified response, or a reduced, or buffered response in the response of the system. A simple every-day example of this could involve restaurants or diners along a highway. The unfamiliar traveler looking for a good meal may observe several establishments. Some may have no cars in the parking lots, while one may be crowded with cars with local license plates. The traveler may assume that others know more than he/she does, so will go to the more crowded place, expecting a great meal, thus increasing the crowd further. Alternatively, consider two nightclubs next door to each other. The one that has more people in it (up to a point) will attract more people. This is a positive feedback that leads to a non-linear response between population and influx of people. Similarly, a negative feedback can be observed in the choices of grocery store checkout lines, or highway tollbooths, where the more people there are in a line, the more likely it is to be avoided, thus evening out the lines in the various counters or booths.

In the Earth System, there are non-linearities caused by both positive and negative feedbacks (Pielke et al., 2003; Schellenhuber 2011). A very obvious non-linearity involves floating Arctic sea ice and the ice-albedo positive feedback (Figure 2.5).

FIGURE 2.5 An iceberg amidst Arctic sea ice

If atmospheric warming leads to some ice melting, the highly reflective ice converts to highly sunlight-absorbing water, thus leading to warming of the water and more ice melting, leading to more absorption, and more melting. ... When such a process is triggered by even a small amount of initial warming, the run-away effect can lead to very rapid ice loss, far out of proportion to the initial warming. This is the essence of non-linear behavior. For the case of non-linear behavior caused by negative feedback, consider the population of squirrels that is right at its carrying capacity (meaning they are all at the brink of starvation, with just enough food to survive). A doubling of birth rate of the squirrels will not lead to a doubling of the population, as a strong negative feedback will cause many more baby squirrels to die of starvation before they reach maturity. In both positive and negative feedback cases, the response of the system to a change in input is disproportional (greater or lesser) than the change in input.

Language Barriers

Scientists have historically had trouble effectively communicating with the general public in part because the specialized language they use is often confused or taken out of context (Table 2.1). Even when a scientist is careful not to use esoteric jargon or acronyms, common words can have contrasting meanings. Theory is only one of these. A number of such examples are in the table below.

TABLE 2.1

Scientific Term	Public Understanding	What the Scientist Really Meant
Theory	Hunch; speculation	Highest level of scientific understanding
Aerosol	Spray can	Tiny atmospheric particle
Enhance	Improve	Intensify; increase
Positive feedback	Good response; praise	Vicious cycle; runaway growth
Values	Ethics; monetary worth	Numbers; quantities
Uncertainty	Ignorance	Range of possible amounts
Error	Mistake; wrong; incorrect	Quantified difference from actual value
Bias	Distortion; political motive	Offset from an observation
Sign	Posted notice; indication; astrological constellation	Arithmetic plus or minus sign
Manipulation	Illicit tampering	Scientific data processing
Scheme	Devious plot	Systematic approach
Anomaly	Abnormal occurrence	Difference from broader average

Different meanings of terms used in science and by the general public. In communicating scientific results to the public, it is important to use terms as they would be understood, rather than used within the scientific community. (Modified after "Communicating the Science of Climate Change," Richard C. Somerville and Susan Joy Hassol, *Phys. Today* 64, no. 10 2011: 48. doi: 10.1063/PT.3.1296)

References

Pielke, R., Schellnhuber, H. j., & Sahagian, D. (2003). Nonlinearities in the Earth system. IGBP Newsletters, 55, 12–15.

Schellnhuber, J. (2010). Tipping elements in the Earth system. Proceedings of the National Academies of Science of the United States of America, 107(3), 1254–1258.

Somerville, R. C., & Hassol, S. J. (2011). Communicating the science of climate change. Physics Today, 64(10), 48. doi:10.1063/PT.3.1296

Figure Credits

Fig. 2.1: Copyright © 2014 Depositphotos/ w20er.

Fig. 2.2: Copyright © Cornelia Schrauf, Josep Call, Koki Fuwa and Satoshi Hirata (CC BY-SA 2.5) at https://commons.wikimedia.org/wiki/File:Common_Chimpanzee_uses_cuboid_tool_in_the_lab.png.

Fig. 2.3: Copyright © Toony (CC BY-SA 3.0) at https://commons.wikimedia.org/wiki/File:-Global_Warming_Predictions_Map.jpg.

Fig. 2.5: Copyright © Aweith (CC BY-SA 4.0) at https://commons.wikimedia.org/wiki/File:Iceberg_in_the_Arctic_with_its_underside_exposed.jpg.

CHAPTER 3

Environment as a System

A system is simple, or so we once thought.
Complexity emerged, so models were sought
That revealed interactions,
Between human actions
And Earth system changes society wrought.

A SYSTEM IS AN entity composed of a number of parts, or components, each of which contributes to the function of the whole. Consider a car, a cell phone, or the human body—all systems. So is the Earth. Some systems are open, with matter and energy flowing in and out of them, while others are closed—completely self-contained. The Earth System is closed with respect to matter—nothing significant flows in or out (except the occasional spaceship going out, meteorites coming in, etc.), but it is open with respect to energy, as solar radiation enters, and infrared leaves, thus maintaining a steady temperature through radiative energy balance.

The components within a system can interact in numerous ways, and a key concept in these interactions is positive and negative feedbacks within the system.

FIGURE 3.1 Skiers

Downhill skiers exhibit both kinds of feedback. (a) On powder days when fresh snow has fallen, they tend to prefer untracked slopes, leading to a negative feedback, since once a slope is skied, other skiers will avoid it (or at least that part of it) so as not to cross tracks and disturb the smooth feel of the fresh snow. (b) Alternatively, on popular slopes, many skiers making the same turns create moguls that attract other "bump skiers" who build the moguls further, attracting more, in a positive feedback that persists until the system collapses due to scraping down to the ground (or an ice layer) between the moguls.

Feedback is the interaction between the driver and the response within a system. In some cases, the processes that occur within a system may take an input driver and reduce its magnitude, thus leading to a **negative feedback** (Figure 3.1a). In other cases, the output from the system may be amplified and serve as an increased input for the system, in which case it is a **positive feedback** (Figure 3.1b). Negative feedback serves to stabilize the behavior of a system, wherein any externally imposed alterations are reduced in magnitude by the system, which tends to maintain a steady-state equilibrium. An example of negative feedback is automobile traffic. If congestion acts to delay drivers, they choose to travel at different times or by different routes, thus reducing the severity of traffic congestion.

Negative Feedback: A damping or reduction in the magnitude of a system input by processes within the system. This tends to stabilize the system.

Positive Feedback: An amplification of the magnitude of a system input by processes within the system. This tends to destabilize the system.

In the Earth System, CO_2 fertilization of terrestrial ecosystems is a negative feedback, in which the additional CO_2 in the atmosphere leads to accelerated plant growth, which removes some CO_2 from the atmosphere, thus acting to partially stabilize atmospheric concentration of CO_2. Positive feedbacks are somewhat more worrisome in most cases, as the system serves to amplify perturbations, thus leading to instability. A prime example of this is Arctic sea ice. Because ice has a very high **albedo** (high reflectivity of the Sun's energy), it does not absorb solar energy, thus keeping the polar region very cold, and the Arctic ocean water covered by ice. However, if some of the ice should melt, exposing ocean water to the sun, the Sun's rays are almost completely absorbed by the water, thus serving to heat it up. By warming the water, more ice is melted, and more open water exposed to heat further, thus melting more ice, exposing more water, and starting a vicious cycle of ice melting. This is happening right now and we are seeing unprecedented low amounts of ice in the Arctic each summer.

Describe one positive and one negative feedback you encounter in your daily life.

Albedo: The reflectivity of a surface. Albedo of 1 reflects all light. Albedo of 0 absorbs all light.

A classic example of negative feedback is the case of **Daisyworld** (Figure 3.2). This is a world occupied by black and white daisies. The white daisies reflect the Sun's rays, and thus serve to cool the planet, but white daisies thrive in warm weather. If there are too many white daisies, they will die off from cold. Black daisies, on the other hand, absorb solar radiation and warm the planet, but they like it cold. With too many black daisies, the planet will warm to the point at which they cannot live any longer. With both kinds of daisies on Daisyworld, temperature is regulated. If, for some reason, the temperature warms a bit (due to changes in solar output, or greenhouse gases, for example), black daisies will die off and white daisies will thrive, thus serving to cool the planet back down to a stable equilibrium. The opposite would counteract an externally imposed cooling. In this way, the temperature of Daisyworld is kept as a steady equilibrium between coverage by black and white daisies. This is negative feedback at its finest.

Daisyworld: A fictitious planet populated by black and white daisies. The black ones thrive in cold temperatures but, with their low albedo, warm the planet. The white ones like it hot, but cool things with their high albedo. This stabilizes the temperature, because if outside perturbations cause warming, for example, more white daisies will thrive and tend to cool things back down.

Uniformitarianism and Catastrophism

There are two contrasting views of natural processes in the Earth System. The first is **uniformitarianism**, the basis of which is the premise that normal, day-to-day processes dominate the behavior of the system as a whole, and that extreme events may occur, but are so rare that their effects are eventually smoothed out by the ongoing normal processes. A famous uniformitarian maxim is "The present is the key to the past," meaning that if you can understand modern processes, then you can assume that these same processes were ongoing in the past and that thus you understand past processes as well. While understanding past processes is interesting to geologists and paleoecologists, most environmental scientists are more concerned about understanding future responses to environmental perturbations. In this case, the uniformitarian principles can be turned around so that "The past is the key to the future."

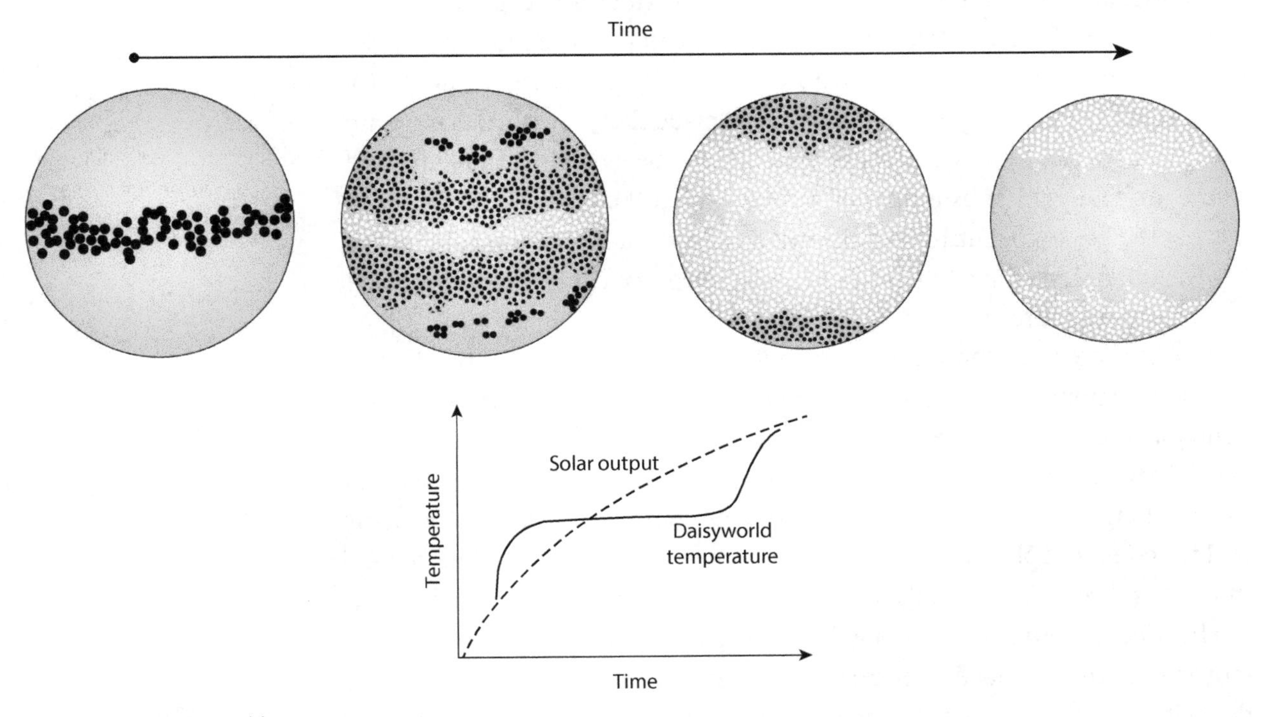

FIGURE 3.2 Daisyworld
Daisyworld stabilizes temperature due to negative feedbacks, even in the case of a warming Sun. Starting too cold for any daisies, black daisies are first established at the equator, and spread as the planet quickly warms to ideal black daisy temperatures due to the absorption of energy by the low-albedo black daisies. As the Sun warms further, the equator gets too hot for black daisies (that generally like it cool), and white daisies move in (that like it warmer), cooling their regions by reflecting incoming sunlight. As the Sun gets more intense, white daisies proliferate, keeping temperatures within tolerable limits. Eventually the overwhelming heat of the warming Sun is too much even for the white daisies, even at the poles, and all daisies are lost. In the intervening time, the negative feedback stabilized the temperature.

Suggesting that the past be used as the key to the future can be dangerous, however, in that humanity has recently (geologically) become a major force in altering the state of the planet and the environment it offers to all living organisms. Humans have made such a large impact that several key scientists have suggested establishing a new geologic epoch, the Anthropocene, to indicate that we have a whole new ballgame, geologically and environmentally.

As a result of human activities (mostly land use and emissions), the earth is in a **no-analogue state** because we have artificially altered the balance between atmospheric chemistry, ocean chemistry and temperature, land cover, species distribution, and many other key global parameters. With these changes, it is difficult to predict how the Earth System will respond with different parts of the system responding to perturbations in other parts. Global changes in the past have lacked saws and fossil fuel burning, so the various parts of the system were able to stay in quasi-equilibrium with each other. For example, there was never a time during the Phanerozoic (the last 542 million years, when multicellular life evolved) when atmospheric CO_2 concentration was as high as it is now at the same time as ocean temperature was as low. With such a cold ocean, the water can dissolve much more CO_2 from the abnormally CO_2-rich atmosphere, making it more acidic, to the detriment of marine organisms that need to make their calcite and aragonite shells in an ocean from which these minerals can be biologically precipitated. There are many other examples of ways in which the past can no longer be used as the key to the future, but the more understanding we can gain regarding the operation of the Earth System, the better we can predict how it will respond to past, present, and future human activities.

No-Analogue State: There is no time in Earth history that is like the environmental conditions of today and that are expected in the coming centuries, so we cannot find a comparable example from geologic studies of the past because atmospheric and ocean temperature, chemistry, and circulation, biome distributions, ice cover and many other physical, chemical, and biological conditions have been suddenly knocked out of equilibrium with each other, primarily due to human land use and emissions.

In contrast to uniformitarianism, **catastrophism** is the thought that history is driven by unusual, extreme events that determine subsequent events, and obliterate the record of day-to-day, more mundane processes. Consider a sandy beach in which waves and tides move sand grains up and down the beach, as well as along the beach by longshore drift. One could study these processes in real time (a most pleasant study in the summer), measuring the flux of sand in various directions, as well as deposition and erosion at any specific place. If a hurricane were to come along, however, a great deal more sand would be moved than the normal flux. Sand would be deeply eroded in some places and deposited in others. Any thin layers of sand deposited by normal waves and tides would be wiped out.

Neither uniformitarianism nor catastrophism can be used as a complete worldview. Both concepts are useful when studying the environment, as both are relevant to the Earth System. Feel free to think up shorter words for them.

Stocks and Fluxes

A **stock** is the amount of a certain entity in a reservoir within a system. There is a stock of water in a lake, gas in a tank, CO_2 in the atmosphere, and money in your bank. The amount can grow or decrease depending on the balance of input to and output from the reservoir. **Flux** is the rate of exchange of material

Stock: The amount of a certain entity in a reservoir within a system.

Flux: The rate of exchange of material into or out of a reservoir.

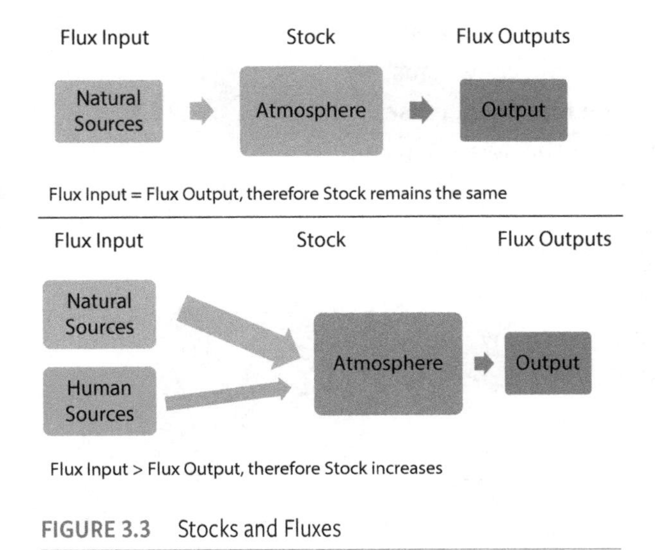

FIGURE 3.3 Stocks and Fluxes

into or out of a reservoir. A positive flux increases a stock, while a negative flux decreases it. The direction of the flux is linked to the reservoir being discussed (Figure 3.3). The flux of CO_2 as it leaves the atmosphere to enter the ocean is negative for the atmosphere and positive for the ocean.

In a system with a reservoir, the residence time of material in the reservoir depends on the magnitude of the reservoir divided by the flux through the reservoir (assuming the input flux and output flux are equal, such as in a dammed lake. If they are unequal, the stock within the reservoir would grow or shrink.). In some systems, this is an important factor. For example, a river can be polluted by contaminants in runoff from industry, agricultural fields, etc. However, in a river, the flux is large and the reservoir small, so if contamination were to end, the river would become clean very quickly (a matter of days, usually). A large lake, on the other hand, may have a very small flux in and out, so that the residence time is many years. This would therefore take longer to clean up. The extreme example is that of groundwater, which has a very large stock and very small flux. When groundwater is contaminated, because of the very long residence time (sometimes thousands of years), it would take a very long time to clean up by natural cleansing, and contaminated aquifers are sometimes cleaned up actively by very expensive and laborious means.

It is a common mistake to confuse stocks and fluxes. The Kyoto Protocol called for a reduction of the rate of greenhouse gas emissions to 1990 levels. Emissions rates are a flux, so in order to comply with the protocol, nations would slightly reduce the rate at which greenhouse gases were emitted to be the same as they were in 1990. In 1990, the flux was not zero. Thus, the rate of increase of the stock of greenhouse gases in the reservoir of the atmosphere would be the same as it was in 1990 (very great, but a little slower than it is today). Many throughout the world assumed that such a protocol would put a halt to climate change. As you know by now, it would merely reduce the rate of increase of greenhouse gases to that of 1990. Climate change would continue to accelerate in the twenty-first century. In order to stabilize the climate, the reservoir of greenhouse gases in the atmosphere would need to be stabilized at a constant value, meaning the flux would need to be zero, implying zero emissions. In any case, some nations, most notably the U.S., didn't even agree to stabilize the rate of emissions at 1990 levels, never mind reducing them to zero. In considering fluxes,

A lake with volume 3×10^8 m³ is fed and drained by a river with a flux of 6×10^6 m³/yr. What is the residence time of the water?

Answer: Looking at the units of the numbers, to get years, you need to divide the lake volume by river flux. So

$$\frac{3 \times 10^8 \text{ m}^3}{6 \times 10^6 \text{ m}^3/\text{yr}} = 50 \text{ yrs}$$

there can be both inputs and outputs, and in the atmosphere, there are fluxes of greenhouse gases into the ocean and into terrestrial ecosystems. These will be discussed later, but they serve to reduce the rate of increase of atmospheric CO_2, for instance to less than it would be otherwise. This will be considered in more detail in Chapters 12 and 13.

Stock: The amount of something stored in a "reservoir."
Flux: The rate at which something enters or leaves a reservoir.
Equilibrium: When input and output are balanced.
Residence time: stock/flux

Stock constant	output = input	Beach sand Conveyor belt Earth's radiation
Stock shrinks	output > input	Tropical forests Fish Lobsters? Mineral ores Fossil fuels Ground water
Stock grows	output < input	Atmospheric CO_2 People

An interesting socio-economic example of an actual and continuing mismatch between stocks and fluxes is property taxes. Property is a stock and assessed a value, whether it be a house, a car, or pair of shoes. In many places, an annual tax is levied on the assessed value. Annual taxes are a flux, as they are measured as money per year (or quarter, or other unit of time). As such, a flux is being applied to a stock. The property does not produce money in order to pay for the flux of taxes and sometimes the stock (say, a pleasure boat, or a widow's farm) must be sold in order to pay for the flux of tax levied against it. Personal or corporate income on the other hand is a flux, measured in dollars per year. When income taxes are levied, a flux is applied to a flux. If there is no income, there is no tax, and as income grows, so does the tax. No one ever has to quit their job in order to pay for the income tax. Conversely, sales tax balances a stock against a stock in a one-time transaction.

In the natural environment, there are numerous reservoirs that have little positive flux, and yet are tapped for human consumption, creating a negative flux, thus depleting the reservoir (like the farmer's widow selling a little piece of the farm every year to pay for the property tax). A prime example of this is groundwater from aquifers with low recharge rates (small positive fluxes), such as the famous Ogallala, or High Plains Regional, Aquifer in the American Midwest, which is being depleted at an alarming rate.

Exponential Growth

Exponential growth: The increase of a stock whose growth rate is proportional to the amount of stock in a reservoir.

A special case of positive feedback is **exponential growth**, in which the rate of growth flux is proportional to the amount that exists (stock) (Figure 3.4). The usual example is that of money in an interest-bearing bank account. With an interest (growth) rate of 3%, for example, an additional 3% of the amount in the account will be added to the account. Because it is added to the account, the amount in the account increases, so the next period, the 3% is calculated on a greater base, and more is added than the previous time. As time goes by, the amount of each increase increases because the balance in the account increases as a result of previous increases. This makes exponential growth. A more relevant example of positive feedback is human population. If each set of parents has four children per generation, the population grows, so the rate of increase of population also increases (as measured in units of people per year). The reason we have seen extra-exponential (faster than exponential) population growth in recent decades is that there is an additional factor of an increasing fraction of children surviving to create grandchildren.

> If a group of 100 bacteria reproduces to increase its population at a rate of 3%/day, how many there be after a year? Answer: 3% is 0.03 in real numbers. There are 365 days in a year. So
>
> $N = N_o (e^{kt}) = 100 (e^{03t}) = 100 (e^{10.95})$
> $N = 5{,}695{,}400$
>
> or, using only the significant figures given in the problem, $N = 5.7 \times 10^6$ bacteria.

The relation that describes exponential growth is

$$N = N_o e^{kt}$$

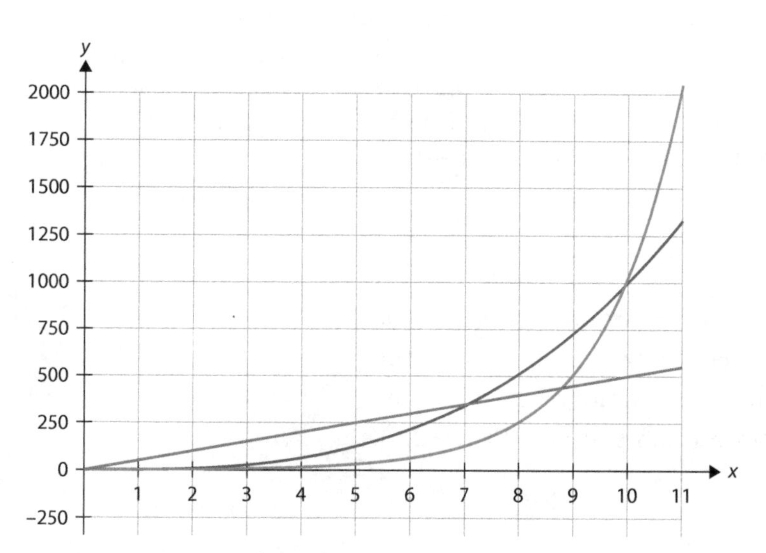

FIGURE 3.4 Exponential growth graph
Exponential growth occurs when the rate of growth depends on the amount already there. As such, the flux into the reservoir depends on the stock already in the reservoir. Compound interest is a common financial example, but BEWARE! Business students commonly use discrete compounding (quarterly, daily, etc.) which is not strictly exponential. In science, we use what would be continuous compounding.

Where N is the stock (amount) at any given time, t; N_o is the initial stock; e is the base of natural logarithms (about 2.718); and k is the growth rate expressed in fraction of the total stock per time. If the growth rate is 5% per year, then k is 0.05 per year, and you would express t in years to get a non-dimensional exponent (exponents are always non-dimensional). It is important to note that this exponential relation is based on continuous growth. The additional stock is added continuously to the reservoir. This is in contrast to usual business accounting methods, in which the accrued interest is added to the reservoir (e.g., bank account) only at periodic intervals, be they annually, quarterly, or daily. In natural systems, growth is continuous, not compounded at discrete intervals.

Using this relation, it is easy to determine the stock within a reservoir given a growth rate and initial amount N_0. A special case of this is the characteristic time it takes to double an exponentially growing quantity (e.g., your money in the bank bearing interest). In this case, $N = 2N_0$, so using this simple relation, dividing by N_0, and taking the natural log, you get $kt = \ln(2)$. Thus $t = .693/k$ with k as a fraction. If you express k as a percentage increase per year, K (commonly done), then you get a simple approximation of **doubling time** as about 70/K. So with a growth rate of 7%/yr, the doubling time is about ten years. This is the origin of the commonly used and very practical doubling-time formula because $.693 \times 100$ is about 70.

Doubling time: The length of time required to double the stock in a reservoir. For exponential growth, this can be approximated as 70/K, where K is the growth rate in percent.

THE CLEVER PEASANT

A poor peasant in a great kingdom performed a heroic deed and saved the lovely princess, the king's only daughter, from certain death at the hands of a ferocious dragon. In an overly emotional offer of appreciation, the king promised the peasant anything he could possibly want, including the princess's hand in marriage. The peasant thought for a moment, and made a rather unusual request. He asked that a single grain of rice be placed on the corner square of a chessboard (chess being the noblest game in the kingdom). On the adjacent square, two grains of rice should be placed. On each subsequent square, the number of grains should be sequentially doubled until all 64 squares were accounted for. The king had not studied exponential growth and thought that the peasant was asking for too little (thinking jewels and riches were more appropriate), but the peasant insisted that this is exactly what he wanted—no more and no less. It turned out, of course, that there was not enough rice in the entire kingdom or enough gold in the treasury to buy enough rice to fulfill the peasant's request. To do so would have meant how many grains of rice? Just one grain less than 2^{64}, or 18,446,744,073,709,551,615 grains of rice (1.84×10^{19}). If there are 30,000 grains of rice in a pound, and 45 pounds in a bushel, there are 60,750,000 grains in a bushel of rice. Dividing the total number of grains by the number of grains per bushel, the peasant was asking for over 300 billion bushels of rice.

The Earth System

The point of discussing systems in the context of environmental science is to promote the System-level thinking necessary to begin to understand the operation of planet Earth and the environment it provides for all life, including humans. The Earth system is composed of several sub-systems (Figure 3.5), and one way to categorize these is:

- ocean physics, chemistry, and biology
- terrestrial ecosystems, including soils
- atmospheric processes and climate
- glaciers and the cryosphere
- water and global hydrology

There are many other ways one can subdivide the structure and processes of the Earth System, depending on regions, processes, or functions of interest.

However, considering any part in isolation necessarily assumes something about interactions or lack of interactions with other parts. The behavior of the atmosphere, for example, can be considered without joint consideration of the ocean, but it would be necessary to impose "known" marine conditions of temperature distribution, heat flux, mass flux (of water, carbon, oxygen, etc.) and the evolution of their exchange with the atmosphere in order to correctly determine atmospheric behavior. However, the feedback between ocean and atmosphere makes it exceedingly difficult to know these attributes a priori. Consequently, it is typically best to consider the Earth System as a whole with internal processes and interactions that determine the behavior of the system itself, not merely a summation of the behavior of the various parts.

The Earth System can be considered closed for environmentally practical purposes in terms of matter exchange with the exterior (space), but it is not closed in terms of energy. It receives energy from the sun (mostly in the frequency of visible light), and emits energy to space (in the infrared). Atmospheric chemistry influences the balance between incoming and outgoing radiation, and thus is a critical component of the Earth System.

People are a peculiar part of the Earth System, in that they affect and are affected by their environment, but are the first (that we know of) to alter it on purpose to suit near-term needs at the expense of long-term needs. Any species that may ever have done this in the past are apparently now extinct. The

FIGURE 3.5 Ocean Floor, Terrestrial ecosystem, Atmosphere, Cryosphere, Water

traditional view of humanity and social systems residing outside the natural environment, and thus insensitive to Earth System dynamics, has now given way to a more accurate view of people as an integral part of the system, responding to and being affected by the behavior of the rest of the system, in addition to causing perturbations in the system through activities such as emissions and land use.

The Gaia Hypothesis

It has been known for some time that life affects its physical environment. Key examples are the rise of atmospheric oxygen caused by photosynthetic organisms billions of years ago, or even just the shade and protection from wind provided to seeds by mature plants all around them. Life affects its environment.

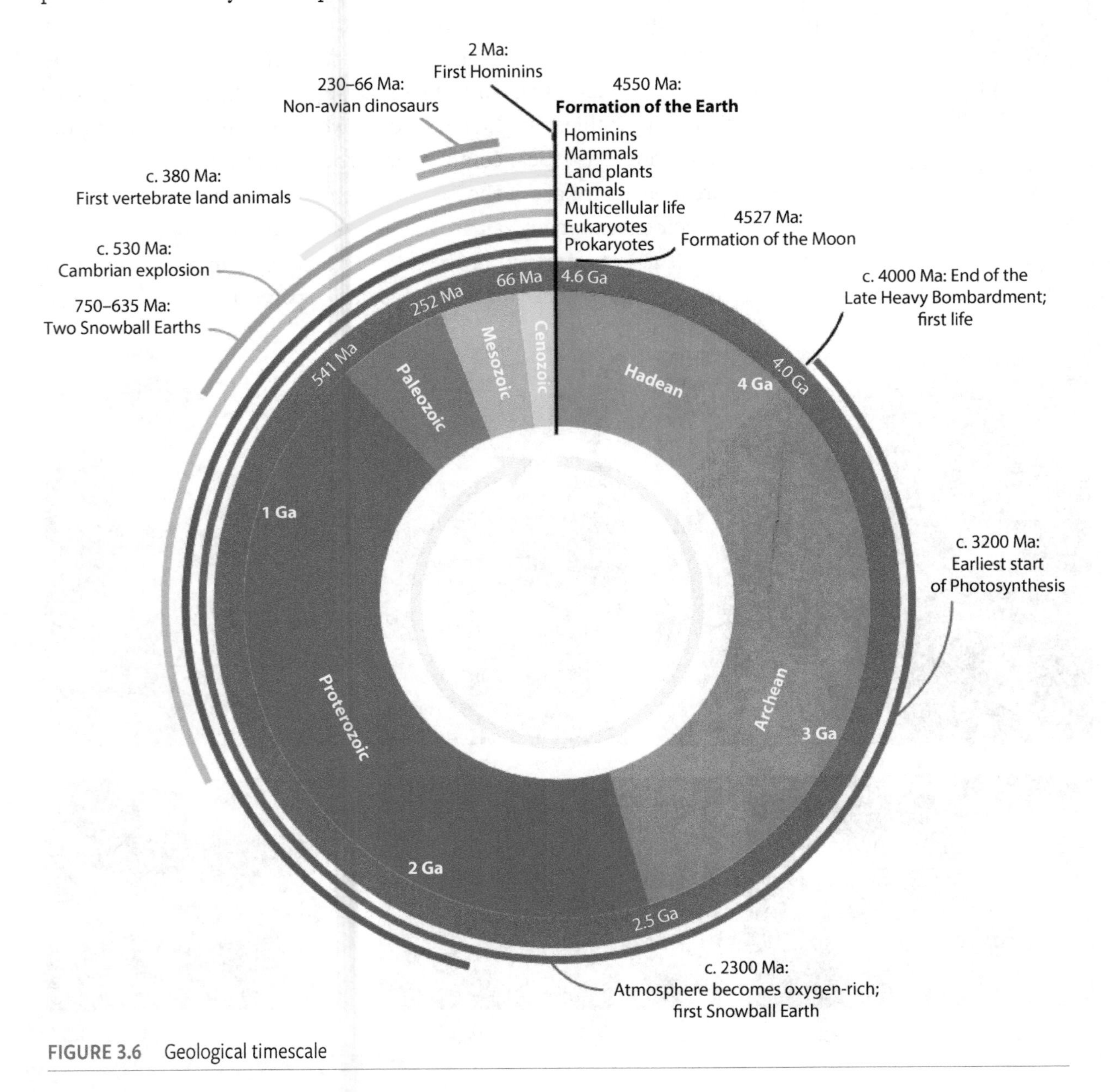

FIGURE 3.6 Geological timescale

Over geologic time (Figure 3.6), the global environment has been evolving in such a way that life has persisted at the planetary level. The Gaia Hypothesis is founded on these two key observations (Lovelock and Margulis, 1974; Moody, 2012). As such, the hypothesis can be stated as follows:

- Life affects its physical environment
- Life has altered its planetary environment such that it persists
- Life controls the global environment *on purpose.*

The third point is what causes the Gaia Hypothesis to diverge from science. There is no indication that the Earth's living systems have, over time, collaborated *on purpose* to alter or maintain a specific planetary environment. This is the stuff of science fiction and fantasy stories in which the "Mother Earth" does things to protect itself as a thinking being. It is not environmental science, or science of any kind.

Obstacles to Finding Simple Solutions to Environmental Problems

It would seem a simple matter to solve environmental problems once they are identified. Just study the system, alter human behavior, apply appropriate remediation measures, and enjoy the result, no? No! Because of feedbacks within the Earth System (both positive and negative), **delayed responses** of system behavior to perturbations caused by a component within the system, and **non-linearities** that involve critical thresholds that, once crossed, alter system behavior in an irreversible way, many environmental problems are exceedingly difficult to solve. In fact in many cases, it is not even clear exactly what should be done, much less how much and how soon.

Delayed response: The reaction of a system or component to forcing or perturbation, when the response does not occur until after the time required to overcome some thresholds and buffers within the system.

Non-linearities: Responses of a system that are disproportionate to the changes in input either by amplifying (positive feedbacks) or buffering (negative feedbacks) the inputs.

The difficulties posed by feedbacks, including but not limited to exponential growth, make it difficult to isolate cause and effect within the Earth System. For example, an examination of paleoclimate from ice core records shows atmospheric CO_2 and temperature rising and falling together in glacial cycles of the Pleistocene. Sometimes CO_2 changes preceded temperature changes and sometimes not. Did CO_2 cause warming to increase, or did warming cause CO_2 to increase? Because of the positive feedback between ocean temperature, climate, and solubility of CO_2 in water, the cause and effect cannot be uniquely identified as a single direction. They act together. While this is a complication in interpreting the natural paleoclimate record, the addition of geologically stored carbon into the atmosphere by fossil fuel burning is a simple trigger for warming, and this is explored in more detail in Chapter 13.

Sometimes, when a component within the system changes, other parts of the system respond, but not for a while. For example, the CO_2 that was added to the atmosphere in the twentieth century has caused some climate change, but glacial ice has not yet completely responded to that change. While Arctic sea ice is rapidly declining, and Greenland ice has been observed to be melting at an increasing rate, Antarctic ice has not yet responded in a clearly definable way. This will take centuries to millennia to become manifest, with its associated sea level rise (Chapter 9), and will thus continue to cause impacts long after we stop emitting fossil carbon into the atmosphere.

Interactions within the Earth System are often highly non-linear, in that changes in one component do not affect other components significantly until a critical threshold, or choke point, is reached, at which point the entire system switches to a new mode of behavior. After that, changing the "offending" component back to its original condition does not cause the system to revert to its original mode. It may have to be driven far beyond the original condition to cause the system to switch back—this is **hysteresis**. An example of this is North Atlantic bottom water formation, which is cold, salty water that sinks down to the bottom of the ocean in the region northeast of Iceland. This drives the global thermohaline circulation (Chapter 9), but there have been events that caused a decrease in the flux of bottom water formation past a critical threshold, and the heat-transporting global circulation system switched to a different mode, affecting climate around the world.

Hysteresis: The difference between the path a process takes to some endpoint, and the path it must take to return to the starting point. A trivial example is driving across town on one-way streets. The return trip must take a different path.

References

Lovelock, J. E., & Margulis, L. (1974). Atmospheric homeostasis by and for biosphere—Gaia hypothesis. *Tellus*, 26(1–2), 2–10.

Moody, D. E. (2012). Seven misconceptions regarding the Gaia hypothesis. *Climatic Change*, 113(2), 227–284.

Figure Credits

Fig. 3.4: https://commons.wikimedia.org/wiki/File:Exponential.svg

Fig. 3.6a: Copyright © 2012 Depositphotos/Rostislavv.

Fig. 3.6b: Copyright © Alan Silvester (CC BY-SA 2.0) at https://commons.wikimedia.org/wiki/File:Mount_Revelstoke_National_Park_1.jpg.

Fig. 3.6c: https://commons.wikimedia.org/wiki/File:ISS-40_Thunderheads_near_Borneo.jpg

Fig. 3.6d: Copyright © Liam Quinn (CC BY-SA 2.0) at https://commons.wikimedia.org/wiki/Category:Ice_blocks_in_the_ocean#/media/File:Icebergs_in_Pl%C3%A9neau_Bay,_Antarctica_(6058724537).jpg.

Fig. 3.6e: Copyright © Thomas Gensler (CC BY-SA 2.0 DE) at https://commons.wikimedia.org/wiki/File:Manavgat_waterfall_by_tomgensler.JPG.

Fig. 3.7: https://commons.wikimedia.org/wiki/File:Geologic_Clock_with_events_and_periods.svg

CHAPTER 4

Human Population

So many young people inhabit the land
Our proliferation was somewhat unplanned.
Our numbers keep growing
Increase hardly slowing
As resource imbalance gets way out of hand.

IN 2021, THERE are over 7.9 billion people on the planet. This is remarkable in a number of ways. Most notable is the rate of growth, which is about 3 people per second, or an additional equivalent to the entire U.S. population every three years. This explosive rate of population growth (Figure 4.1), were it any other species on the earth, would be considered an infection or a plague (depending on organism size), and antibiotics or other sterilization measures would be applied immediately. Of this growth, 90% is in the developing world, or the poorer nations globally, where about 80% of the population already lives. As such, the spatial distribution is skewed, as is the distribution of wealth and security, shifting ever more toward an increasing fraction of the world population who are under food, water, health, and other stresses. Perhaps one of the most alarming observations is that about 50 to 100 billion modern humans have ever lived on Earth (depending on whether you count infant mortality during times of life expectancy of ten years or less over most of the last 50,000 years). Of those, 7.9 billion, or about 7% to 14% are alive today.

FIGURE 4.1 Large family

The history of humanity and precursors shows that there was a small (fewer than a couple hundred million) and stable population for the last 200,000 years or so, until the most recent thousand years or so (Figure 4.2). Population increase accelerated, and then exploded starting about 1850, or the time of the industrial revolution. In part, the stability of the early population was due (as for any species) to a balance between births and deaths. High mortality rates made it necessary to maintain high fertility rates (children born per woman) in order to maintain the population. These high fertility rates became the norm, and without them, the species would have been extinct long ago. About 10,000 years ago, a stabilizing climate enabled ecosystems to remain in one place for long periods, and this quickly led to the concept of agriculture, enabling the purposeful growth of crops and animals specifically (and increasingly) bred for human consumption. This enabled the onset of specialization, stationary communities, and the support of larger populations. It was not until the advent of antibiotics, uses for fossil fuels, and industrialization that the greatest imbalance between fertility and mortality began. With increasing global access to medicine and medical care, an unprecedented fraction of babies lived to reproduce. This caused a huge increase in the number of grandchildren born (Figure 4.2). (It is not the children born, but the grandchildren who determine growth or decline of a population. You could have twelve children, but if only one of them survives to bear you grandchildren, the population could still decline.)

Two hundred thousand years of "having as many children as possible" is a difficult habit to break, and even after most children born survived to maturity, cultural norms maintained high fertility. Eventually, in the most industrialized, wealthiest countries where medicine, food, water, sanitation, technology, and personal safety enabled almost every child to survive, people began to feel the confidence to bear fewer children and fertility declined to closer to constant population-maintenance levels. The delay between the ability for children to survive and the confidence of parents to bear fewer children is called the "**demographic transition,**" which is affected by both poverty and education (Figure 4.3) (Bottencourt, 2018; Snopkoski et al., 2016).

Before the transition, population is held stable by high death rates matching high birth rates. After the transition, it is stabilized by low birth rates finally matching low death rates. In between, there is population explosion, and this is what happened in the previous century in Europe and America, and is happening in this century in the rest of the world (Figure 4.4).

If there were 100 million people on Earth in the year 1, then 500 million in the year 1500, then 7 billion in 2000, was the growth exponential? If so, what is the constant growth rate, k?

Answer: Let's find the growth rate from years 1 to 1500.

$N = N^{\circ} (e^{kt})$ so

$5 \times 10^8 = 1 \times 10^8 (e^{1500\,k})$. Now divide by 100 million

$5 = (e^{1500\,k})$. Now take the natural log of both sides to isolate the unknown k

$\ln 5 = 1500k$. so

$k = \ln 5/1500 = 1.61/1500 = .001$, or a growth rate of 1%.

Now let's see how this growth rate compares to that from 1500 to 2000.

$N = N_o (e^{kt})$. Follow the same steps as above ...

$7 \times 10^9 = 5 \times 10^8 (e^{500\,k})$

$14 = (e^{500\,k})$

$2.64 = 500k$

$k = .005$, or growth rate of 5%, WAY faster than previously. Therefore, human population has been FASTER than exponential, or "super-exponential," during the population explosion of the last several hundred years.

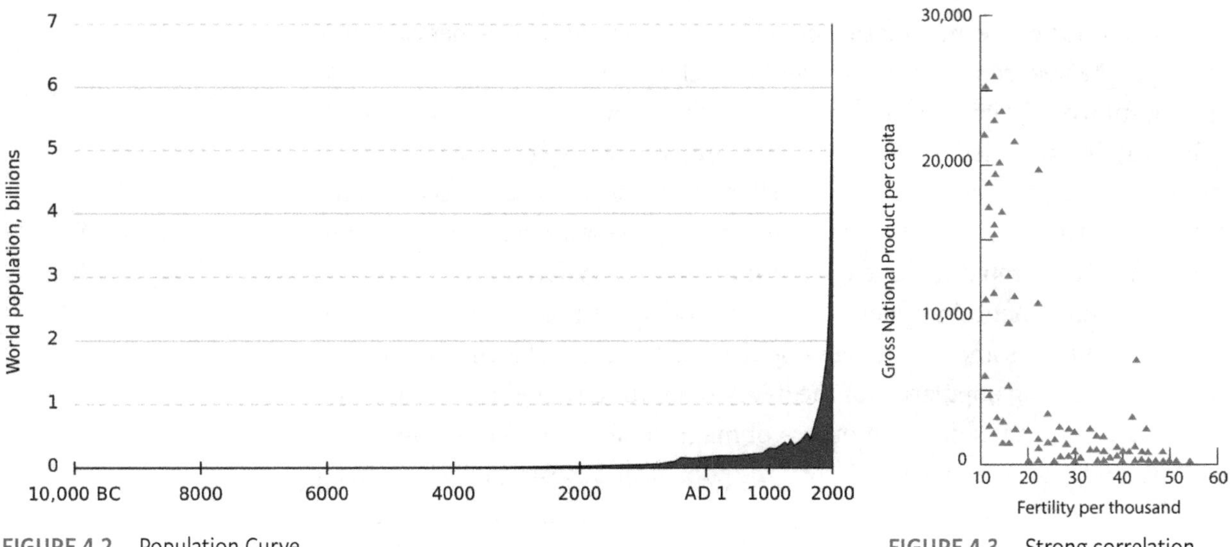

FIGURE 4.2 Population Curve
Human population remained stable for thousands of years until a recent explosion, surpassing 7 billion in 2012, and 7.9 billion in 2021.

FIGURE 4.3 Strong correlation with poverty and birth rates

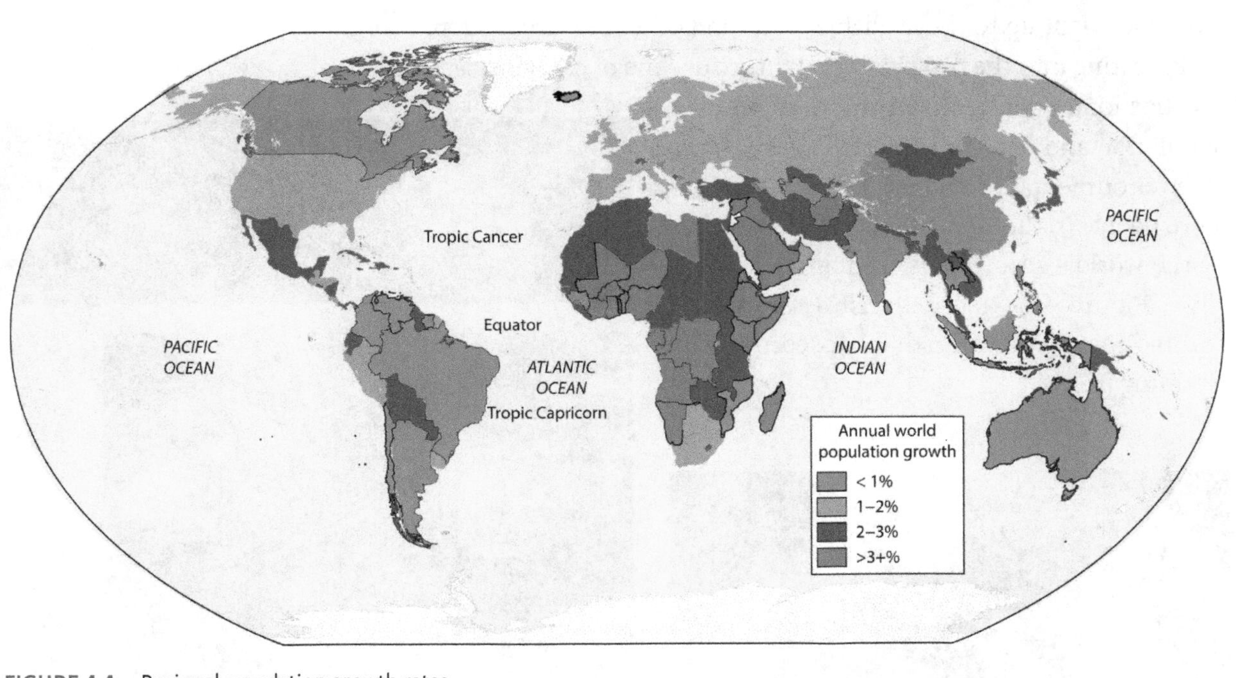

FIGURE 4.4 Regional population growth rates

What Is the Carrying Capacity of the Earth?

Carrying capacity is the limit of population that an ecosystem can sustainably support. For every other species (besides human), the carrying capacity is determined by the balance of reproductive rate, food, and habitat on the one hand, and mortality and predation on the other. Organisms use only as much food and other resources as needed to survive. With humans, however, there is a large range of resource consumption. Some, in very poor regions, consume a small fraction of the amount that others in rich areas do. The planet can carry more people who consume very little than rich people who consume a lot. So the question of carrying capacity comes down to "How do people want to live?" While this is not a scientific question, it bears on the science of the environment in that human populations are responsible for the degradation of ecosystems throughout the world, and the more people there are, and the more each consumes, the more rapid the rate of environmental degradation. In general the overall environmental impact is just the product of the average impact of individual humans and the human population. The general feeling globally is that the exploding population of the developing world wishes to live and consume like Americans do, with orders of magnitude greater consumption of energy, food, and other resources than the poorest and most rapidly increasing populations of the world. This being the case, the carrying capacity declines dramatically. In effect, humans are the first species (that we know of) that can determine its own means of population control. It does not need to be starvation, disease, and predation, as it is for other species. If we choose not to reproduce right up to the available food (and other resource) supply, we can avoid "eating into the principal" and future decline of carrying capacity.

It has long been realized that human overpopulation and overconsumption is the root of all environmental problems. There are already more people in the world than can be sustained by the world's ecosystems, and degradation of soils (Figure 4.5), aquifers, fish (Figure 4.6), and biodiversity is already proceeding at an alarming pace.

If it took 30 days for an exponentially increasing population of bacteria to completely fill a test beaker, when would the beaker be only half full of bacteria? Answer: Day 29.

FIGURE 4.5 Soil loss

FIGURE 4.6 Overfishing

The usual concern regarding this is the continued availability of food for human consumption. There are billions of people in the world who are undernourished, and technologic and economic advances could help to provide more food and better living conditions for a large segment of the human population. The problem is that if these populations are still within the demographic transition, prior to reducing fertility voluntarily (Battacharya and Chakraborty, 2017), more food and better conditions will merely enable more children to survive and will lead to further population explosion, potentially getting right back to where they started, but now with more mouths to feed. This same concept applies to all environmental problems caused by human activities. Essentially, "There is no solution to an environmental problem that cannot be negated by a concomitant increase in human population." With an increasing portion of the world getting through the demographic transition, the rate of population increase appears to be slowing down a bit, but population is still increasing rapidly (Figure 4.7).

The global ecosystem can be considered as a bank account that bears interest in the form of food, water, and a wide variety of goods and services that support our human population. The interest is provided at a certain rate that depends on the health of the ecosystem and the extent to which the functions it performs are permitted to operate in the face of human perturbations such as land use for agriculture. However, in recent centuries, we have been extracting resources from our environment at a rate much greater than that at which they are restored. As such, we have been eating into the principal, a concept abhorrent to any financial manager. The result of this is that with more limited ecosystem function, the global environment bears less "interest" for us to use, and thus the carrying capacity actually declines. So the more people there are, the faster they use up the remaining resources, and the fewer people who can ultimately be supported in the long run. What happens in typical populations when they exceed their environment's carrying capacity is that they increase for a while, during which they deplete the resources upon which they depend, and then when the resources are stressed to a critical limit, the population crashes to a new and very low number, determined by the few remaining resources available (Figure 4.8).

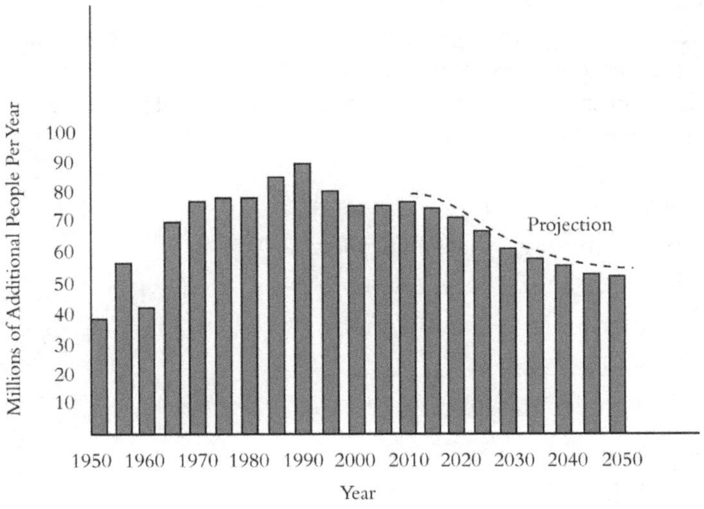

FIGURE 4.7 Population has slowed
The rate of population growth has slowed already and is projected to be further reduced. This should not be confused with a reduction in population. It is merely projected to increase as a slightly slower rate than it has to date. In order to population to decrease, the growth rate would need to become negative.

About half of the world's primary productivity (the amount of biomass produced by plants) is already used by humans. This is used in an unsustainable manner that leads to soil loss, decline of fisheries, and reduction of biodiversity. If the population were to double (in just a few decades), the entire planet would be used for agriculture to support human life. This would quickly lead to reduction in the rate of provision of ecosystem goods and services, and subsequent reduction of carrying capacity, as humans need other resources besides food (e.g., clean fresh water, clean air, proper sanitation, etc.).

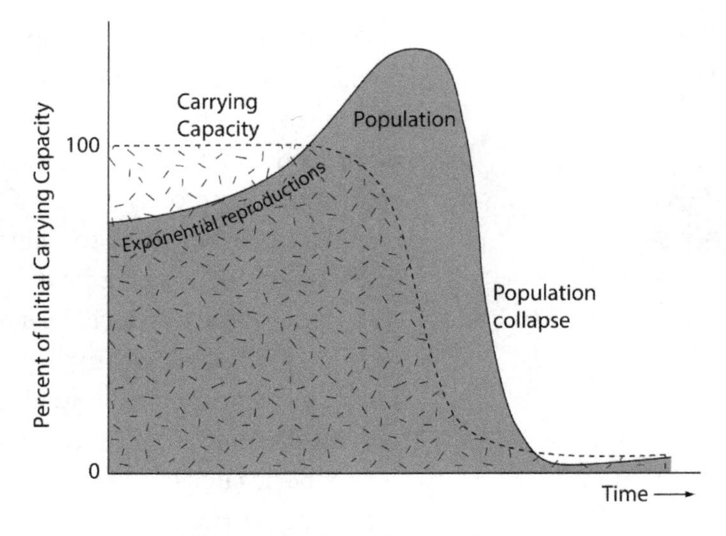

FIGURE 4.8 Population exceeds carrying capacity
A population can exceed its carrying capacity for a while. As long as it is beneath its carrying capacity, it is supported by the rate of provision of environmental goods and services (e.g., food) provided by the ecosystem, much like bearing interest on a bank account. The population can exceed the carrying capacity, at which point it reduces the amount of the ecosystem that provides support, like "eating into the principal" of a bank account. This causes fewer goods and services to be provided, reducing carrying capacity. When the ecosystem is exhausted in what the population needs, the population crashes to a new, much lower (maybe zero) level, according to the rate at which the remnants of the ecosystem can provide the necessary goods and services.

Population by the Numbers

The difference between the number of deaths in a population and the number of births (per year) is the growth rate. At present, the global birth rate is about twenty-one people per thousand (per year), and the death rate is about nine people per thousand (per year). This makes a global growth rate of twelve per thousand per year, or 1.2% per year. While this may not sound like a very large number, it is. In 10 years the population increases by over 13%. In fact, one can easily calculate the time it would take the entire population to double—roughly 70/1.2 = 58 years. At a 3% growth rate, it would double in 23 years.

The history of birth and death rates of a segment of the population determines its age structure. If there are a great many babies born, but life expectancy is short, there is a very young average population. If, on the other hand, there are very few births and long life expectancy, but there were more births in the past, there would be a very old population with declining numbers. If the case of the young population gains medicine, security, and longer life expectancy, it would enter the demographic transition and population would explode. The current population of the world is unevenly distributed, with most in Asia, yet with the greatest rate of growth in Africa. In general, high birth rate is related to poverty (Figure 4.9). Historically, it has been advantageous for old people to be supported by a large number of young people. As our life expectancy increases, with more and older people, an inverted pyramid age structure would lead to a greater burden on younger,

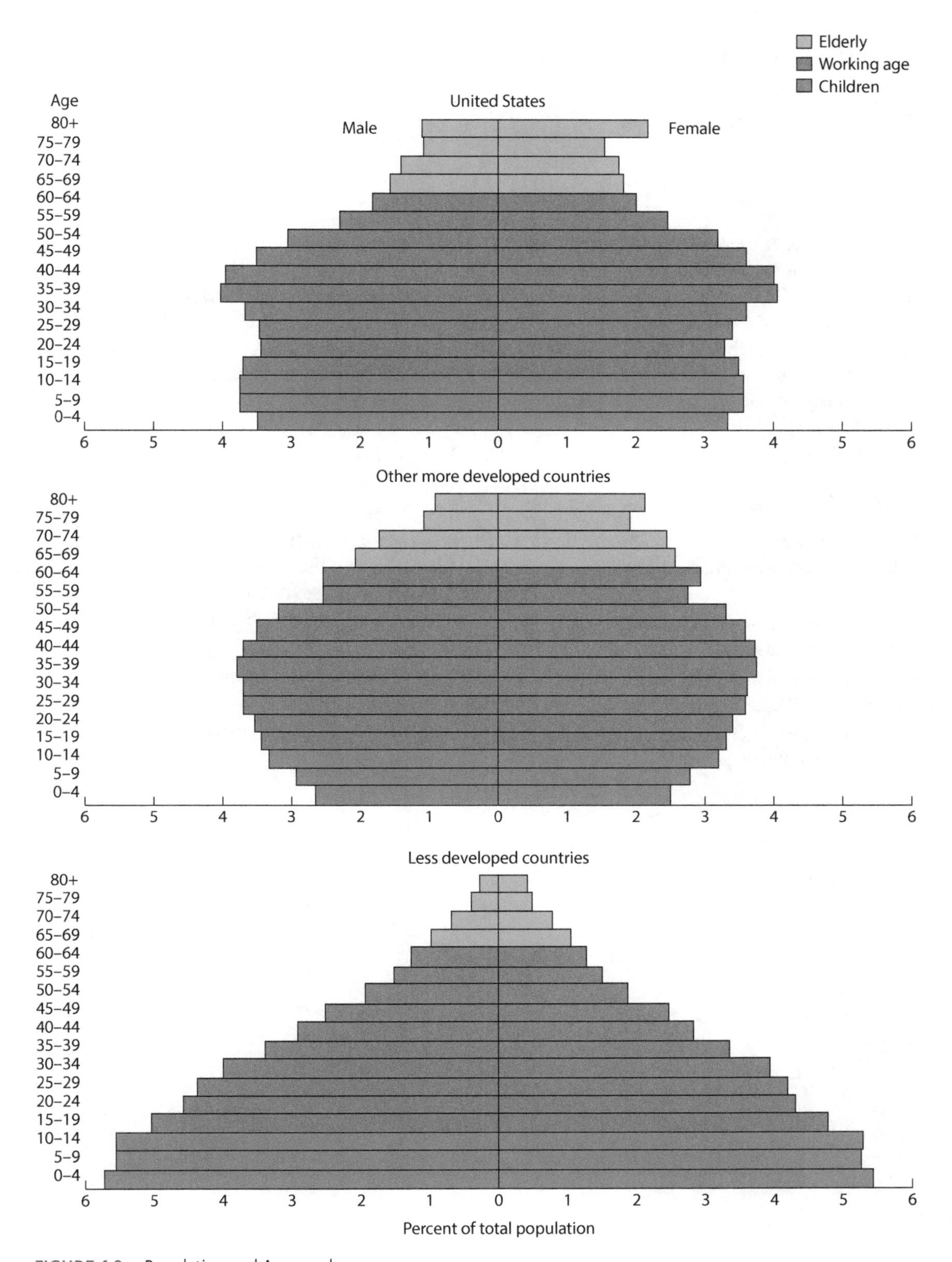

FIGURE 4.9 Population and Age graph

There is a strong correlation between poverty and birth rates. Wealthy nations are generally past the demographic transition, while parents in poorer nations are still unsure of children's survival. The cause and effect between having many children and poverty is not clear, and feedbacks may be involved.

working people to support the aged. This is an inevitable and necessary, yet temporary, price to be paid for stabilizing populations that grew quickly in the previous generation.

References

Bhattacharya, J., & Chakraborty, S. (2017). Contraception and the demographic transition. *Economic Journal*, 127(606), 2263–2301.

Bittencourt, M. (2018). Primary education and fertility rates: Evidence from Southern Africa. *Economics of Transition*, 26(2), 2` 283–302.

Snopkowski, K., Towner, M. C., Shenk, M. K., & Colleran, H. Pathways from education to fertility decline: a multi-site comparative study. (2016). *Philosophical Transactions of the Royal Society B-Biological Sciences*, 371(1692), 20150156

Figure Credits

Fig. 4.1: Copyright © Ojedamd (CC BY-SA 3.0) at https://commons.wikimedia.org/wiki/File:-FamiliaOjeda.JPG.

Fig. 4.2: https://commons.wikimedia.org/wiki/File:Population_curve.svg

Fig. 4.4: https://commons.wikimedia.org/wiki/File:BlankMap-World6.svg

Fig. 4.5: https://commons.wikimedia.org/wiki/File:STS007-03-0058.jpg

Fig. 4.6: Copyright © CSIRO (CC by 3.0) at https://commons.wikimedia.org/wiki/File:CSIRO_ScienceImage_3691_Fishermen_haul_a_catch_of_orange_roughy_aboard_a_fishing_vessel.jpg

Fig. 4.9: http://nationalatlas.gov/articles/people/IMAGES/int-fig5.gif

Fig. 14.4f: Copyright © 2012 Depositphotos/iofoto.

CHAPTER 5

Ecosystems

Life on the Earth is spread far and wide
Plants and animals live side by side.
They make trophic levels.
Ecology revels!
All from energy the sun can provide.

ECOLOGY IS THE study of communities of living organisms and the environment in which they interact, or the "study of the house" by its Greek origin. A **community** is simply a group of organisms that interact with each other (Figure 5.1).

FIGURE 5.1 Community

Ecosystem: Short for "Ecological System." A set of interacting organisms (community) and the physical resources that sustain it, including light, water, nutrients, and other needs that enable the community to function as a unit in marine and terrestrial environments.

Indeed no species can survive in isolation—all need a community in which to thrive. As such, the persistence of life, and evolution of species, has depended on the interactions between species within communities and with the physical environment in which they are immersed. An **ecosystem** consists of both living (biotic) and non-living or physical (abiotic) components. The latter is the physical environment that includes sunlight, temperature precipitation, soil substrate, geomorphology, etc. Some basic components of both the physical and biological parts of ecosystems must be considered in order to understand ecosystem function. First and foremost, and the source of virtually all energy in the world's (near surface) ecosystems, is photosynthesis. (Chemosynthetic organisms obtain energy from chemical compounds rather than sunlight—see discussion below.)

Photosynthesis

Plants (and other organisms such as algae) have evolved a remarkable mechanism for using sunlight to create biomass through photosynthesis (Figure 5.2). Through a complicated set of biochemical reactions involving large molecules, plants achieve a simple net chemical reaction:

$$6CO_2 + 6H_2O + \text{solar energy} \rightarrow C_6H_{12}O_6 + 6O_2$$

How many grams of carbon dioxide would be removed from the atmosphere by a plant to make 180 grams of plant biomass?
Answer: $C_6H_{12}O_6$ has a molecular molar weight of 180 gms ($6 \times C^{12}$; $12xH^1$; $6 \times O^{16}$). It requires 6 C atoms to make one of these, so 6 CO_2 would be required, each with a molecular weight of 44 grams, making a total of 264 grams of CO_2 removed from the atmosphere.

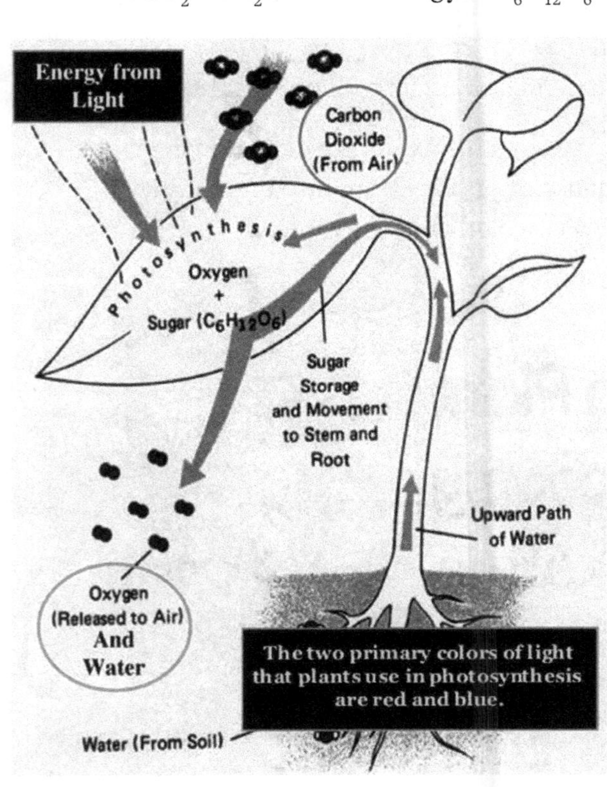

FIGURE 5.2 Photosynthesis
Photosynthesis is the basis for virtually all life on Earth. Energy from the sun (mostly in the blue and red wavelengths) is absorbed by chloroplasts in leaves (for example) and used to combine CO_2 and H_2O to make sugars $C_6H_{12}O_6$ that then go to build plants and perform various biological functions. Photosynthesis and cellular respiration are the inverse of each other.

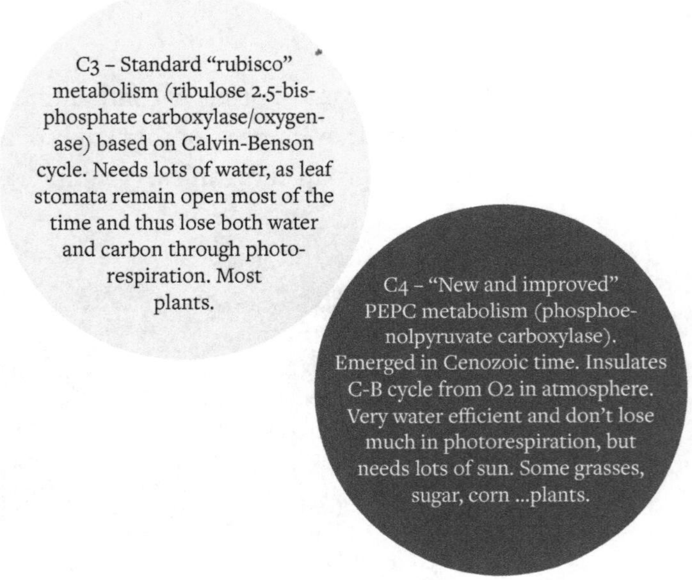

C3 – Standard "rubisco" metabolism (ribulose 2.5-bisphosphate carboxylase/oxygenase) based on Calvin-Benson cycle. Needs lots of water, as leaf stomata remain open most of the time and thus lose both water and carbon through photorespiration. Most plants.

C4 – "New and improved" PEPC metabolism (phosphoenolpyruvate carboxylase). Emerged in Cenozoic time. Insulates C-B cycle from O2 in atmosphere. Very water efficient and don't lose much in photorespiration, but needs lots of sun. Some grasses, sugar, corn ...plants.

FIGURE 5.3 C3 and C4
There are two metabolic pathways plants can take to create sugar from CO_2 and water. The first is the "traditional" C3 "rubisco" metabolism (ribulose 1,5-bisphosphate carboxylase/oxygenase) based on Calvin-Benson cycle (Collatz et al., 1990). This is what originally evolved on Earth a few billion years ago, and most plants do this. This pathway needs lots of water, as leaf stomata remain open most of the time and thus lose both water and carbon through photorespiration. The second pathway (C4) evolved only about 65 million years ago, and adds another carbon atom to the mix through a "new and improved" PEPC metabolism (phosphoenolpyruvate carboxylase). This insulates the Calvin-Benson cycle from O_2 in the atmosphere. It is very water-efficient but needs lots of sunlight. Common C4 plants include many grasses, sugar, corn, and bamboo.

This converts CO_2 and water into sugar and free oxygen. This is fortunate for us animals, as without photosynthesis, there would otherwise be no molecular oxygen in the atmosphere for us to breathe. It would all combine inorganically with something else through oxidation (rusting or burning). In fact, breathing, or respiration of animals, is just the inverse of photosynthesis:

$$C_6H_{12}O_6 + 6O_2 \rightarrow 6CO_2 + 6H_2O + \text{energy}$$

In animals the resulting energy is used for many things, including locomotion.

In the process of photosynthesis, the energy of sunlight incident on chlorophyll in leaves is stored in Adenosine Triphosphate (ATP), the universal energy storage molecule in cells (for animals, too), which is then used to convert CO_2 and water into sugar for the plant to grow. There are two kinds of plants, depending on how they do this—C3 plants use the Calvin cycle that involves a 3-carbon molecule, while C4 plants, evolved to be more water-efficient, use a 4-carbon molecule before providing the CO_2 to the Calvin cycle (Figure 5.3). In either case, the net effect is the same—CO_2 and water are converted to plant biomass. Plants also respire to convert the sugars into energy they use for a number of metabolic processes.

Follow the Energy

In ecosystems, and even individual organisms, the primary currency is energy. The way organisms obtain and use energy, and the ways ecosystems cycle energy through, between, and throughout communities determine overall ecosystem function. Consider energy at the most basic level.

Energy is the ability to do work. From an ecosystem or organism perspective, work involves moving, growing, and all the processes involved in life. There are two basic types of energy—kinetic (energy of motion) and potential (stored energy). Energy can be stored in numerous ways, but there are only three fundamental forms linked to the three fundamental forces of nature. These are gravitational, electrical, and nuclear. (There is actually a fourth, the "weak" force, but we will ignore that for now.) When a mass is lifted against gravity, energy is stored in that the mass can be allowed to sink again while doing work (performing a function). Nuclear energy is the very strong force that binds protons and neutrons in an atomic nucleus, and can store great quantities of energy. Electrical energy is the force between protons and electrons, and electric (and electromagnetic) fields perform a great many functions, including all aspects of chemistry, light and the entire electromagnetic spectrum, and all the electric currents involved in organisms and modern technology. Energy cannot be created or destroyed, but it can be converted from one form to another. (Actually, mass can be converted to energy and back again, but we will not delve into special or general relativity for the purposes of this book.)

Sometimes the different aspects of electrical potential energy are given their own names, such as "chemical energy," which involves the interactions between atoms. When wood or coal is burned, for example, energy is released when atoms are recombined into electrically more strongly bound molecules (e.g., hydrocarbons converted to carbon dioxide and water). When animals eat food, the same process occurs and the animal is able to gain energy from chemically altering the food. "Radiant energy" is the energy carried by electromagnetic waves such as visible light, x-rays, and radio waves. These waves can be absorbed by electrons in atoms, and move the electrons into higher energy levels. This is how the ozone layer of the atmosphere shields the Earth's surface from harmful ultraviolet rays coming from the sun, for example.

"Thermal energy" is actually a form of kinetic energy at the molecular level. When molecules move or oscillate, they carry kinetic energy, which, when aggregated over a material, is called heat. When two substances of different temperatures are put in contact with each other, heat can flow from the higher temperature substance to the lower, causing them to equilibrate at an intermediate temperature.

Each of these forms of energy will be explored further in Chapter 11 in the context of energy sources for modern society, but energy is also the key to ecosystem function, so is explored in that context here.

Autotrophs: Organisms that can produce organic compounds (biomass) from inorganic materials, such as CO_2 and water. They typically use energy from sunlight (photosynthesis) to accomplish this. They use CO_2 and water and produce oxygen and sugars. Chemautotrophs use the energy stored in chemical bonds (such as sulfur compounds) instead of sunlight, typically at deep-sea hydrothermal vents along mid-ocean ridges.

Trophic Levels

The web of life, including the food chain, is all about who eats whom (Figure 5.4). (As humans, we try to stay at the top of the food chain.) Plants are our primary producers, and are at the bottom. This is because they have the unique ability, through photosynthesis, to produce biomass directly from CO_2, water, and sunlight. This makes them **autotrophs**. One of the most important measures of an ecosystem is **net primary productivity** (NPP), completely determined by plants. NPP is the amount of carbon incorporated into biomass by photosynthesis minus the amount of biomass converted back to CO_2 during cellular respiration. Plants and animals both respire to maintain metabolic functions. Animals can spend as long as they like in the sun, and will not gain any energy at all from the experience (although one may get a sunburn ...). Animals have to eat plants to gain the energy they need for survival, and are thus **heterotrophs**. As such, these animals, herbivores (vegetarians), are the second trophic level. Other animals (carnivores) eat herbivores. This is a third trophic level. A few species of terrestrial animals (super-carnivores) even eat other carnivores, placing them at yet a fourth trophic level. By eating biomass, animals gain energy from the food. However, this is a very inefficient process and only 10% of the energy stored in biomass (at any trophic level) is transferred to the animal eating it; 90% is lost. Consequently, in an ecosystem of predators and prey, there needs to be ten times as much prey as there are predators (Figure 5.5). This is

Heterotrophs: Organisms that eat organic materials (plants, animals, fungi, etc.) to produce the energy needed for metabolic function and growth. They consume biomass and produce CO_2 and water through respiration. Animals and Fungi, for instance, are heterotrophs.

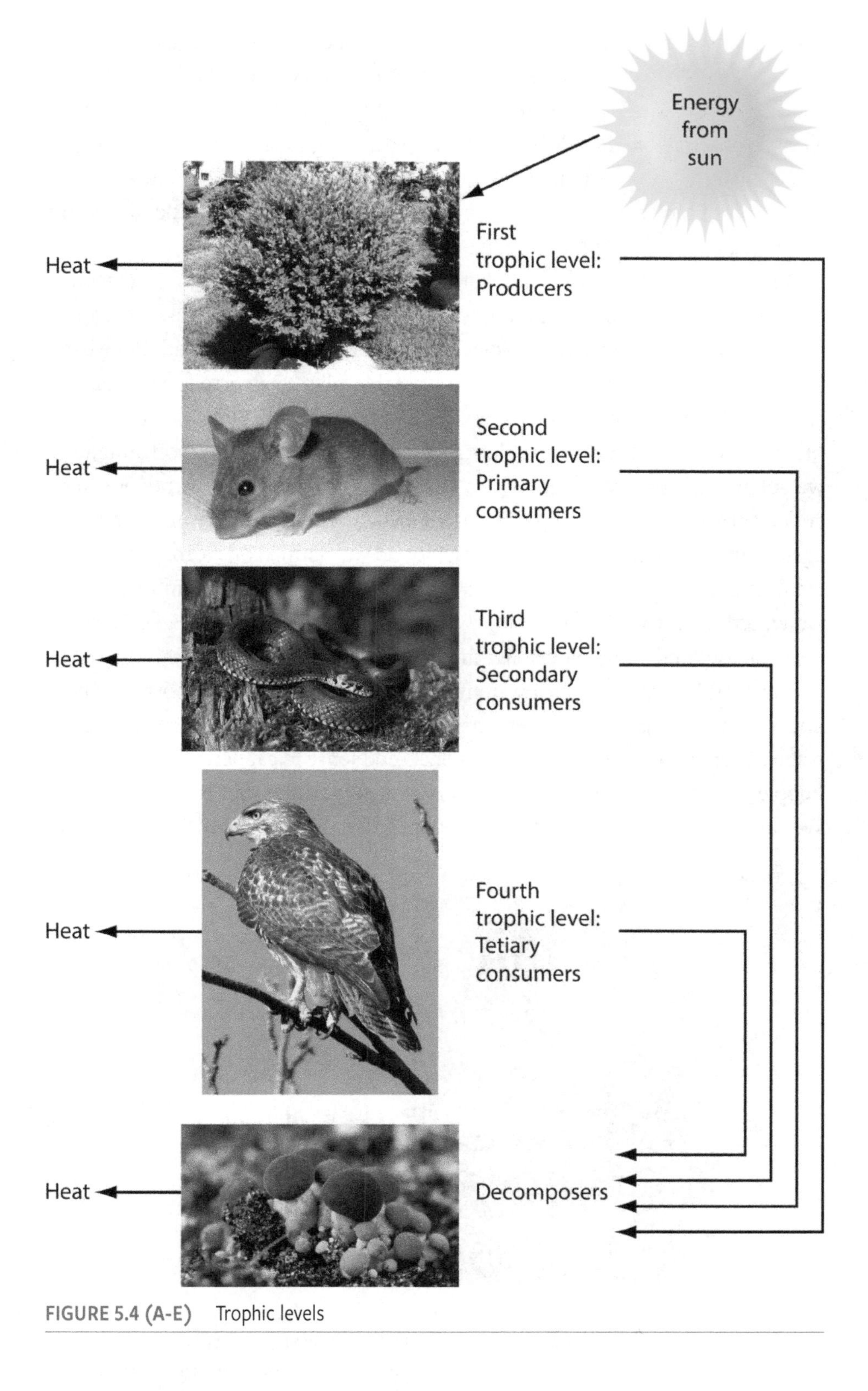

FIGURE 5.4 (A-E) Trophic levels

why in any ecosystem there are lots of plants, some herbivores, a few carnivores, and very few super-carnivores (Figure 5.5).

Once any organism dies, it decomposes by bacterial and fungal action. It could be argued that because decomposers eat everyone in the end, they are the highest trophic level (but eating dead things doesn't count in that sense). They play an important role by eliminating dead organisms from an ecosystem and returning the nutrients back into circulation (Figure 5.6). They require oxygen to operate, and in extreme cases, can deplete an aquatic system in oxygen, causing eutrophication, described in Chapter 7.

While many animals are specialists, eating at a specific **trophic level**, some, such as humans, are more generalists. We can eat plants, herbivores, carnivores, and even super-carnivores. However, consider the energy involved. If we eat a plant, we get about 10% of the solar energy that plant converted to biomass. If we eat an herbivore (e.g., cow), we get 10% of the herbivore, who got 10% of the plant. So we get only 1% of the plant. If we eat a carnivore (e.g., wolf, alligator) we get 10% of that, which leaves us only 0.1% of the original plants. Eating a super-carnivore would bring that down to 0.01%. Thus if we were to plan an agricultural system for feeding humanity, and we planned on eating herbivores, we would need to plant ten times as much plant biomass for them to eat than we would need to plant if we ate the plants ourselves (Figure 5.5). As such, agricultural land can support ten times as many vegetarian humans as it can carnivorous humans. Eating lower on the food chain is thus an important mechanism for reducing environmental degradation caused by agriculture. There are some circumstances in which it makes energetic sense to eat herbivores, however. People cannot eat grass (small children try it sometimes, but it does no good). We can't digest it. Thus, vast areas of global grassland ecosystems do us no good

Trophic level: Where an organism is on the "who eats whom" food chain. Based on how an organism obtains energy needed to sustain its metabolism, trophic levels start with primary producers who typically use photosynthesis to generate organic molecules from inorganic CO_2 and water (e.g., plants) as the lowest trophic level. Herbivores who eat plants are the next, carnivores who eat herbivores are the next, and supercarnivores who eat other carnivores are the next. It is all about what you eat.

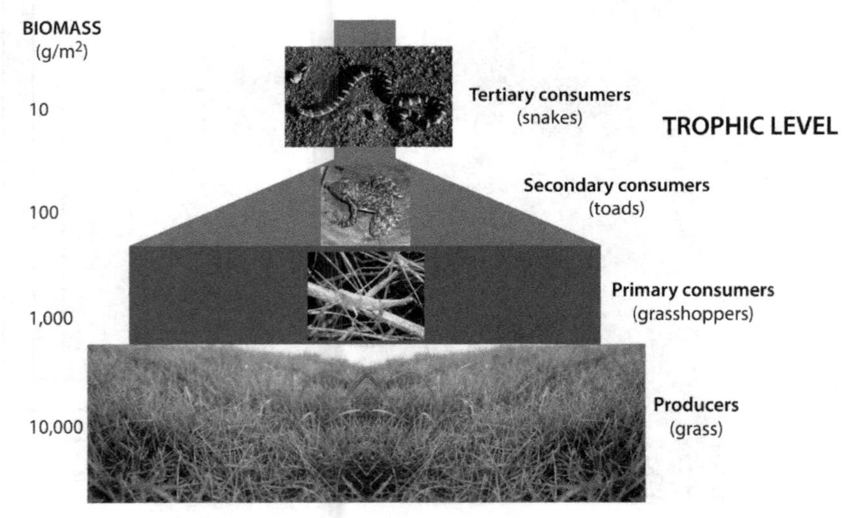

FIGURE 5.5 Energy

Energy derives only from the sun. Plants (and algae) are primary producers that can create biomass from inorganic materials using photosynthesis. Herbivores are primary consumers. Carnivores and supercarnivores follow. Finally, all are consumed by decomposers. Only 10% of the energy transfers from one trophic level to the next.

at all for food. However, cows and other herbivores CAN digest grass, and indeed it is their primary diet, so if we eat the cows who eat grass, we at least get 1% of the energy stored in grass. It's better than nothing. Enter the modern beef industry: If a cow is fed corn instead of grazing on grass, we would have been energetically better off eating the corn ourselves. We would obtain ten times more energy than we can get from the cow who eats the corn.

> If ten acres of farmland can grow 5000 kg of corn, how many kg of corn-fed beef can be produced from that? Answer: 500 kg.

In the ocean, trophic levels are a bit more complicated. As opposed to a mere four levels, there are about eight trophic levels in the marine ecosystem. Again, it all starts with primary producers who use solar energy to convert CO_2 and water into biomass. These are phytoplankton, which float around in the photic zone at the top of the ocean where sunlight reaches their chlorophyll. Herbivores such as zooplankton (e.g., copepods, small shrimp-like crustaceans) eat the phytoplankton, and carnivores (e.g., small fish) eat the zooplankton. Larger fish eat smaller fish, and so on, up the trophic levels. Sometimes very large organisms, such as many whales, eat lower trophic level organisms such as copepods. This is very advantageous due to the need for lots of food for large animals and the abundance of biomass at low trophic levels (recall the 10% rule).

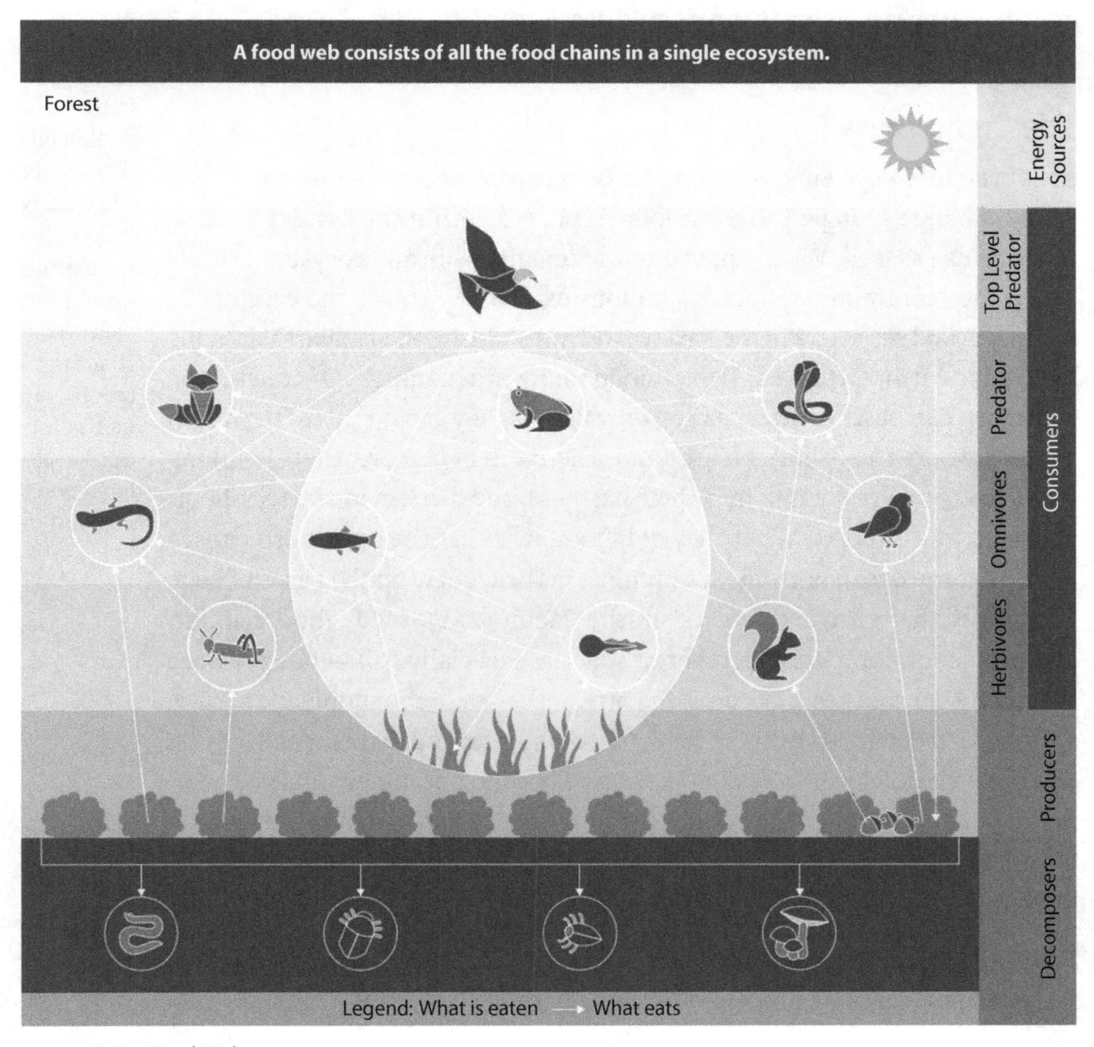

FIGURE 5.6 Food Web

People can eat any marine trophic level. However, we seem to prefer the highest trophic levels, such as swordfish and salmon. This is what there is the least of, and we are depleting fisheries world-wide in these top predators. Consequently, in many cases, we settle for fish from lower trophic levels. This is called **fishing down the food web** (Pauly et al., 2000; Pauly et al., 1998), and while there is more biomass at successively lower trophic levels, ocean fishing is by no means sustainable, and we have already destroyed many of the world's most productive marine ecosystems (and fisheries), such as the Grand Banks off New England.

Fishing down the food web: The practice of catching fish from lower trophic levels, because fishermen have already caught so many higher trophic level fish as to deplete their populations below what is practical for commercial fishing.

In 2013, the maximum allowed catch of haddock and cod, two key market fish, was drastically reduced in an attempt to restore and hopefully preserve this fishery in New England. In order to maintain a diet of top trophic level fish, some fishermen have resorted to mariculture, which is basically farming in the ocean. Salmon, for example, are kept in huge netted enclosures, and are fed until they grow to marketable size. They are fed lower trophic level fish that are caught in the ocean for that purpose, of course, and as a result, rather than eating the lower trophic fish ourselves for 10% of their energy, we eat the salmon and gain only 1% of the energy of the lower trophic fish. This is called **farming up the food web** (Stergiou et al., 2009), and serves to quickly deplete the ocean in fish that we could eat ourselves.

Farming up the food web: The feeding of lower trophic fish to commercially preferable higher trophic level fish in the course of fish farming (aquaculture and mariculture). Because about 90% of the energy is lost between trophic levels, this greatly accelerates the rate of overfishing the ocean, because we could have eaten the fish that are fed to the higher trophic level fish. So, farming up the food web actually accelerates the rate of fishing down the food web.

The Community Effect

As is true in all systems, interactions between components mean that you cannot change a single thing without responses within the system altering other things as well. When applied to the organisms in an ecosystem, this is called the "community effect." A famous example of this is the case of the fishermen and sea otters in the Pacific Northwest. Fishermen make their living by catching fish. Otters eat fish, among many other things. Of course, the fishermen can observe otters eating only fish, as they can do this at the ocean surface, where the fishermen are (and where otters can breathe). Thinking that the otters are cutting into their business, the fishermen started shooting sea otters as pests. What they didn't know is that the otters also eat sea urchins at the ocean bottom. Sea urchins, in turn, gnaw on the bases of kelp, huge algae that grow in "forests" in the Pacific and provide food, habitat, and protection for a great number of species, especially fish—the very ones that the fishermen seek. Without the otters, the sea urchin population grew, and the kelp declined. Without kelp, there was a rapid decline in fish stocks, and the fishermen were left with nothing. They did not understand the community effect, and the important role the sea otters had in making it possible to maintain their fishery. This same community effect applies to all ecosystems in both marine and terrestrial realms. Consequently, it is critical to consider the roles of each species before hunting, harvesting, or otherwise disturbing the population structure and balance within the communities in an ecosystem.

Net Primary Productivity (NPP)

Net Primary Productivity (NPP): The rate or flux of carbon (in the form of CO_2) drawn out of the atmosphere and converted to biomass (e.g., simple sugars) by photosynthesis in plants, after subtraction of the nighttime respiration that "burns" these sugars during plant metabolic processes and growth.

Gross Primary Productivity (GPP): The rate or flux of carbon (in the form of CO_2) drawn out of the atmosphere and converted to biomass (e.g., simple sugars) by photosynthesis in plants.

Net Ecosystem Productivity (NEP): NPP minus soil respiration from the decomposition of dead organic matter in the soil.

Net Ecosystem Exchange (NEE): NEP minus the loss of carbon from soil by leaching, removal, and transport out of the system by moving water.

Net primary productivity (NPP) is the rate at which plants convert CO_2 into biomass. The productivity of an ecosystem is controlled by the **autotrophs**, primarily plants and algae. By day, they engage in photosynthesis, incorporating carbon, but they respire all the time, including at night, thus using up about half of the sugar biomass made by photosynthesis. As such, net primary productivity is **gross primary productivity** (just photosynthesis) minus respiration (NPP = GPP—R). This sometimes small net amount is the amount that the plant grows. When applied to an entire ecosystem, it reflects the rate at which carbon is sequestered in the living biomass of the ecosystem. Different ecosystems have different average NPP (Figure 5.7). **Net ecosystem productivity** (NEP) includes the effect of decomposition of dead organic matter (usually in leaf litter and in the soil), and is invariably less than NPP. Finally, **net ecosystem exchange** (NEE) includes the effect of leaching and water-borne removal of carbon into and out of the ecosystem in question. Each of these entities is a critical consideration in determining carbon budgets of an ecosystem as well as the global carbon cycle.

Climate determines the nature of any ecosystem, and thus land cover and NPP. However, humans also determine the nature of ecosystems, as we have altered and in most cases completely removed ecosystems in favor of agriculture. As such, we have appropriated much of the planet's NPP for human consumption (Krausmann et al., 2013). By two independent calculations,

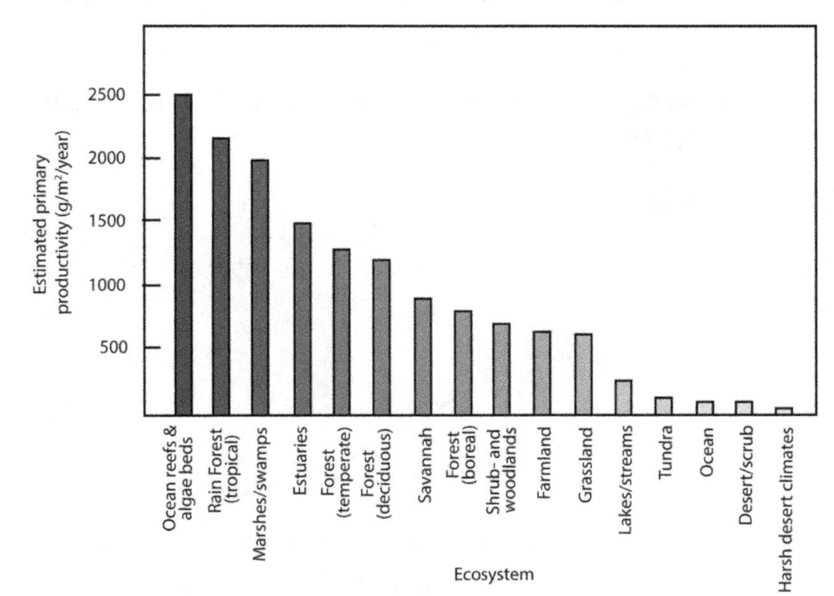

FIGURE 5.7 Typical NPP for various ecosystems, measured as grams of carbon per square meter per year. Note how NPP generally correlates to water availability.

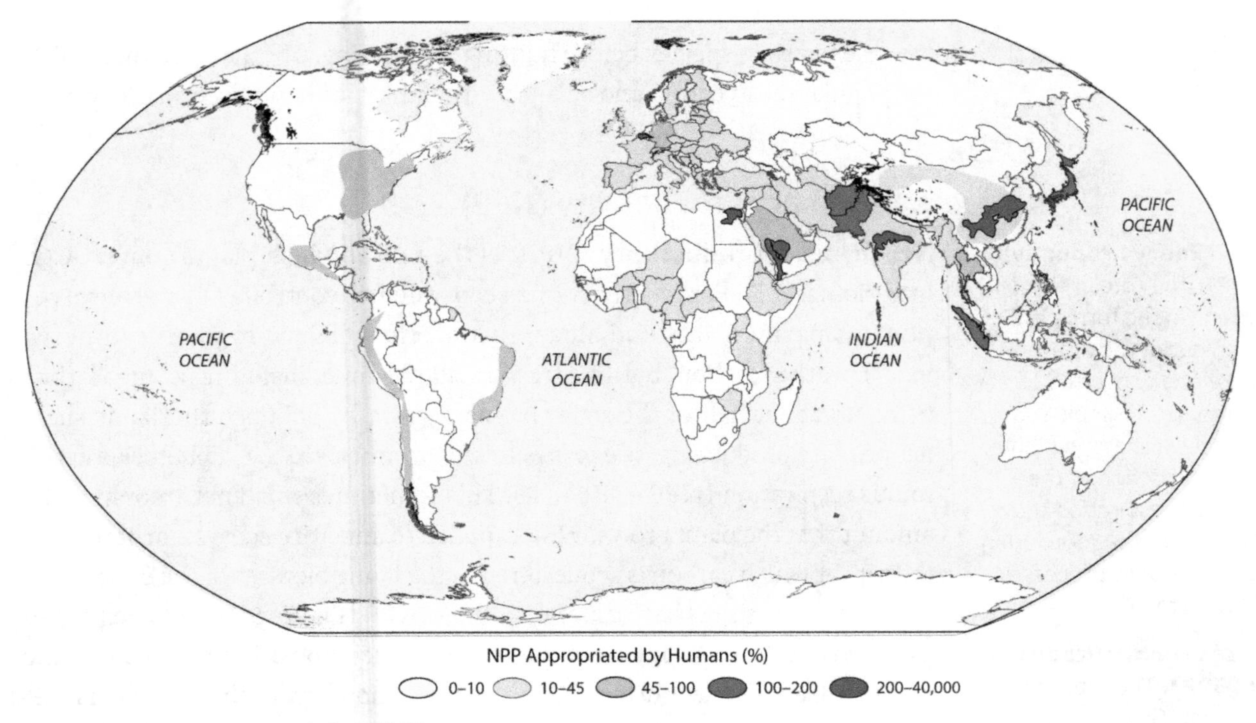

FIGURE 5.8 Human appropriation of NPP
Appropriation of global Net Primary Productivity by people is the percentage of potential NPP of the natural ecosystem used for food, wood, and other purposes (Krausmann et al., 2013). Note that there are areas in which human appropriate far MORE than 100% of NPP. This is due to irrigation of food crops in deserts and other areas of otherwise low NPP.

humans have appropriated about 32%, or a third of global NPP (Figure 5.8). In some regions, humans appropriate MORE NPP than the ecosystem provides. This is done through irrigated agriculture in otherwise less-productive areas, such as northern India and Saudi Arabia.

TABLE 5.1 LINNAEAN CLASSIFICATION OF LIFE (WITH EXAMPLE OF HUMANS). THE KINGDOMS ARE THEMSELVES GROUPED INTO DOMAINS (FIGURE 5.9).

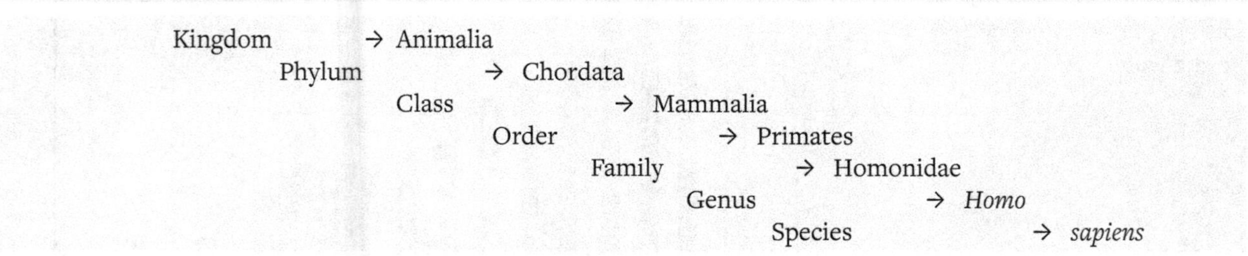

Kingdom	→ Animalia
Phylum	→ Chordata
Class	→ Mammalia
Order	→ Primates
Family	→ Homonidae
Genus	→ *Homo*
Species	→ *sapiens*

Biodiversity

Biodiversity is defined as all hereditarily based variation at all levels of organization (Table 5.1). This practically amounts to the number and variety of species, genera, and all forms of life (Figure 5.9) that constitute a community within an ecosystem. These levels can be genes within a single species (consider people or dogs), different species within a community, and differences between communities within a broader ecosystem. The usual measure of diversity is the number of different species within a community, or within a part of the ecosystem of concern. There has been a great deal of discussion about preserving biodiversity, with considerably less being done to actually preserve it. (As is so often the case, when all is said and done, a lot more has been said than done.) The number of species in a community depends on a number of factors such as climate, soil, topography, and general geography. At the global level, major groups of communities, as characterized by their predominant plants, constitute the world's **biomes.**

Biomes: Major groups of communities at the global level. Examples are deserts, grasslands, and tundra. Various researchers have made different global definitions and lists of biomes, each grouping the world's communities in different ways with finer or courser resolution. There is no single accepted list of the world's biomes.

In order to determine how many species there are, one needs to first define "species," which can be done in terms of a biological concept, in which a species is defined as one that can interbreed freely under natural conditions, and produce fertile offspring. In some cases, such as the geologic record, there is no knowing which dinosaurs could breed with which, so a morphological definition is used, in which organisms that looked more like each other (in fossils, usually) are defined as a species. Using modern techniques, one could define species on the basis of genetic similarities, but this is yet to be done effectively or comprehensively.

DOMAIN ARCHAEA Kingdom Archaea	DOMAIN BACTERIA Kingdom Bacteria	DOMAIN EUKARYA Kingdom Protista	DOMAIN EUKARYA Kingdom Fungi	DOMAIN EUKARYA Kingdom Plantae	DOMAIN EUKARYA Kingdom Animalia
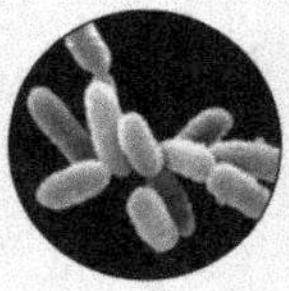	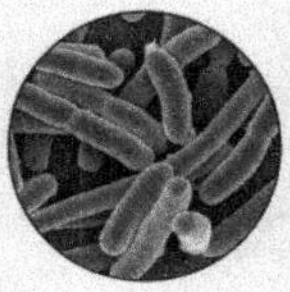	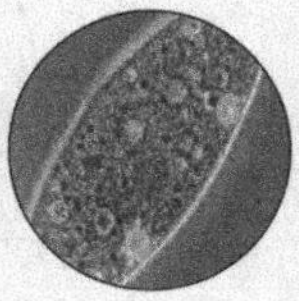			
Frequently live in oxygen-deficient environments; often adapted to harsh conditions, such as hot spring, salt ponds, and hydrothemal vents in deep ocean floor.	All other prokaryotes; thousands of species; most are decomposers, some are parasites; some cause diease; some are photosynthetic; important in biogeochemical cycles.	Eukaryotes that are unicellular or relatively simple multicellular organisms, such as algae, protozoa, slime molds, and water molds; important in aquatic food chains; algae are important producer.	Most are complex milticellular eukaryhotes that secrete digestive enzymes into their food and then absorb the predigested nutrients; decomposers; some are parasites; some cause disease.	Complex milticellular eukaryotes; most use radiant energy to manufacture food molecules by photosynthesis; play important role as producers and as source of atmospheric oxygen.	Complex milticellular eukaryotes that ingest their food and then digest it inside their bodies; consumers-herbivores, carnivores, omnivores, and detritivores.

FIGURE 5.9 Kingdoms

The six kingdoms of life on Earth, grouped in three domains. Protista (all eukaryotes that are not plants, animals, or fungi) are just a convenient grouping that is actually divided between five supergroups of eukaryotes, one of which also includes animals and fungi, and another of which includes plants.

The number of species may not fully account for the full functions of diversity because a community could be diverse, but not even. Evenness relates to the relative proportions of the numbers of the species within the community. Some communities are dominated by a single species, although they contain many others in very small numbers. As such, whatever the dominant species does will be done a lot, and this may be sustainable or not.

Each species in a community occupies a specific **ecological niche**. A niche is determined by the food available to eat and the places available to live and reproduce. Two species that have the exact same needs for food and habitat cannot coexist, as one will outcompete the other. This is the basis of the **competitive exclusion principle**. Some species occupy very narrow niches, such as the koalas, that eat almost only eucalyptus leaves. In part, this is because this relatively inactive mammal (sleeping most of the day) is resistant to the toxins in the leaves, so has few other species to compete with for this source of food. Other species are generalists, such as humans, and cockroaches even more so, which can eat a wide variety of foods and obtain their needed nutrition. They occupy broader niches, and partially compete in various ways with a wider variety of species than does the koala.

Ecological niche: The occupation and living needs of a species. It is the specific environmental needs, including food, shelter, conditions for reproduction, and protection from predation. Only one species can occupy each ecological niche.

Competitive exclusion principle: Two species cannot occupy the exact same ecological niche.

If a niche becomes available, it will not take long (geologically) before an organism evolves to fill it. They do this by altering their habits as well as their morphology by the usual evolutionary processes of genetic mutation (or genetic drift, more generally) and natural selection. Because of this, a single species can split into several, taking different evolutionary paths driven by the variability of the niches available. This is called diversification, and has occurred at various times throughout geological history, with major radiations occurring in the wake of mass extinctions such as after the Permian or Cretaceous periods (Figure 5.10). A famous example of more modern diversification is Darwin's finches, which migrated to an island and took advantage of the numerous separate niches available there by evolving differing beaks to draw nectar from different kinds of flowers, or at the other extreme, crack seeds and nuts.

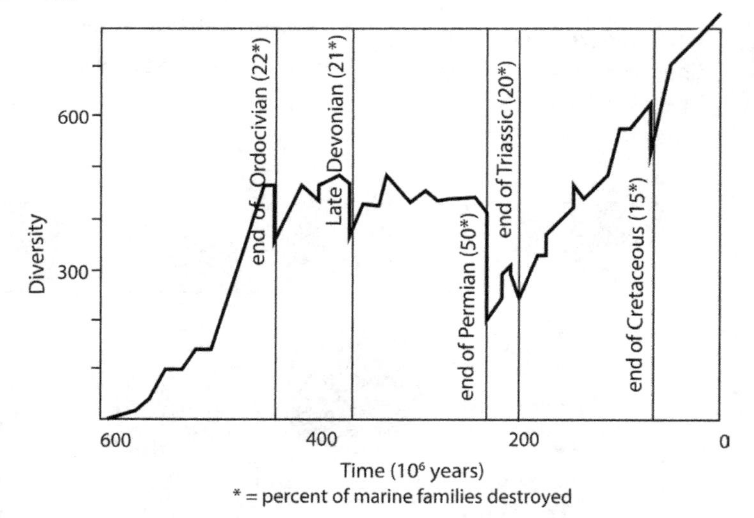

FIGURE 5.10 History of global marine biodiversity
History of global marine biodiversity at the family level. In order for a family to become extinct, all the genera in the family must become extinct, and in order for a genus to become extinct, all of the species within it must become extinct. Note the major mass extinctions at the end of the Permian and end of the Cretaceous. Despite numerous mass extinctions, overall biodiversity has increased over geologic time. This means that ecological niches have narrowed, increasing specialization. This specialization makes more efficient use of resources and can increase NPP, but would cause individual species to become more vulnerable to environmental changes. However, narrow ecological niches also means that once a species is gone, species in adjacent niches are so close in behavior and needs that they may find it easier to evolve to fill the empty niche than if there was lower diversity with broader niches. An example of this in the modern day is the growth of coyotes in the absence of wolves in much of North America (Fener et al., 2005; Otis et al., 2017). We may see more of this during the sixth mass extinction now being caused by anthropogenic land use and emissions.

New niches form as a result of climate change, extinction of a particular food or predator, and more recently, human activities. In some environments, there are relatively few niches. Examples are the extreme environments of glaciers and hot springs. These niches are filled by a few species that are able to take advantage of all of what little these environments have to offer. Other environments, most notably tropical rain forests, have a great many niches, with so much biological productivity (fundamentally driven by abundant sunlight and enabled by copious water) that species can become very specialized and occupy a very narrow niche, leaving room for many other species with only minor differences in what they eat or do. (It is important to note that diversity applies to all species, regardless of kingdom—plant, animal, fungi, or otherwise.) The ecosystems that have evolved the narrowest niches and greatest biodiversity are those that occur in the most stable environments. These typically occur in the tropics in rain forests and coral reefs, where there is sufficient time of stability for organisms to specialize to fill increasingly narrow niches.

Another factor that controls the number of niches is the size of the area available to a community. In the case of islands, the terrestrial ecosystem is clearly limited such that the larger the island, the greater the diversity. More niches will be available on an island with a mountain as well as coastal plain than on a flat, featureless island such as a coral atoll, for example. This is because the variations in climate caused by the mountain (colder at higher altitudes, and drier in the lee of prevailing winds) offer a broader variety of conditions under which to grow, and thus a greater variety of niches to be filled.

An important question regarding biodiversity is how much biodiversity there is on Earth. The answer is that we do not really know. There are some preliminary guesses as to the number of species of organisms, and about 1.7 million have been described to date, but these are skewed toward the birds and mammals that we can readily observe, especially in North America and Western Europe, where naturalists have been cataloging them for centuries (Larsen et al., 2017). On the basis of likely diversity of niches in the ecosystems of the world, estimates of true global diversity fall into the range of four to 40 million species, with a most likely value of about million. It is unlikely that the majority of these will be described and cataloged, as the niches they occupy are being destroyed at an alarming rate through deforestation, agriculture, invasion by foreign species moved by human activities (such as the zebra mussel, kudzu, or purple loosestrife), and climate change. Further, the number of bacterial species is largely unknown, and may well dominate the number of species globally.

Succession

An ecosystem is not a static entity, but one that evolves over time. All ecosystems are disturbed by natural (and anthropogenic) factors throughout their existence. Fire, lightning, storms, floods, insect invasions, and of course, chainsaws and plows can disturb or completely eliminate a fully functioning ecosystem. After such a disturbance, the ecosystem recovers, but it does not start out by replacing the same community of organisms that it lost. In general, they are not suited to colonizing bare ground, for example. A different set of organisms called colonizers include a group of "pioneer species" that make the first inroads into a formerly occupied patch of land (Figure 5.11).

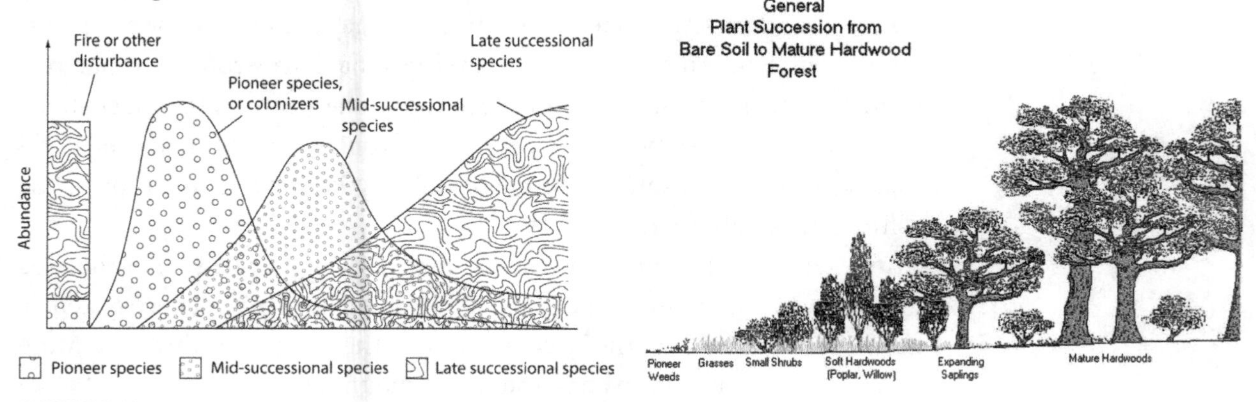

FIGURE 5.11

After disturbance, rapidly germinating and growing pioneer species colonize a landscape, fix nitrogen and prepare the soil with carbon and other nutrients, facilitating succession of later stages. Taller, longer-lived plants (e.g., trees) overshadow and overtake the pioneer species, to be subsequently overtaken by the late successional plants (e.g., tall hardwoods). The late successional plants deplete nutrients from the soil, making themselves increasingly vulnerable to future disturbance, after which the cycle begins anew.

If there has not been an ecosystem in a newly available part of the land surface, perhaps because a glacier retreated, or a large parking lot was removed, **primary succession** commences where there was no ecosystem before. If a functioning ecosystem is disturbed by fire or chainsaw, for instance, and is immediately able to recover, **secondary succession** occurs (Figure 5.12).

Primary succession: The establishment of an ecosystem where there was no ecosystem before.

Secondary succession: The reestablishment of an ecosystem that has been recently destroyed by natural or anthropogenic processes, such as fire, land use, or other disturbances. Designation as "Primary" and "Secondary" depends on what was there before—otherwise the stages of succession are the same for both.

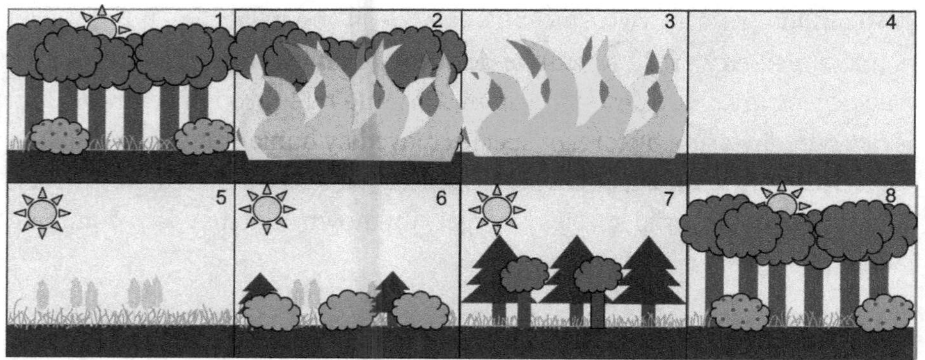

FIGURE 5.12 Secondary succession

In both cases, early successional species pave the way for later successional species. (Do not confuse primary and secondary with early and late!) Early successional plants are generally quick to germinate and grow, but short-lived. Depending on climate, soil, and other environmental conditions, they could include grasses, small shrubs, or other small plants. Early successional plants serve some very important functions. They cover the soil, providing shade that moderates temperature of the surface of the soil, and help retain moisture that can help seeds of larger, longer-lived plants to germinate and grow. In addition, early successional plants serve to add nitrogen and other nutrients to the soil as well as carbon, making it better suited for later successional species that need more fertile soil to grow. Once later successional species take root and grow, the shade they create eliminates the early successional species. Still later, even longer-lived and larger plants can move in. These stages of succession are a natural and critical function and enables ecosystems to survive many disturbances and bounce back on a regular basis.

During the stages of succession, biomass and species diversity increase up to a steady state that is maintained until the next disturbance. Nutrients in the soil and organic matter in leaf litter and soil carbon also increase until the latest stages of succession. The late-stage plants use up nutrients and they begin to decline in, for instance, an old forest. As the ecosystem ages, the plants become increasingly stressed in nutrients and vulnerable to disturbance such as fire, pests, or storms. When a disturbance occurs, the successional cycle begins anew with pioneer species. In most ecosystems, successional cycles are very variable in length, as they depend on the changing relation of vulnerability and disturbance severity. Periods of decades to centuries are not uncommon.

Facilitation: The process of soil alteration (chemistry, shade, moisture) by early successional species to provide nutrients and physical environment needed by later successional species.

Interference: The process of forming impenetrable ground cover or other means to prevent later successional species from taking root.

The stages of succession described so far are based on early successional species facilitating the establishment of later successional species. **Facilitators** accomplish this by altering soil chemistry and providing shade and moisture for seed germination. However, in some ecosystems, early successional species are less welcoming. If early successional species prevent later successional species from moving in, **interference** occurs. Some grasses make a dense mat on top of the soil so that seeds from other species cannot reach the ground and needed moisture to germinate. This enables the grasses to persist for long periods, until some disturbance (fire, disease, animals) creates a break in the continuous grass cover, providing the opportunity for later successional seeds to germinate and grow, casting shade on the grasses, which then weaken and die out, giving later successional species more opportunities. Once a break in the grassy mat occurs, the establishment of late successional species occur with a positive feedback. The more breaks, the more tall late successional plants, the more shade, the more breaks ...

Interactions between Organisms

Symbiosis: A relationship between species in which each depends on the other for critical life functions.

Within a community, organisms can interact in a number of ways. Some are **symbiotic**, which means that two species rely on each other through their life cycles. For example, bees and flowering plants evolved together and depend on each other. The bees collect pollen to make nectar, and in doing so, pollinate the plants in their reproductive cycle. Neither could live without the other.

In some cases, two species help each other, but do not require each other. This is called **commensalism** and an example is the fungi (mycorrhizae) that are associated with most plant roots (Figure 5.13). The fungi provide phosphorus and other nutrients for the plant roots, while the roots give back to the fungi biomass produced by photosynthesis. (Actually the plants could struggle along without the fungi, but the fungi would not last long without the plant roots.) Egrets commonly stand on the backs of water buffalo and pick lice, ticks, and other insects from the backs of the large animals. Both benefit, as one gets food and protection, while the other gets rid of bothersome bugs. In cases of commensalism, the organisms could live without each other, but live better in each other's company.

Commensalism: A relationship between species in which each helps the other in ways that are not immediately critical for survival.

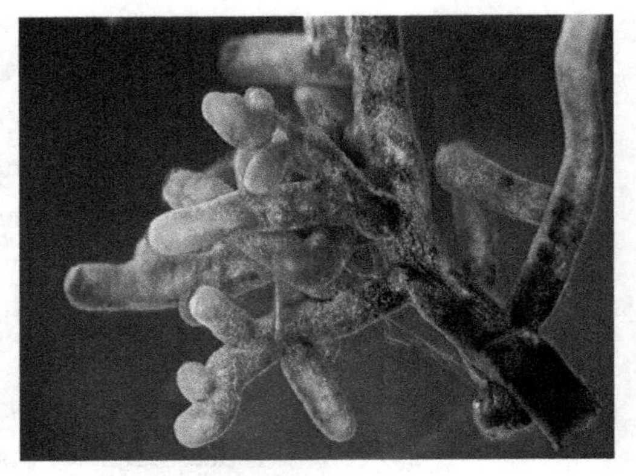

FIGURE 5.13 Mycorrhiza
Mycorrhizae are fungi that attach to plant roots and provide to the plant water and nutrients obtained from the soil in far greater quantities than the plant's roots could accomplish alone. In return, the fungi obtain energy from the plant for their own metabolism.

In many species' interactions, one depends on the other but does nothing in return. This is **parasitism**, where the parasite lives off its host and may weaken it a bit or cost it some energy, but doesn't usually kill it.

Predation, on the other hand, is a case in which the predator kills and eats its prey in order to survive. This is the foundation of the various trophic levels in terrestrial and marine ecosystems. Prey animals have evolved various and seemingly imaginative defense mechanisms, including toxic poisons (monarch butterflies), irritating odors (skunks), speed (rabbits), and many others. Plants also have defense mechanisms, and poison is the most common (milkweed). In the case of milkweed, monarch butterflies coevolved to tolerate the poison, and by eating milkweed, the butterflies become poisonous to their own would-be predators. Other defenses include thorns (roses), waxy leaf surfaces (holly), mimicry (passion flowers), and many others.

Parasitism: A relationship between species in which one lives off the energy from the other but does not kill the host (e.g., mosquito, tapeworm, etc.).

Predation: A relationship between species in which one organism eats another (killing it in the process).

Ecosystem Restoration

In response to ecosystem loss due to land mismanagement, agriculture, and other anthropogenic disturbances, restoration efforts are being made for numerous ecosystems (Wagner et al., 2016; Rosenfield et al., 2017). Most notable in the U.S. are forests for their carbon cycling and sequestration, wetlands for their hydrologic functions as well as key biogeochemical cycling, and prairies for the vast areas of grassland habitat they once represented. The question is—"What should be restored?" For that matter, "What is natural?" Given the process of succession, there is no sense trying to plant seedlings of late successional species on bare ground. They will not grow. Knowledge of the appropriate communities at each stage of succession enables restoration workers to plant the appropriate early successional species that will facilitate the stages of succession (depending on their objectives), after which the ecosystem will take care of itself.

The key to successful restoration is to ensure that critical ecosystem functions are put into place and set into motion so that no further intervention is necessary. These functions are different in detail for every ecosystem. Some of the most critical functions include water and nutrient cycling, pollination, seed dispersal and germination, appropriate balance of trophic levels, soil development and maintenance, and appropriate and sustainable interactions between the myriad species (of all kingdoms) that populate the community.

It is often difficult to assess the success of restoration efforts because some processes are not easily monitored, and overall sustainability can only be demonstrated over decades to centuries. In fact, it is sometimes not obvious to the ecologist (much less the general public) what the goals of restoration should be (Table 5.2). One could attempt to bring an ecosystem back to the condition it was in prior to the industrial revolution. Alternatively, one could go back to pre-agricultural conditions, or even pre-human. One could attempt to maximize NPP, or biodiversity, or standing biomass. Depending on the goal, different criteria would be used to measure success. In any event, no goal will be reached without restoration of the key functions by which an ecosystem operates, so these are always the top priority.

The ecosystem of a large agricultural region with degraded soil is being restored. What factors should be considered in determining the distribution of species that should be considered in that restoration?
Answer: Soil chemistry, climate and climate change, invasive species, extinct species, disease (pestilence, blight, etc.), keystone and engineer species, future land use, and many other factors.

TABLE 5.2 RESTORATION OBJECTIVES.

Objectives and Approaches
1. Pre-Industrial—Ecosystems maintained as in AD 1500
2. Presettlement—Ecosystems maintained as in AD 1492
3. Preagriculture—Ecosystems maintained as in 5000 BC
4. Preserve an endangered species—Whatever stage is appropriate to animal
5. Range of variation in history—Future mirrors the known past

References

Collatz, G. J., Berry, J. A., Farquhar, G. D., & Pierce, J. (1990). The relationship between the Rubisco reaction-mechanism and models of photosynthesis. *Plant Cell and Environment*, 13(3), 219–225.

Fener, H. M., Ginsberg, J. R., Sanderson, E., & Gompper, M. E. (2005). Chronology of range expansion of the coyote Canis latrans in New York. *Canadian Field Nationalist*, 119(1), 1–5.

Krausmann, F., Karl-Heinz, E., Gingrich, S., Haberl, H., Bondeau, A., Gaube, V., & . . . Searchinger, T. D. (2013). Global human appropriation of net primary production doubled in the 20th century. *Proceedings of the National Academies of Science of the United States of America*, 110(25), 10324–10329.

Larsen, B. B., Miller, E. C., Rhodes, M. K., & Wiens, J. J. (2017). Inordinate fondness multiplied and redistributed: the number of species on earth and the new pie of life. *Quarterly Review of Biology*, 92(3), 229¬–265.

Otis, J., Thornton, D., Rutledge, L., & Murray, D. (2017). Ecological niche differentiation across a wolf-coyote hybrid zone in eastern North America. *Diversity and Distributions*, 23(5), 529–549.

Pauly, D., Christensen, V., Froese, R., & Palomores, L. M. (2000). *Fishing down aquatic food webs. American Scientist*, 88(1), 46–51.

Pauly, D., Christensen, V., Dalsgaard, J., Froese, R., & Torres, F., Jr. (1998). Fishing down marine food webs. *Science*, 279(5353), 860–863.

Rosenfield, M. F., & Mueller, S. C. (2017). Predicting restored communities based on reference ecosystems using a trait-based approach. Forest Ecology and Management, 391, 176–183.

Stergiou, K. I., Tsikliras, A. C., & Pauly, D. (2009). Farming up Mediterranean food webs. *Conservation Biology*, 23(1), 230–232.

Wagner, A. M., Larson, D. L., DalSoglio, J. A., Harris, J. A., Laubus, P., Rosi-Marshall, E. J., & Skrabis, K. E. (2016). A framework for establishing restoration goals for contaminated ecosystems. *Integrated Environmental Assessment and Management*, 12(2), 264–272.

Figure Credits

Fig. 5.1: Copyright © Bernard DUPONT (CC BY-SA 2.0) at https://commons.wikimedia.org/wiki/File:Roan_Antilopes_with_Buffaloes,_Zebras_and_one_Eland_at_water_hole_..._(30033578573).jpg.

Fig. 5.2: http://www.nasa.gov/centers/goddard/news/topstory/2007/spectrum_plants.html

Fig. 5.4a: Copyright © Jerzy Opiola (CC BY-SA 3.0) at: https://commons.wikimedia.org/wiki/File:Buxus_sempervirens_2.jpg.

Fig. 5.4b: Copyright © Roger McLassus (CC BY-SA 3.0) at: https://commons.wikimedia.org/wiki/File:Mouse-19-Dec-2004.jpg.

Fig. 5.4c: Copyright © Marek Szczepanek (CC BY-SA 3.0) at: http://commons.wikimedia.org/wiki/File:Natrix_natrix_(Marek_Szczepanek).jpg.

Fig. 5.4d: Copyright © Dori (CC BY-SA 3.0 US) at: https://commons.wikimedia.org/wiki/File:-Hawk_3713.jpg.

Fig. 5.4e: Copyright © Przykuta (CC BY-SA 3.0) at: https://commons.wikimedia.org/wiki/File:Xerocomus_chrysenteron_7840.jpg.

Fig. 5.5a: Copyright © W. A. Djatmiko (CC BY-SA 3.0) at: https://commons.wikimedia.org/wiki/File:-Cyl_ruffus_061212_2021_tdp.jpg.

Fig. 5.5b: Copyright © L. Shyamal (CC BY 2.5) at: http://commons.wikimedia.org/wiki/File:Micrixalus.jpg.

Fig. 5.5c: http://en.wikipedia.org/wiki/File:Grasshopper_%2827%29.JPG

Fig. 5.5d: Copyright © Sugeesh (CC BY-SA 3.0) at: https://commons.wikimedia.org/wiki/File:Carpet_Grass.JPG.

Fig. 5.6: Copyright © Roberta Rosina, DensityDesign Research Lab (CC BY-SA 4.0) at https://commons.wikimedia.org/wiki/File:Food_Web.svg.

Fig. 5.8: https://commons.wikimedia.org/wiki/File:BlankMap-World6.svg

Fig. 5.9a: http://en.wikipedia.org/wiki/File:Halobacteria.jpg

Fig. 5.9b: http://en.wikipedia.org/wiki/File:EscherichiaColi_NIAID.jpg

Fig. 5.9c: Copyright © Kankitsurui (CC BY-SA 3.0) at: http://en.wikipedia.org/wiki/File:Protist_collage.jpg.

Fig. 5.9d: Copyright © Liz West (CC BY 2.0) at: https://commons.wikimedia.org/wiki/File:Baby_mushrooms_.jpg.

Fig. 5.9e: http://en.wikipedia.org/wiki/File:Potato_plant.jpg

Fig. 5.9f: https://commons.wikimedia.org/wiki/File:NSRW_Butterfly.png

Fig. 5.11a: Adapted from Copyright © Jai Li (CC BY-SA 3.0) at: https://commons.wikimedia.org/wiki/File:Fire_ecology_succession.svg.

Fig. 5.11b: http://www.calrecycle.ca.gov/SWFacilities/Closure/Revegetate/Images/successn.gif

Fig. 5.12: Copyright © Katelyn Murphy (CC BY-SA 3.0) at https://commons.wikimedia.org/wiki/File:Secondary_Succession.png.

Fig. 5.13: Copyright © Ellen Larsson (CC by 2.5) at https://commons.wikimedia.org/wiki/File:Mycorrhizal_root_tips_(amanita).jpg.

CHAPTER 6

Global Biogeochemical Cycles

The elements are simply C, N, and P
That serve to set life on a nutrient spree.
It's these that we cherish.
Without them we perish,
For they are for all life the chemical key.

A **BIOGEOCHEMICAL CYCLE** IS the movement of a particular element through linked biological and geological systems. Certain elements that play a key role in biology, such as carbon, nitrogen, sulfur, phosphorus, and a host of others, are cycled through plants, animals, soils, rivers and the ocean, the atmosphere, and rocks, in a continuing set of chemical reactions and transfers. Of particular interest to environmental science is the way key elements cycle through ecosystems. Plants can incorporate these elements from the air, water, and soil; animals get them by eating the plants; decay of the plants returns them to the soil, atmosphere, and water, where they are exchanged between the ocean, land, and atmosphere; and then cycled again.

Some elements, such as calcium, are a major component in rocks, and do not form gaseous molecules. Because of this, calcium cycles slowly through ecosystems. Other elements, such as carbon or sulfur, form gases when oxidized, and can rapidly enter and leave plants and ecosystems in general. Plant growth in most ecosystems is limited by any one of the following: water, sunlight, temperature, or one of the critical nutrients that support life by cycling through ecosystems. When a limiting factor is added either naturally or artificially, it is called **fertilization**. This is commonly done on grassy lawns, for example, by the addition of nitrogen, a common limitation to plant growth.

Fertilization: The making available of critical nutrients that are in short supply to plants to accelerate growth and metabolic processes.

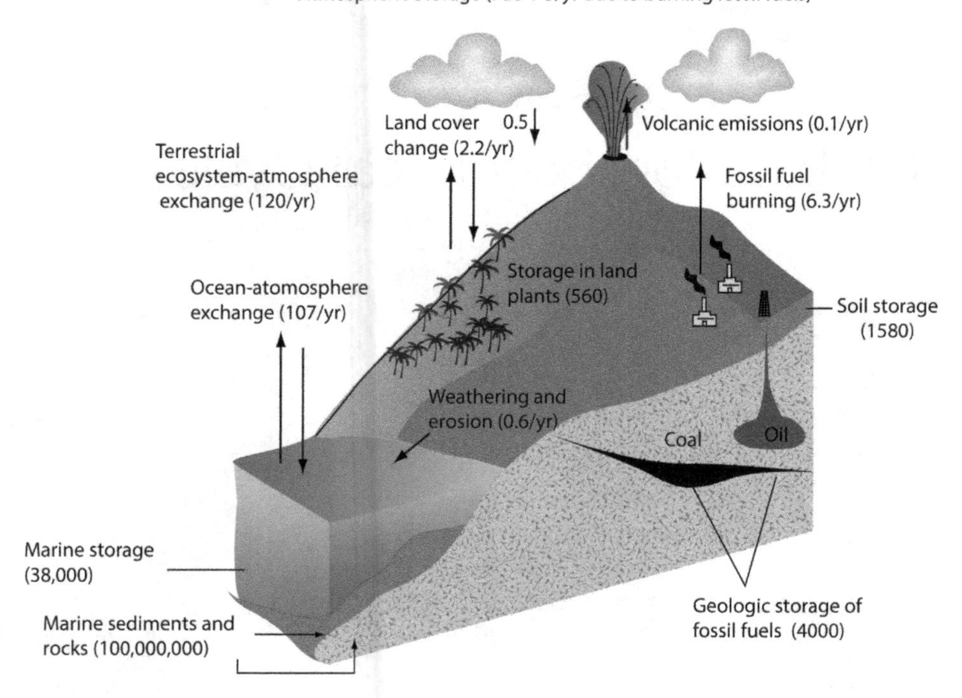

FIGURE 6.1 Global carbon cycle

Idealized view of the global carbon cycle, as of the late twentieth century. Values are in gigatons carbon for storage reservoirs, and gigatons per year for fluxes between reservoirs. To convert to weight of CO_2, multiply by molecular weight ratio 44/12 =3.67.

The Carbon Cycle

The most important of all biogeochemical cycles is the carbon cycle. (Figures 6.1 and 6.2) This is in part because we are carbon-based life forms, and in part because our manipulation of parts of the carbon cycle is leading to global climate and other changes in real time. Carbon is present in rocks, in soils, dissolved in water, and as CO_2 (carbon dioxide) and CH_4 (methane) in the atmosphere, in addition to being the foundation for all life on Earth. It moves through and between each of these reservoirs by a number of processes that are beginning to become understood in some detail (Le Quere et al., 2018).

CO_2 in the atmosphere is absorbed by plants through photosynthesis to produce biomass. This is sometimes eaten by herbivores, but most of it dies and decays in the soil or oceans, adding carbon to these reservoirs. Aside from crustal rocks and sediments, the ocean is the largest reservoir for carbon. Precipitation of dissolved carbon in the ocean, either by biologic processes that make carbonate shells or chemical processes that make carbonate rocks, transfers carbon from the ocean to the crust of the earth as rock. It remains on the ocean bottom until, in the course of plate tectonics, it is subducted into the mantle at a subduction zone. Because carbonate sediments have relatively low melting temperatures, further depressed by the presence of water, heating as the subducted material plunges deeper into the mantle causes them to melt first, ensuring that magmas so generated contain almost all of the carbon deposited at the bottom of the ocean. These magmas are erupted from volcanoes that overlie the subduction zones, and release the carbon as CO_2 back into the atmosphere, where it is available for plants to use again. This entire cycle takes tens to hundreds of millions of years, but is a constant "carbon-processing machine" that keeps the quantity of carbon residing in each of the reservoirs involved at a reasonably constant level on human timescales. Over geologic time, however (100s of millions of years), an increasing amount of carbon has been stored in rocks, with decreasing concentration in the atmosphere (until the last years of fossil fuel burning).

One reservoir of carbon in the crust is hydrocarbons formed by concentration and organic maturation of dead marine microorganisms. When these form on the continental shelves, shallow epicontinental seas, or other marine environments that are not subject to subduction, they become permanently stored in the crust rather than continuing their global cycle. Over a century ago, we discovered that if we just combine these organic molecules with oxygen (burn them), we can extract great quantities of energy to drive an increasingly industrializing human society. In this way, fossil fuel burning has artificially transferred carbon out from permanent geologic storage (where it resided for tens to hundreds of millions of years) and into the atmosphere as CO_2, an important greenhouse gas that affects global climate.

Solubility pump: The uptake of atmospheric CO_2 by ocean surface water in response to increasing atmospheric CO_2 concentration in order to maintain chemical equilibrium between the partial pressure of CO_2 in the air and adjacent water at the interface.

Biological pump: The incorporation of dissolved CO_2 in surface ocean water by phytoplankton and other marine organisms to make their shells. When the organisms die, their shells settle to the deep ocean where they dissolve for long-term storage or to the ocean bottom where they become carbonate sediments for even longer-term storage.

The global carbon cycle is being perturbed by human activities, both from emissions of CO_2 from fossil fuel burning and by deforestation (usually for increasing agricultural lands). The fate of anthropogenic carbon released from fossil fuel burning (coal, oil, natural gas) is not as simple as merely adding to the atmospheric carbon reservoir. Over the twentieth century, about a third of emitted CO_2 remained in the atmosphere, a third was absorbed by the ocean, and a third by the terrestrial biosphere. In this way, the ocean and terrestrial ecosystems serve as sinks for the uptake of anthropogenic atmospheric carbon. (However, the sinks of carbon may be weakening, and at present, about half of the carbon being released into the atmosphere remains there, adding to the greenhouse effect.) As the concentration of atmospheric CO_2 increases, more of it dissolves in the ocean in order to maintain chemical equilibrium at the ocean–atmosphere interface. This is called the **solubility pump** (DeVries et al, 2009). One quarter of emitted carbon is now taken up by the global ocean. The carbon absorbed into the surface water is then incorporated into phytoplankton and other marine organisms, which ultimately die and deliver that carbon into the deep ocean. This is the **biological pump** (Thomsen et al., 2017). This combination of the solubility pump and biological pump still provides a large discount for the

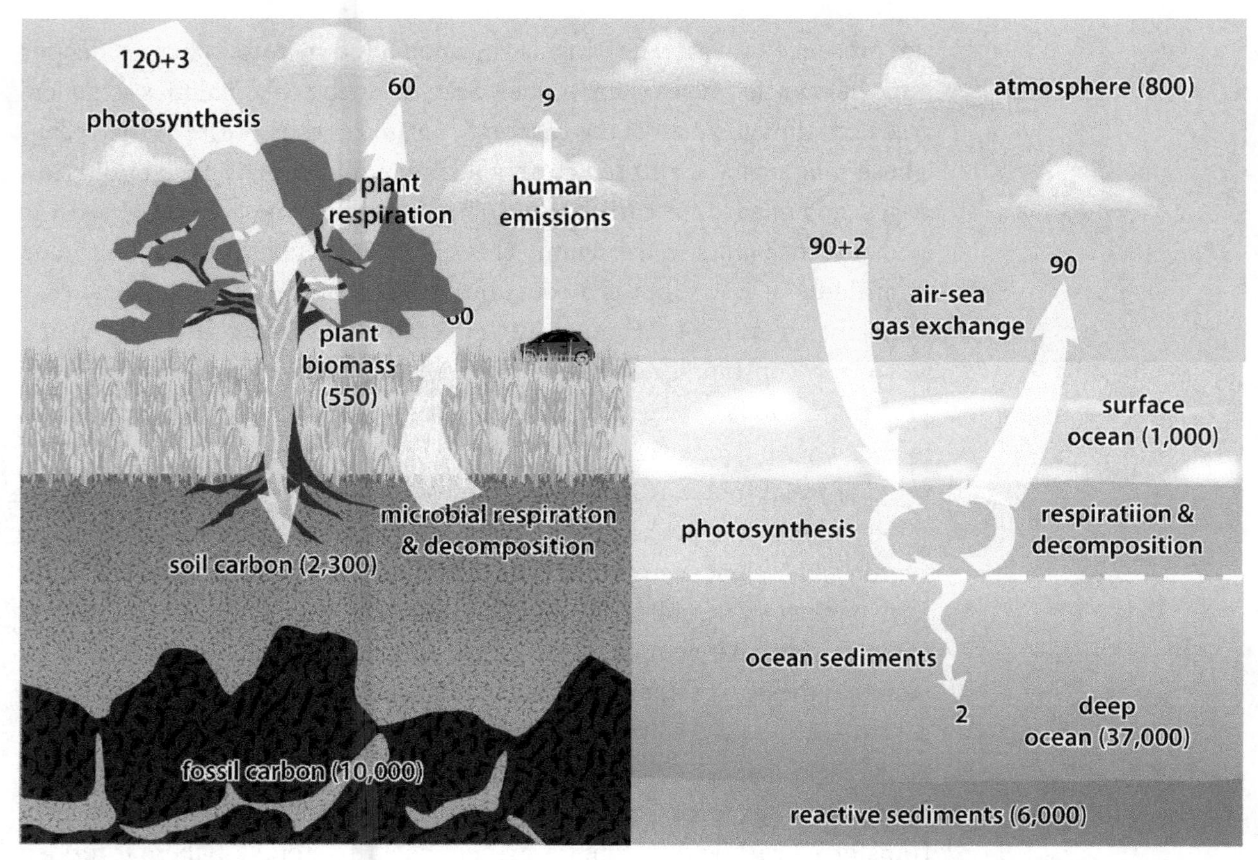

FIGURE 6.2 Global carbon flux
More recent values of global carbon flux. Values of fluxes in gigatons/yr, and reservoirs (in parentheses) in gigatons.

impact of carbon emissions on global climate, but the discount is shrinking.

Another carbon sink is the regrowth of northern hemisphere forests cut down in the last century or two during the days of subsistence agriculture. These forests are growing back, now that agriculture has become more centralized in the plains of the U.S., for example. This terrestrial sink is estimated to have drawn down another third of the fossil fuel CO_2. (Of course this does not account for the fact that these forests were cut down, and thus released their carbon into the atmosphere through decomposition, so that current forest uptake is just replacing what was lost in the previous century.) The result of these carbon sinks is that we burn fossil fuels at a large discount with only a third of our emissions contributing

What weight of CO_2 is produced if a power plant burns 1 ton of coal (pure C)?
Answer: When coal is burned, oxygen is added to make CO_2. Carbon has an atomic weight of 12, while Oxygen has an atomic weight of 16. CO_2 thus has a weight of 12 + 16 + 16 = 44. So the weight of the coal is increased by 44/12 by combining with oxygen. Each ton of burned coal produces 11/3 tons, or 3.67 tons of CO_2. Try doing the same calculation for the burning of natural gas (methane), CH_4. In measuring CO_2 and other greenhouse gas weight, scientists sometimes use the gross weight as calculated here, and sometimes the carbon equivalent, including only the weight of the carbon itself (not oxygen). So be careful to note the units used in reports of emissions rates, reservoir amounts, etc.

to greenhouse warming of the planet so far, but an increasing fraction of our emissions are remaining in the atmosphere as the sinks weaken.

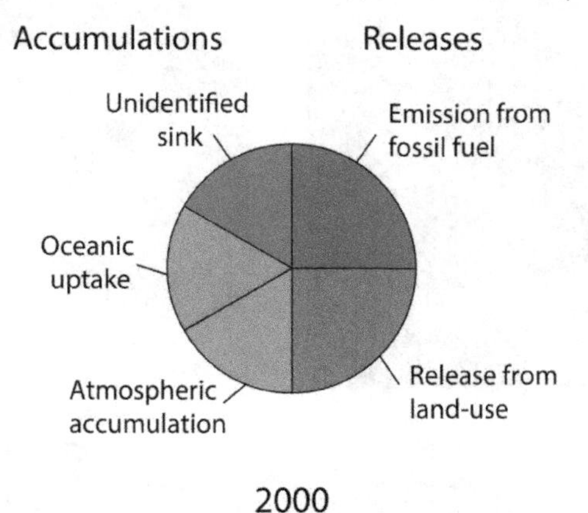

FIGURE 6.3 Carbon at turn of millennium
At the turn of the millennium, some carbon known to have been emitted into the atmosphere from fossil fuel burning was left unaccounted for, so a "missing sink" was postulated. More recently, it was "found" in the terrestrial ecosystem, whose carbon uptake has been faster than had been previously thought. However, the terrestrial and marine sinks are vulnerable to reduction due to saturation of the terrestrial sink, and loss of the solubility and biologic "pumps" in the marine sink.

The ability of the ocean and terrestrial ecosystems to continue to take up carbon from the atmosphere is a subject of serious scientific inquiry. If the oceans become so acidic (due to additional solution of CO_2) that marine organisms cannot readily make calcium carbonate shells, then the biological pump could be severely impacted. Also, once northern forests regrow to equilibrium, and if tropical deforestation is allowed to continue, terrestrial uptake could be reduced. As such, the discount we have been getting regarding the climate impacts of carbon emissions could be reduced, and global climate could become more sensitive to anthropogenic CO_2 emissions in the future.

For many years, the fate of emitted carbon was not clear. While the solubility of CO_2 in the ocean was reasonably well understood, after accounting for that and forest regrowth, there was still a significant flux of carbon left over. It wasn't in the atmosphere—we could measure that. It wasn't in the ocean. It was just missing. Thus commenced the search for the **"missing sink"** (Figure 6.3). Discussion ensued in the scientific community regarding the role of forests and other ecosystems in taking up the remainder of the emitted carbon, and model and observational data have since demonstrated that a combination of temperate forest regrowth and growth in intact tropical forests can account for the missing sink. As such, the missing sink was found in terrestrial ecosystems, and the carbon cycle was closed. In more recent years, the measurements of atmospheric CO_2 on the ground and at various heights have begun to suggest that while the regrowth of northern hemisphere forest

Missing sink: The discrepancy between the known global atmospheric carbon emissions and known uptake by terrestrial and marine sinks. The discrepancy was resolved when new calculations of productivity determined that the terrestrial sink (including northern hemisphere forest regrowth and tropical productivity) had been underestimated.

FIGURE 6.4 Carbon emissions and sinks
Upper panels: sources of atmospheric CO2. Lower panels: sinks of atmospheric CO2 in the atmosphere, land and ocean. While Figure 6.3 indicates the relative proportions of total exchange between carbon reservoirs (stock) since the industrial revolution, the current rate (flux) of carbon (2000–2007) is very different in that emissions are a much greater source than land use, and more remains in the atmosphere, leading to enhanced greenhouse effect.

(after major deforestation in previous centuries in North America and Europe) is an important terrestrial sink of carbon, it may not be quite as large as previously thought (or is slowing down), in part due to the effects of thawing of permafrost and accelerated ecosystem respiration in warming northern latitudes. Further, recent atmospheric measurements suggest that despite the high rate of tropical deforestation, atmospheric CO_2 levels within tropical forests are not as high as would have been expected, indicating that the intact areas of tropical forest are even more productive than previously thought. This highlights the importance of preserving tropical forests. The role of terrestrial ecosystems in the carbon cycle and the influences of climate change are active areas of research internationally, and new developments are expected in the coming years.

The Carbonate-Silicate Cycle

A part of the carbon cycle is linked to silicate rocks (the most common rocks in the Earth's crust) in the carbonate-silicate cycle. Because it involves the weathering of silicate rocks, this cycle works at very long timescales—hundreds of millions of years. Atmospheric CO_2 dissolves in rainwater, making rain slightly acidic (carbonic acid—H_2CO_3). This dissociates to hydrogen (H^+)

and bicarbonate (HCO_3^-). The hydrogen ions (mere protons) are chemically very active and alter silicate minerals releasing calcium (Ca^{2+}) that dissolves in water to join the bicarbonate in streams that ultimately reach the ocean. There, marine microorganisms incorporate the calcium and bicarbonate into their shells and when they die, their shells sink into the deep ocean and dissolve there, or make it to the bottom to accumulate as carbonate sediments. This is the natural part of the marine biological pump (in addition to the role it plays in absorbing anthropogenically released fossil carbon). The sediments are subducted and the carbon is delivered back to the atmosphere through volcanoes. It then rains onto rocks as carbonic acid and starts the cycle all over again.

The Nitrogen Cycle

Another very important biogeochemical cycle is that of nitrogen (Figure 6.5), critical for proteins and the formation of DNA of all cells (Fowler et al, 2015). It is good news that the atmosphere is nearly 80% nitrogen, so there is plenty available. The bad news is that animals cannot use it in that form (it needs to be in organic molecules in food), and even plants need it in other forms, combined with hydrogen (reduced) or oxygen (oxidized). Consequently, a chain of biogeochemical events must ensue in order for the much-needed nitrogen to be useful for ecosystems (Figure 6.5).

This cycle begins with bacteria in soil and plant root nodules that **fix** nitrogen by breaking the N_2 molecule and combining the N with hydrogen, using a special enzyme (nitrogenase). This must be done in an anoxic environment

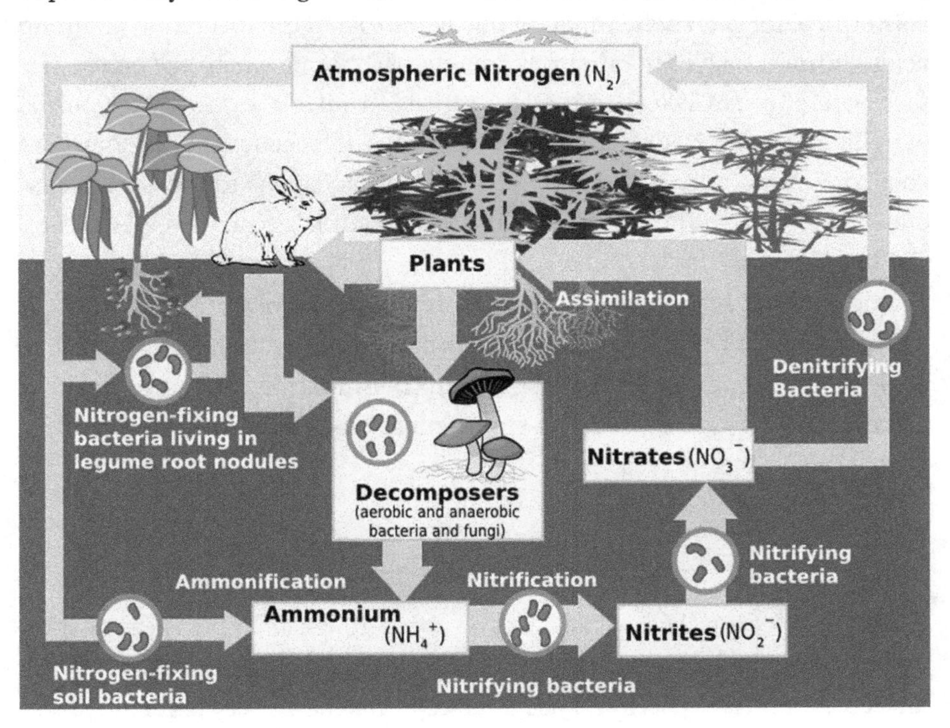

FIGURE 6.5 Nitrogen cycle

Nitrogen fixation, nitrification, and dentrification are all accomplished by bacteria in the soil. This provides plants nitrogen in the form they need to grow.

so that the N does not combine with oxygen instead (oxidize). **Nitrogen fixation** produces ammonia (NH_3), which when combined with water produces ammonium (NH_4^-). Ammonia and ammonium can be taken up by plants and used to form proteins and DNA. The nitrogen cycle continues, however, and in a process called **nitrification**, other bacteria oxidize ammonium to form nitrite (NO_2^-). Yet another set of bacteria further oxidizes nitrite to make nitrate (NO_3^-). Nitrate is also useful for plants. However, plants take up only some of the ammonia, ammonium, and nitrate produced by bacteria. The rest undergoes **denitrification** by yet another set of bacteria that converts nitrate back in to gaseous nitrogen (N_2), also in an anoxic environment, where it is not as difficult to remove the oxygen from the nitrogen. This completes the nitrogen cycle. It also demonstrates the critical role bacteria play in biogeochemical cycling, and indeed, in the Earth System in general. The same process occurs in the oceans, with a similar sequence of nitrogen fixation and denitrification, providing phytoplankton with useful N, a critical nutrient. As will be discussed in Chapter 8, the Haber-Bosch process is an artificial mechanism to fix nitrogen (to make fertilizer), and globally, people surpassed bacteria as the dominant planetary nitrogen-fixers in the mid-1980s.

Nitrogen fixation: The conversion of atmospheric molecular N_2 to ammonia and ammonium, done naturally in soils by bacteria. It is also done artificially through the energy-intensive Haber-Bosch process to make fertilizers.

Denitrification: The conversion of nitrate to gaseous N_2 by bacteria. Nitrogen fixation, nitrification, and denitrification are the component processes of the nitrogen cycle.

The Sulfur Cycle

Sulfur is an abundant non-metallic element that can be incorporated in minerals in rocks, dissolved in water, and oxidized to form a gas in the atmosphere. It is critical for metabolic function due to its role in the formation of amino acids. Sulfur dissolved in raindrops falls out of the atmosphere and enters surface waters (rivers, lakes) and soils, where it is available for use by organisms as sulfate (SO_4^{2-}). Dissolved in water, it eventually reaches the ocean, where some of it is incorporated in marine organisms and sinks to the deep ocean, and some enters the atmosphere from sea spray from ocean waves. However, most oxidized sulfur, sulfur dioxide (SO_2) comes from fossil fuel burning by people. When it is emitted into the atmosphere, it dissolves in water to form sulfuric acid (H_2SO_4), which, when incorporated into raindrops, falls out as acid rain. The addition of "scrubbers" in the smokestacks of coal-burning power plants has greatly reduced sulfur emissions, and reduced the severity of acid rain in recent decades.

The Phosphorus Cycle

Phosphorus is critical for all life, most notably through its role in the formation of **deoxyribonucleic acid** (DNA), and **adenosine triphosphate** (ATP), the molecule that stores energy for use by all cells. It behaves very differently from carbon and nitrogen, in that it does not have a gaseous phase, so resides only

Deoxyribonucleic Acid (DNA): The double helix molecule that carries an organism's genetic information within the nucleus (and also mitochondria) of every cell.

in rocks and dissolved in water. Apatite, a phosphorus-bearing mineral, is present in many kinds of rocks, and when it weathers (mostly physically by wearing away, rather than dissolving or chemically reacting), phosphorus dissolves in water as phosphate (PO_4^{3-}), a very useful form that is taken up by plant roots for the production of ATP. At that point, the phosphorus remains in organisms, being eaten by herbivores, which are eaten by carnivores, etc., transferring the phosphorus from one to another. It is passed through trophic levels in both terrestrial and marine environments in this way, and decomposers finally release it as phosphate back into the water. (Figure 6.6).

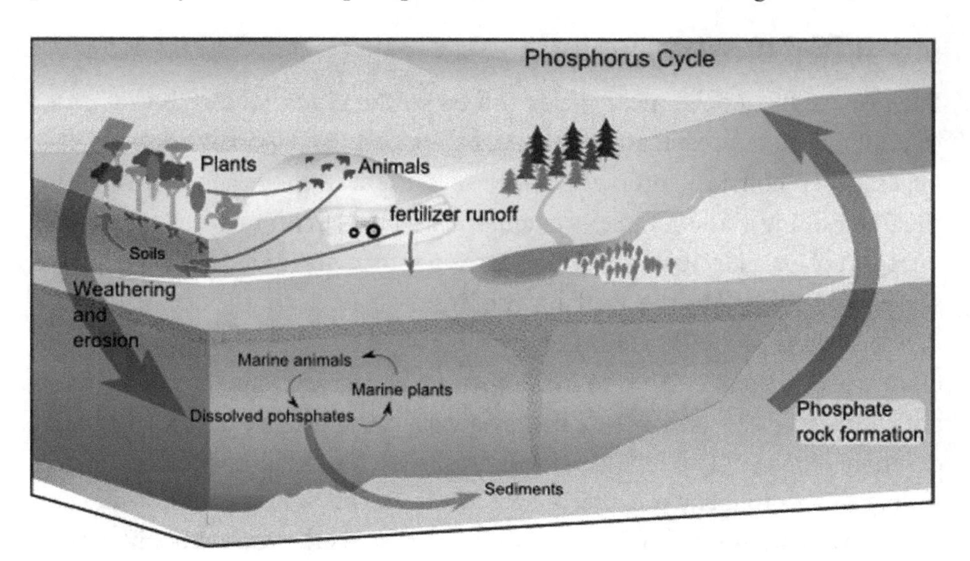

FIGURE 6.6 Phosphorous cycle

All living cells need P to make ATP, the energy storage molecule made within mitochondria. P behaves as a metal, with no gaseous phase, so it resides in rocks, soils, and various compounds in solid form. We mine P from phosphorite deposits, but the problem is that we are not really sure how phosphorite forms. What we DO know is that we are mining it out, and we are facing a "phosphorous crisis" that may affect agriculture globally. Soil erosion removes applied (and existing) P along with the soil so it is all lost to the ocean. Virtually ALL phosphorite is concentrated in Morocco, China, South Africa, and Jordan, so there are political implications as well.

Phosphorite: A rock with high concentration (up to 40%, sometimes) of P2O5. It is not clear how the existing deposits were made in the geological past, but it is apparent that there is no modern analog for the environmental conditions that formed them.

Phosphorus is an important nutrient for natural ecosystems, but also is critical for agriculture (Yuan et al., 2018; Reinhard et al., 2017). Consequently, it is mined extensively for the production of fertilizer. The global deposits of phosphorus, however, are becoming depleted. Like fossil fuels, these deposits took millions of years to form, and we are depleting them in a few centuries. A troubling issue is that it is not obvious how **phosphorite**, the rock that is deposited in ocean bottoms, forms or what special conditions are necessary (Stolper et al., 2016). In any case, we are depleting a global resource that is very unevenly distributed (with almost all of it concentrated in Morocco, China, South Africa, and Jordan). Without the phosphorus in fertilizers, much of the intensive agriculture of the world will be compromised. While phosphorus is not destroyed (it is an element, P), by spreading it around as fertilizer, it is diluted over the landscape such that it cannot be recovered to make more fertilizer. Although plants do not absorb as much P as nitrogen (N) or potassium (K), it is nevertheless a critical nutrient. If soil erosion is eliminated, P applied to soil could remain for decades, but once spread out over the planet, there is no feasible way to concentrate it to produce new fertilizer. This is becoming a concern in the scientific and agricultural communities, and will be revealed in the public realm initially as an increase in food prices.

Integrated Biogeochemical Cycles

Each of the biogeochemical cycles can be studied individually, but in order to understand how each affects the local, regional, and global processes, it is necessary to consider interactions between them as they each contribute to the functioning of the complex system of the global environment. For example, nitrogen plays a key role in plant growth, but nitrogen fixation by bacteria is strongly controlled by the availability of iron, especially in the ocean. Growth of some phytoplankton is limited by their ability to combine dissolved calcium with carbonate CO_3^{2-} to make their calcium carbonate shells. As more anthropogenic CO_2 is dissolved in seawater by the solubility pump, it converts carbonate ions (CO_3^{2-}) into bicarbonate ions (HCO_3^-) making it more difficult for organisms to make their carbonate shells. In this way, the carbon and calcium cycles are related. (This is also the mechanism of ocean acidification, discussed in Chapter 9). In general, because living organisms need a variety of elements in specific forms to support metabolism and growth, each of the biogeochemical cycles described here as well as others must work in concert to support the ecosystem.

References

DeVries, T., & Primeau, F. (2009). Atmospheric pCO(2) sensitivity to the solubility pump: Role of the low-latitude ocean. *Global Biogeochemical Cycles*, 23(GB4020).

Fowler, D., Steadman, C. E., Stevenson, D., Coyle, M., Rees, R. M., Skiba, U. M., . . . & J. N. Galloway. (2015). Effects of global change during the 21st century on the nitrogen cycle. *Atmospheric Chemistry and Physics*, 15(24), 13849–13893.

Le Quere, C., Robbie, A. M., Friedlingstein, P., Sitch, S., Pongratz, J., Manning, A. C., . . . & Zhu, D.(2018). Global carbon budget 2017. Earth System Science Data, 10(1), 405–448.

Reinhard, C. T., Planavsky, N. J., Gill, B. C., Ozaki, K., Robbins, L. J., Lyons, T. J., . . . & Konhauser, K. O. (2017). Evolution of the global phosphorus cycle. *Nature*, 541(7637), 386.

Stolper, D. A., & Eiler, J. M. (2016). Constraints on the formation and diagenesis of phosphorites using carbonate clumped isotopes. *Geochimica et Cosmochimica Acta*, 181, 238–259.

Thomsen, L., Aguzzi, J., Corrado, C., Leo, F. D., Ogston, A., & Purser, A. The oceanic biological pump: Rapid carbon transfer to depth at continental margins during winter. *Scientific Reports*, 7(10763).

Yuan, Z., Jiang, S., Sheng, H., Liu, X., Hua, H., Liu, X., & Zhang, Y. (2018). Human perturbation of the global phosphorus cycle: Changes and consequences. *Environmental Science & Technology*, 52(5), 2438–2450.

Figure Credits

Fig. 6.4a: http://earthobservatory.nasa.gov/Features/Deforestation/

Fig. 6.4b: Copyright © Dori (CC by 3.0) at http://commons.wikimedia.org/wiki/File:Smokestacks_3958.jpg.

Fig. 6.4c: Copyright © 2012 Depositphotos Inc./iofoto.

Fig. 6.4d: Copyright © 2014 Depositphotos Inc./Natashamam.

Fig. 6.4e: Copyright © 2012 Depositphotos Inc./ozaiachinn.

Fig. 6.5: Copyright © Raeky (CC BY-SA 3.0) at https://commons.wikimedia.org/wiki/File:Nitrogen_Cycle.svg.

Fig. 6.6: Copyright © Bonniemf (CC by 3.0) at http://commons.wikimedia.org/wiki/File:Phosphorus_cycle.png.

CHAPTER 7

Water

Thirsty they were so they went and they looted
Water resources, some really unsuited.
Demand by and by
Left aquifers dry.
And what's more, the runoff became all polluted.

WE LIVE ON a water planet, and while most of the surface is covered by ocean, as land animals, humans are most concerned with the availability of fresh water on land. Water is a critical resource, as it not only quenches our thirst, but also enables all of our food to grow as well as supporting all of the world's terrestrial ecosystems. The wisdom with which we manage local, regional, and global water supplies determines the extent to which they will be available for use by us and future generations.

Properties of Water

Latent heat of fusion: The energy required to melt a solid at its melting point without changing its temperature. It is equal to the energy released by freezing the liquid.

Latent heat of vaporization: The energy required to evaporate or boil a liquid without changing its temperature. Unlike latent heat of fusion, evaporation can take place at temperatures lower than the boiling point. It is equal to the energy released by condensation from gas to liquid state.

Water is a very special substance. This is in part why life started in water, evolved, and continues to exist only in the presence of water.

- Water is the only common material on Earth that **expands when frozen**. So when lakes, rivers, or parts of the ocean freeze, the ice floats and the bottom of the water body remains liquid to support the ecosystem that depends on it. If ice sank in water, ponds and lakes would freeze solid and life would be very different on Earth.
- Water has a very large **heat capacity**. That is, it requires a lot of heat energy to raise the temperature of water by a degree centigrade. This property enables water to moderate the Earth's climate in time (day/night; summer/winter), as well as to transport heat great distances between the tropics and poles.
- Water has very large **latent heat of fusion (melting) and vaporization (evaporation)**—much more than most common substances. This means it requires a great deal of heat to melt ice, or to evaporate or boil water. This further contributes to water's ability to moderate climate. Evaporation also

provides the latent heat energy that drives hurricanes and controls weather of all kinds.

- Water, because of its **polarized** molecular structure, can dissolve many solids (e.g., salts), liquids (e.g., alcohol), and gases (e.g., CO_2). As a result, it carries nutrients, salts, and many compounds essential for living organisms.
- Water is a **clear liquid** (between 0°C and 100°C), even when salt is dissolved in it. Consequently, incident sunlight penetrates to considerable depth as it is slowly absorbed. This enables marine photosynthetic organisms to thrive some distance from surface turbulence and waves. It also causes the ocean to absorb almost all of the solar energy that penetrates the surface, giving an albedo of close to zero. This is in sharp contrast to solid water in the form of snow, which has an albedo of very close to one, reflecting virtually all incident solar radiation.
- Water has high **surface tension**, caused by the same polar molecular structure that makes it a good solvent (Figure 7.1). At the surface between water and air, for example, the polarized water molecules line up and arrange themselves facing the same way, with the positive side of one sticking to the negative side of the next. This enables insects to walk on the surface and enhances capillary action that helps blood to flow and plant roots to take up water.

Surface tension: A cohesion of the molecules at the interface between two fluids (usually liquid and gas) caused by the arrangement of polarized molecules at the surface of the liquid.

These properties give water a special and very critical role in the environment. Humans have taken the availability of clean, fresh water for granted for millennia, as there has been plenty relative to the human population. However, our population has exploded to the point that we can no longer assume that there is enough water to provide food for everyone all the time. Thus, water is one of the most critical areas of concern for environmental policy, international treaties, and potential future conflict.

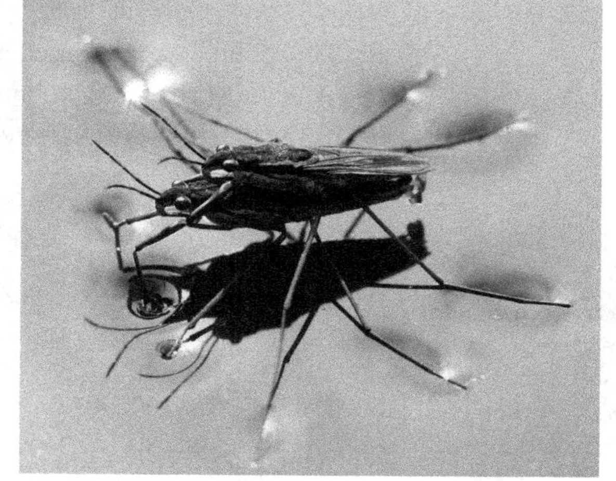

FIGURE 7.1 Surface tension

The Hydrologic Cycle and Water Budgets

The availability of water for consumption, irrigation, industry, recreation, and other uses depends completely on the hydrologic cycle driven by solar radiation, atmospheric transport, precipitation, runoff, infiltration, evaporation, and transpiration (Figure 7.2). As with most processes on Earth, the hydrologic cycle is powered by the sun. Heating of the land and water surface causes evaporation, transferring water vapor and latent heat into the atmosphere. Winds (also driven by the sun through differential heating of the planet) transport the water vapor from sites of evaporation to other places, where it reaches supersaturation in the air, and precipitates as rain or snow.

Almost all of the world's water is found in the ocean (Table 7.1 and Figure 7.4), but it is salty—not useful for terrestrial organisms. Most of the fresh water on Earth is locked up in glaciers and ice sheets. The remaining 1% of the world's water is available to us in groundwater, soils, lakes, rivers, and the atmosphere.

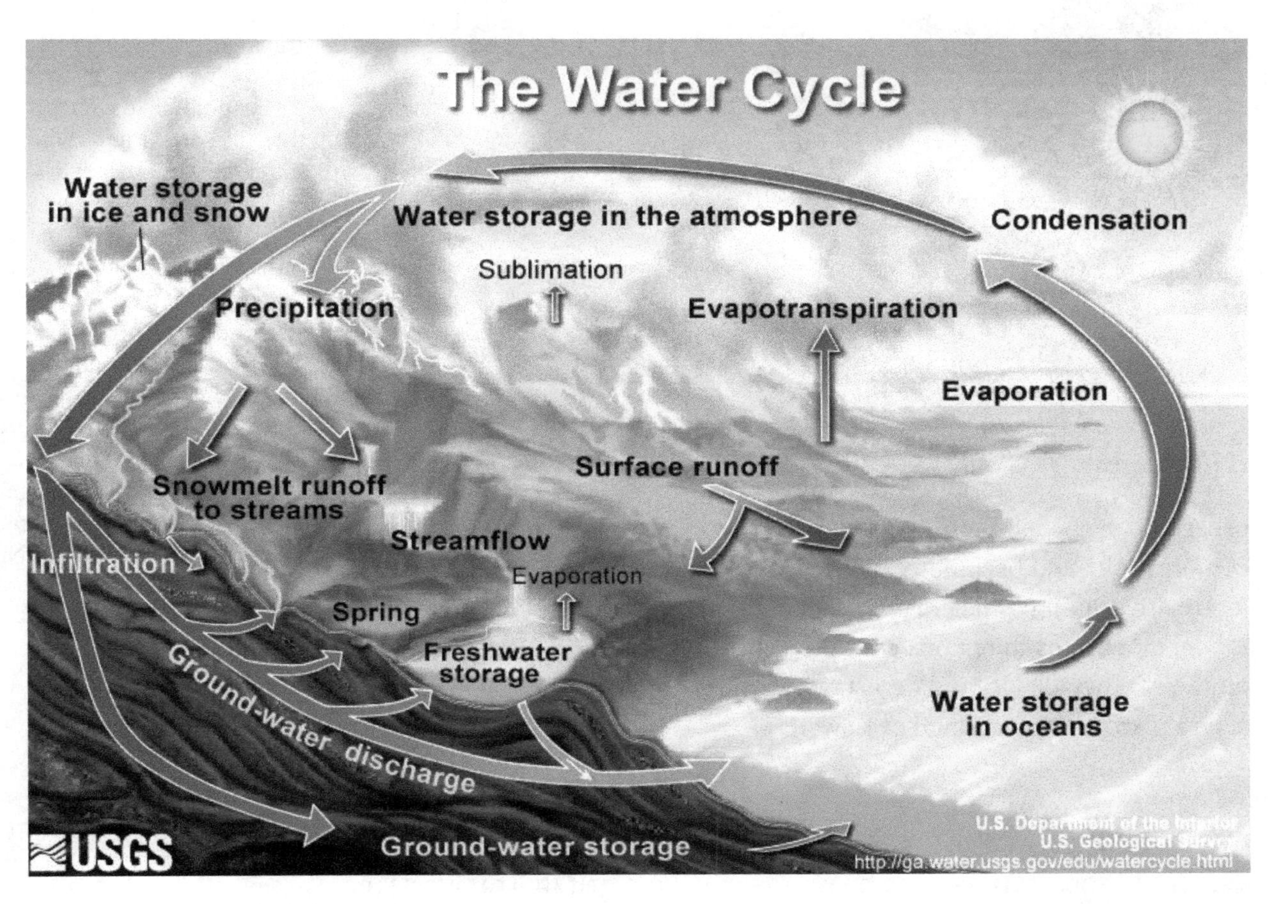

FIGURE 7.2 The basic hydrologic cycle

Supersaturation: The solution of a solute in a fluid at concentrations greater than the maximum possible for chemical equilibrium at a given temperature, pressure, and chemical composition. It often leads to sudden precipitation of the solute when perturbed.
Convection: The vertical movement of fluids due to density differences, usually caused by a thermal gradient with the less dense fluid below the more dense fluid, under the influence of gravity. Convection would not work without gravity.

Supersaturation that leads to rain can occur in a few ways, and they generally involve cooling of the air (Figure 7.3). One mechanism is the rising of a parcel of air over land that is heated by the sun (hot air rises). As it rises by **convection** to the upper part of the troposphere, it cools due to decompression (the air is thin at high altitude), and finds itself supersaturated in water, thus nucleating droplets that can interact in clouds, coalesce, grow, and fall out as rain or snow. This simple convective mechanism occurs most commonly on hot summer days when the ground is intensely heated and water evaporates from the surface.

Another way to supersaturate air is to cool it by lifting it over a mountain as the wind blows across it. This is called the **orographic effect.** As the air rises and becomes supersaturated, the water rains out. As it descends on the

FIGURE 7.3 Supersaturation

Orographic effect: The influence of a mountain standing in the path of air that is forced to rise and cool (thus becoming supersaturated in water vapor—cool air cannot hold as much moisture as warm air) such that it causes precipitation on the prevailing upwind side of the mountain. As the cooled, and now dry, air descends on the downwind side of the mountain, it warms, thus reducing relative humidity, making the downwind side of mountains very dry. The big island of Hawaii is a good example of this, with rainforest on one side and desert on the other.

other side of the mountain, the now water-depleted air warms up again (by compression) and becomes very undersaturated in water vapor with a very low relative humidity. There is no chance of rain under those conditions. This is why mountains in regions with prevailing winds have forests and jungles on the windward side, and deserts on the leeward side. A prime example of such a place is the big island of Hawaii.

TABLE 7.1 SOURCES OF THE WORLD'S WATER SUPPLY. NOTE THAT THE PERCENTAGES DO NOT ADD UP EXACTLY TO 100%. THIS IS BECAUSE THE UNCERTAINTY IN THE VOLUME OF THE OCEAN IS GREATER THAN THE TOTAL AMOUNT OF WATER IN THE ATMOSPHERE, RIVERS, AND STREAMS.

Location	% of Total Water	Surface Area (km^2)	Water Volume (km^3)	Avg. Residence Time in Location
Oceans	97.2	361 million	1.23 billion	Years (thousands)
Ice caps and glaciers	2.15	28.2 million	28.6 billion	Years (tens of thousands)
Groundwater	0.31	130 million	4 million	Years (hundreds to thousands)
Lakes	0.01	855,000	123,000	Years (decades)
Atmosphere	0.001	510 million	12,700	~9 days
Rivers and streams	0.0001	Variable	~1,200	~2 weeks

USGS

Air mass: A region of the troposphere that moves as a unit, bringing with it a set of atmospheric conditions (dry, moist, warm, cool). When air masses interact over a point on the ground, a variety of weather conditions can occur (often precipitation).

Yet another way to supersaturate air is to lift it over another **air mass**, thus cooling it and causing precipitation similar to the orographic effect. This often occurs as weather fronts move and air masses of different temperatures and pressures interact, with the warmer one rising over the colder one such that it cools. Each of these mechanisms causes precipitation, which can be considered the start of the hydrologic cycle.

Of the water that rains out of the atmosphere as precipitation, about 60% evaporates again into the atmosphere before reaching the ocean in either surface flow (rivers) or base flow (underground flow of groundwater). The part that runs off the land without evaporating or infiltrating into the soil and groundwater is organized by drainage divides that determine which way the water will flow in a given area. The drainage divides define watersheds, a.k.a. drainage basins, which are the fundamental hydrological unit as well as the basic unit of ecosystems that depend upon the water for their existence. Watersheds can be defined at any scale, from a small catchment only a few square meters in area, to larger, regional watersheds, whose streams converge and may go to a continental-scale river at sub-continental scale. (A **stream** is the general term for any flow of surface water. A river is a special case of a large stream.) The **watershed** itself is defined as the area of land that drains to a specific stream, the mouth of which empties into a larger stream or ultimately, the ocean. The stream that drains a watershed defines the watershed, and examples are the Missouri, which is part of (because it drains into) the Mississippi.

Watershed: An area of land, bordered by drainage divides, in which falling rain flows into a single, common stream. Watersheds can be of any scale, and are nested such that smaller watersheds drain into bigger ones, until ultimately major rivers drain to the ocean.

A water budget can be constructed on the basis of the flux of water into and out of reservoirs of various sizes. It can be done for the whole planet, a country, or a watershed at any level. To make a budget, one balances the fluxes of water into and out of storage in the various reservoirs within the system.

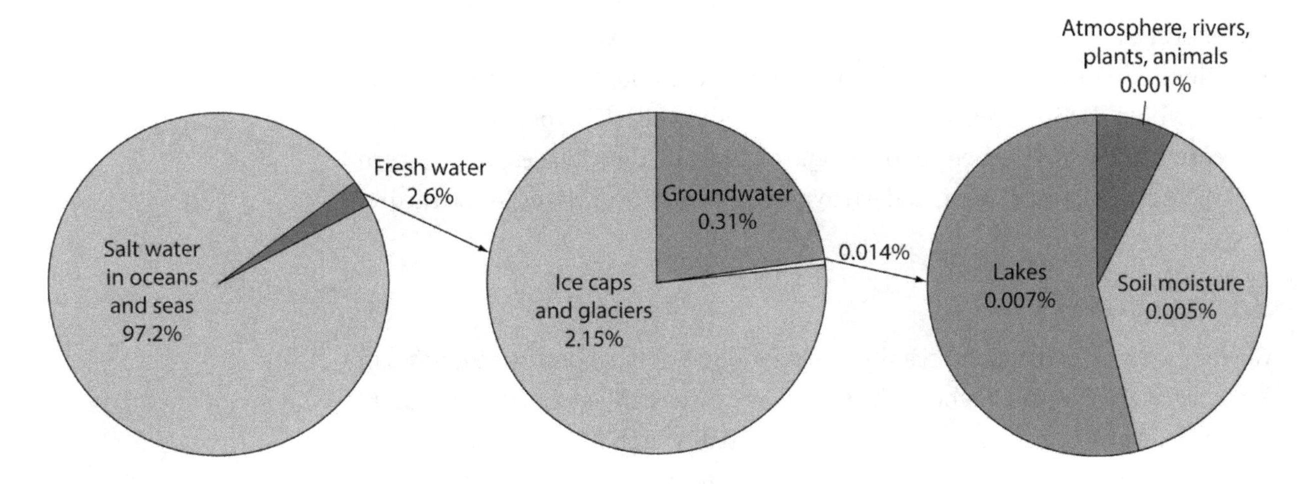

FIGURE 7.4 Distribution of water on Earth
Distribution of water on the earth. Most is in the ocean and is salty. Of the remaining 2.6%, most is frozen as ice, and most of the rest is groundwater. Only 0.014% of the world's water is in the near-surface environment for use by plants, animals, and society.

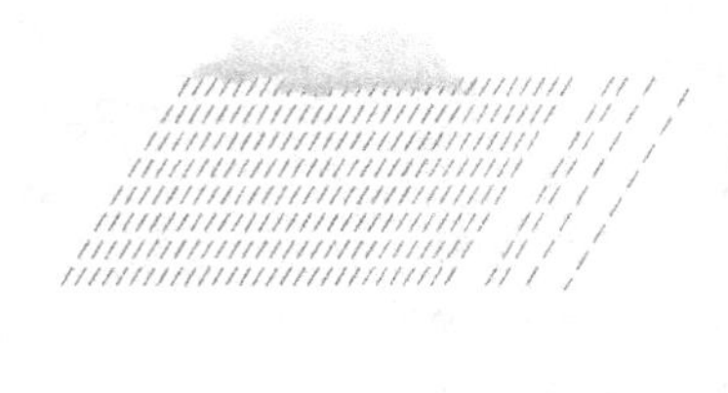

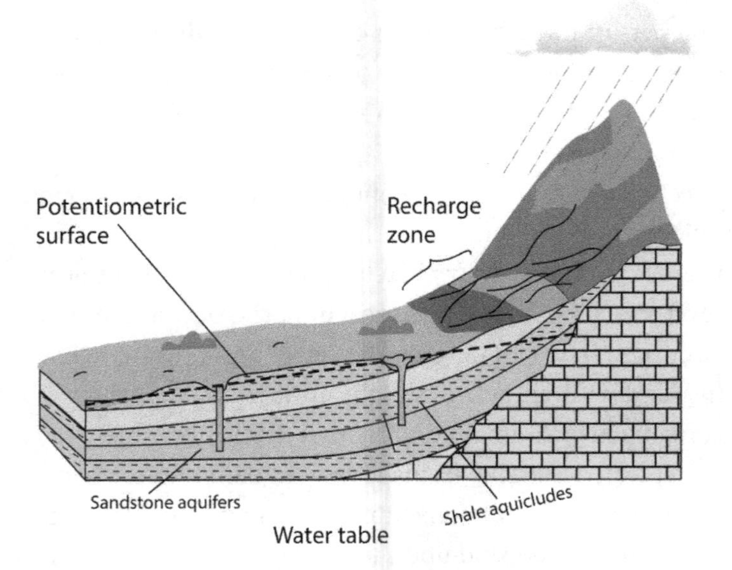

FIGURE 7.5 Artesian wells tap aquifers
Artesian wells tap aquifers that are pressurized by a dipping, or tilted, aquiclude that extends to the areas of recharge. The water pressure pushes water up the well without the need for pumping. A flowing artesian well enables water to reach the ground surface, while a non-flowing artesian well pushes water to below the ground surface, but requires much less pumping than a normal well. When multiple wells tap an aquifer, their cones of depression interact, and a deeper well may cause shallower wells to run dry.

In North America, precipitation and runoff is very heterogeneous, as it is on all continents, with much of its runoff in the Arctic region, far from thickly settled areas or intense agriculture. In the southwest, there is very little runoff, as the flow of any available rivers, such as the Colorado River, is used almost completely for agriculture before it has a chance to reach the sea. In fact, the regulations on state-to-state water use for the Colorado River allocate to each state a certain amount of the flow of the river. If you add up all of these allocations, it sums more than the entire flow of the river. If each state used its full allocation, a dry riverbed would result, with no flow going to Mexico at all (which, by international treaty, is not allowed). The politics of water allocation and use have begun to become an issue in many places, the western U.S. being only one of them, and we may see conflict over water in the future.

Surface Water

Surface water is convenient, readily available, and easy to monitor, as it flows in streams of all sizes throughout the world. Because of this, population centers have arisen on the banks of rivers, as they provide water, food, transportation, and other services. They are also somewhat variable in their flow rates, and are subject to flooding or times of very low flow. Flooding has the obvious impact on buildings and infrastructure, and accounts for huge economic losses throughout the country and the world. Consider the flooding of Houston, TX during Hurricane Harvey, for example. Flooding is also essential for maintaining ecosystems on floodplains. Floodplains are created by the sediments deposited during high flow of a river. So far as a house or town is concerned, there is no "risk" of flooding on a floodplain. There is strictly a guarantee. Floodplains are there because they flood.

For several millennia, a thriving society was maintained along the Nile River, and agriculture was maintained in a desert on the basis of soil whose nutrients were replenished by annual flooding on the Nile. In the 1960s, the Aswan High Dam was built to generate electricity and control the annual flooding that was considered inconvenient for the increasingly urban society of the twentieth century. It did indeed control flooding, and halted the deposition of nutrient-rich sediments on the agricultural land of the Nile Valley, so that farmers now must apply artificial fertilizers to maintain crop productivity. However, the dam also reduced the times of low flow during droughts. Low flow in a river tends to increase the concentration of dissolved material, and in industrial regions, can increase the concentration of pollutants because there is insufficient water volume to dilute contaminants to acceptable levels. Low flow also interferes with the ability of power plants to withdraw sufficient water to flash to steam to drive turbines and then cool in cooling towers enough to return to the stream without raising water temperature and adversely affecting the ecosystem.

Aquifer: A rock that can hold and transmit water in and through the pores and other voids within the rock.

Mining: The removal of a natural resource that does not get replaced or that renews at a rate lower than the rate of removal.

Zone of aeration: The zone of soil and rock near the ground surface that has both air and water in the pores and other voids.

Zone of saturation: The zone of soil and rock below the zone of aeration that contains only water in its voids.

Water table: The surface that divides the zones of aeration and saturation.

Groundwater

Groundwater is the largest reservoir of fresh water accessible to people. Although surface water in lakes and rivers is much easier to obtain in humid climates, in dry areas, groundwater is the only available water, and must be pumped out of the ground through wells drilled into **aquifers** (Figure 7.5). In such arid or semi-arid regions, groundwater is there, having taken millennia to accumulate, or accumulating at times of more humid climate in the geologic past. At present, the aquifers in arid climates do not recharge at appreciable rates, so any water that is withdrawn depletes the aquifer permanently. As such, such groundwater withdrawal is considered **mining** of fossil water.

The soil and rock near the ground surface has voids, or pores, that are filled with air and some water. Hydrologically, this is the **zone of aeration,** and while it is very useful for plant roots, it is not a water resource. However, at a critical depth, the pores between sedimentary particles and rock formations become saturated in water (no more air), and this depth is called the **water**

One form of physical weathering of rocks and soils is ice wedging, in which the expansion of water upon freezing widens cracks and pores, creating larger fractures and smaller particles, respectively. "Frost heaves" encountered on roads in winter are caused by the same process of expansion of a volume of groundwater beneath the road. If the road is lifted by 1 cm, the density of ice is 0.9, and the porosity of the saturated sub-road soil is 0.5, to what depth is the ground frozen beneath the road?
Answer: Water expands 10% when frozen, so to rise by 1 cm, 10 cm of pure water would need to freeze. However, the porosity of the soil is only 0.5, so twice that, or 20 cm of frozen pore water is needed.

table, below which is the **zone of saturation** (Figure 7.6), and potentially the body of an aquifer. An **aquifer** is a rock formation that has a useful level of **porosity** (the volume of void space filled by water) typically between 10% and 30%, and also is **permeable**, meaning that water can move through the aquifer so that it can get to the well that draws it out. A rock that has very low permeability is called an **aquiclude**. Aquicludes separate parts of aquifers, support perched water tables in mountains, create artesian systems from which water shoots without needing to be pumped, and make confined aquifers in which the water partially supports the level of the ground surface above it.

The land surface defines the distinction between surface water and groundwater. Anytime the ground level dips below the water table, surface water is found as either a stream or a lake. In fact, a **spring** is a location where an aquifer intersects the ground surface (usually on a hill, but sometimes in a small pond or lake), and groundwater becomes surface water. As such, "spring water" is the same water as groundwater that comes from a well, (although bottled water salespeople market it as more exotic). In fact, spring water is merely groundwater that has come to the surface and thus gains the opportunity to become contaminated.

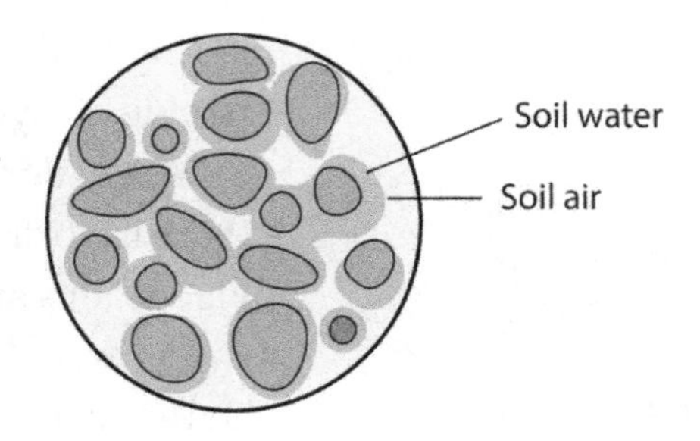

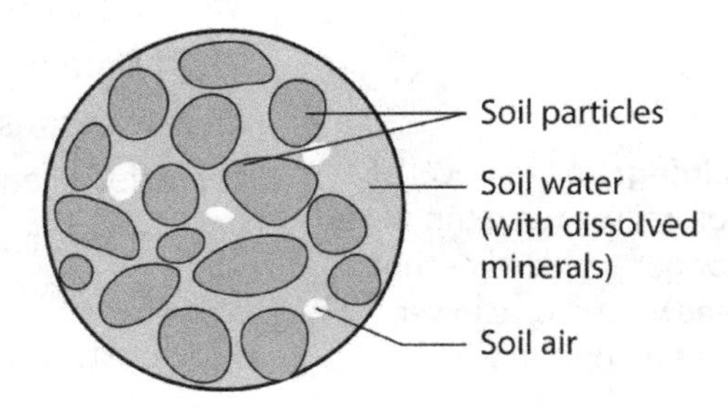

FIGURE 7.6 Intergranular pores in soils
In soils, intergranular pores are variably filled with water. Below the water table, they are completely filled (zone of saturation), while above the water table (zone of aeration), they contain both water and air in varying proportions, depending on how dry the soil is at a given time.

Water Use

Water is used in numerous ways for many purposes. These can be categorized as in-stream use and off-stream use. **In-stream** use does not remove the water from the river or stream and could include navigation, hydroelectric power, fishing, and recreation. Each of these uses has different optimal **hydrographs** (the flow, or level of the river as a function of time over the course of a year), so there could be some conflict, for example, between hydroelectric power generation that needs more flow in summer, and navigation, which needs a constant, but high water level year-round. **Off-stream** use actually removes water from a stream and may or may not return it to the stream after doing something with it. If it does not return the water to the stream, it is considered **consumptive** use. A prime example of consumptive use is irrigation. When crops are watered, almost all of the water evaporates from the air, ground surface, plant surface, or out through the pores (stomata) in plant leaves by transpiration. When the water evaporates, it is transported in the atmosphere to some other location, where it rains and either evaporates, infiltrates, or runs off. Usually, irrigation is done in relatively dry regions with insufficient rainfall to maintain rain-fed agriculture, and so evaporated rain typically leaves the watershed and enters another watershed, with a net loss of water from the irrigated watershed.

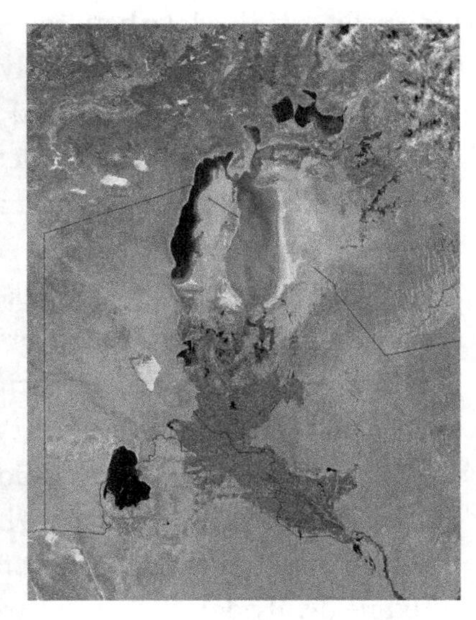

FIGURE 7.7 (A AND B)
(a) The Aral Sea in 1985. (b) The Aral Sea in 2003.

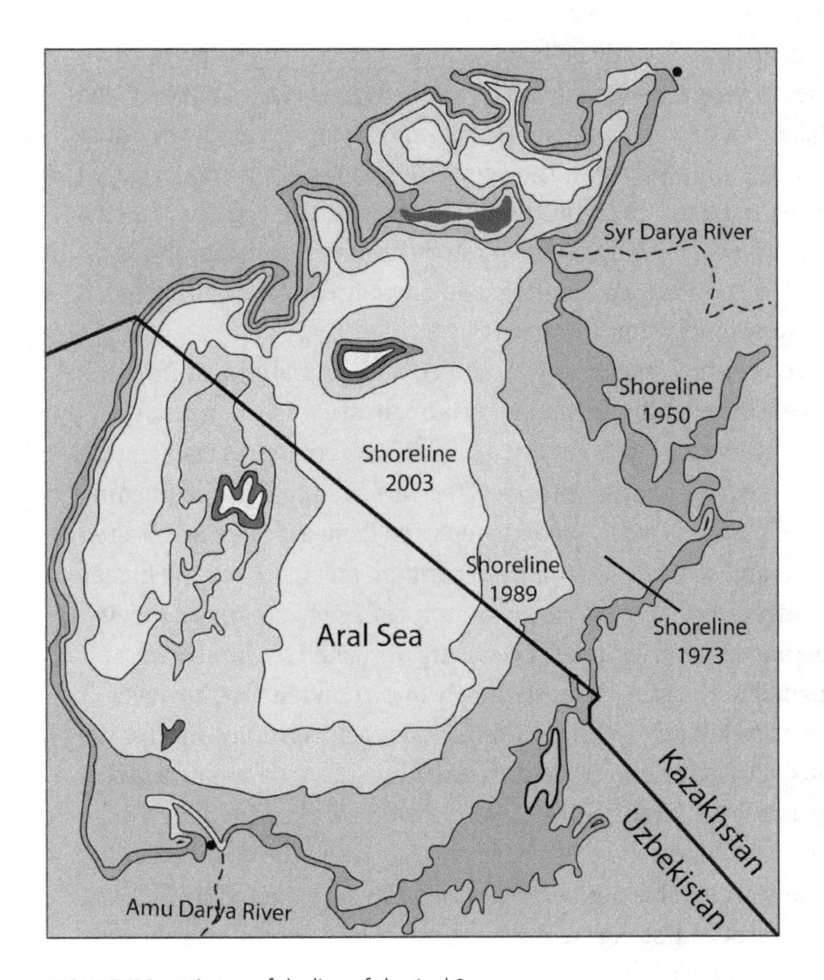

FIGURE 7.8 History of decline of the Aral Sea.

The most extreme case of water loss from a watershed, and one of the greatest environmental catastrophes of all time, involves the Aral Sea, an internally draining lake between Kazakhstan and Uzbekistan in central Asia (Figure 7.7 and 7.8). An internally draining lake has no outlet to the ocean. It is basically a hole in the ground into which water flows but it doesn't fill to capacity to overflow into an outlet river. In any internally draining lake, the lake level is maintained by a balance between the rate of river input and the rate of evaporation from the lake surface (McDermid et al., 2017). Obviously, we only see internally draining lakes in desert environments. Other examples are Lake Chad in the Sahara desert and the Great Salt Lake in Utah. The Aral Sea (not actually a "sea" connected to the ocean) is fed by the Amu Darya and the Syr Darya, two rivers with headwaters in the high mountains to the east. In the 1960s, the Soviet Union began to greatly increase the amount of irrigation of the Karakum desert in order to grow cotton—a cash crop—and began to divert more and more of the water from these two rivers into irrigated fields. By doing this, the area for evaporation greatly increased, and the water did not reach the Aral Sea. Consequently, the Aral began to shrink dramatically, until it is now only a small vestige of its former area (Figures 7.7 and 7.8). While evaporation ensued, all dissolved salts were left behind (only the pure water evaporates), and the lake became increasingly salty, until all the fish died (destroying a previously thriving fishing industry). Further, the dried former lake bed became a large salt flat, and winds created salty dust storms that deposited the salt onto the very crops that were being irrigated by the water, thus reducing their productivity. In addition, the dust storms led to human health problems due to respiratory illnesses. As such, the "Aral Sea Catastrophe" is an infamous example of environmental degradation due to the mismanagement of water resources.

If water is not consumed (meaning evaporated and ultimately lost from the watershed), it is effectively all returned to the stream after use. In a sense, it is merely "borrowed" temporarily. An example of this is electric power generation from coal-burning power plants (although the same applies to gas-burning and nuclear power plants as well). Water is removed from a stream and heated until it boils to steam, so that the rapidly expanding steam drives the blades of a turbine connected to an electric generator, thus generating electricity. The steam cools and condenses and is eventually returned to the stream. If it were to be returned to the stream immediately upon condensing, however, it would be so hot as to kill any aquatic wildlife and would severely disrupt the ecosystem. Thus, it must be cooled before returning it to the stream and this is why power plants have large cooling towers that look like a couple of huge upside-down lampshades on top of each other. Only when the water is cooled to environmentally acceptable temperatures can it be returned to the stream. While some water is, of course, lost to evaporation, it does not necessarily leave the watershed. Although crop irrigation is typically done only in semi-arid or

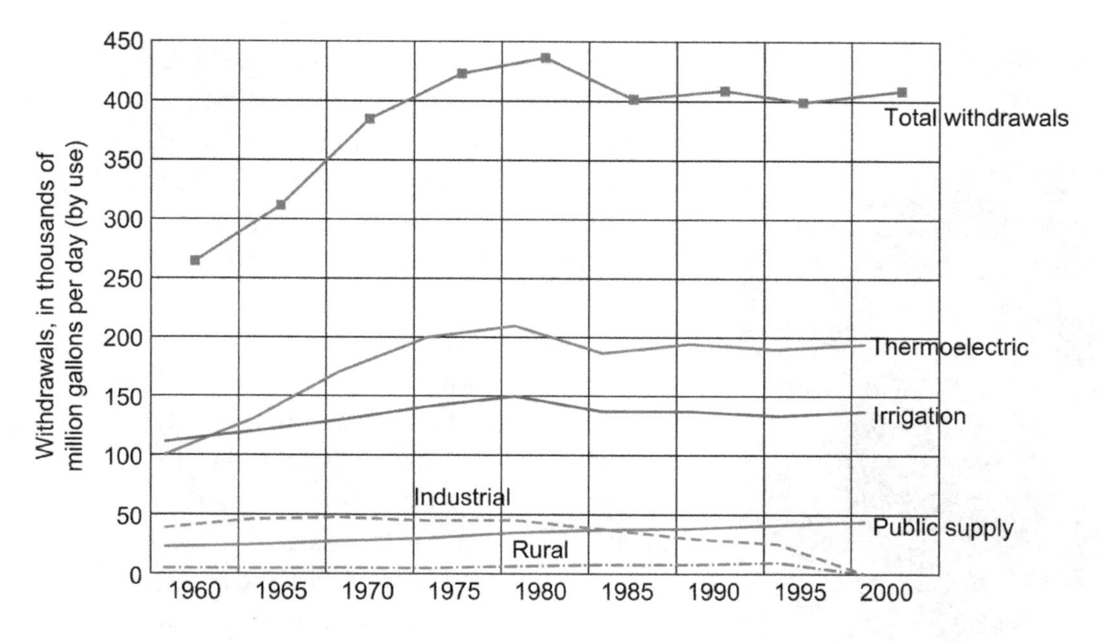

FIGURE 7.9 U.S. Water usage
Note that public supply of domestic drinking, washing, and other water is very small compared to irrigation. Even that is smaller than thermoelectric power, which has dominated since the 1960s. Electricity-generating power plants use water to heat, to steam, to drive turbines, and to generate electricity. It doesn't matter what the source of heat may be—coal, oil, gas, nuclear. They all have the big cooling towers that enable the used and condensed steam to cool down to a temperature that will not kill the biota in the river when returned to the river. This is now the main use of water in the U.S.

very dry regions, power generation is done in humid places, where there are numerous and steadily flowing streams, such that evaporated water has a reasonable chance of raining back down within the same watershed and coming back for an additional round in the steam turbines and cooling towers. In the U.S. electric power generation is the single greatest use of water (Figure 7.9), with irrigation for agriculture coming in second. In many parts of the world, however, agriculture is by far the greatest water user.

Wetlands

Wetlands can be defined as places where **the water table is at or near the ground surface for a significant portion of the growing season** (Figure 7.10). As such, they are partially defined by the plant life they support. The difference between a wetland and a lake, for example, is that plants are rooted in the soil underlying the water of a wetland (e.g., lily pads, seagrass, etc.) and a lake does not have plants rooted on its bottom (not enough light penetrates, or depth is too great). Wetlands do not always have to be covered with water, however, and vernal pools are seasonally wet, supporting a community of plants, fungi, amphibians, and other animals that are unique to wetlands. In fact, a wetland need never be covered with standing water, because if the

FIGURE 7.10 Wetlands

water table is "near the ground surface" it is within the rooting depth of the plants that are adapted to rooting in the zone of saturation of the groundwater. Wetlands do much more than support critical ecosystems, however (Williams et al., 2017). The list of service wetlands perform for human society is long, so just a few will be mentioned here.

Two wetlands with contrasting functional characteristics are illustrated in figure 7.11. This "spider diagram" presents a multi-dimensional spectrum of parameters that describe wetland function and can be used as the basis for a functional classification of wetland types. The two fictitious wetlands shown here would occur in very different environments. Wetland A is warm with a very variable water table and high salinity, as might be found in a salt marsh along the coast of Maine during summer. Wetland B is cold with a stable water table, fresh water, and a steep gradient relative to wetland A, as might be found on a high mountain plateau. If wetlands were to be characterized according to function, the specific impacts of wetland draining or general destruction could be more clearly quantified. At present, this is not done by land developers or regulatory

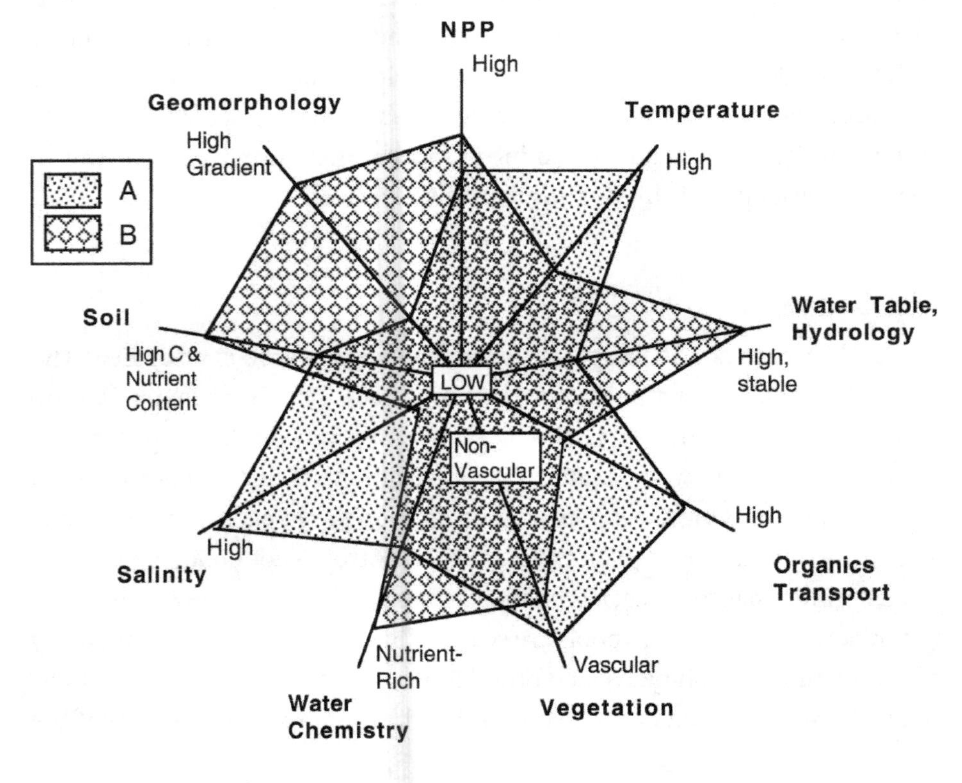

FIGURE 7.11 Spider diagram

agencies. For more details on functional classification of wetlands, see http://gaim.unh.edu/Products/Reports/index.html and choose report #2.

- Wetlands store water in the short term after heavy rains, reducing flooding downstream. Thus, they continue to provide water downstream for an extended time after a rain event, serving to stabilize stream flow that is needed for agriculture, power plants, navigation, and ecosystems within the watershed.
- Wetlands are important areas for the recharge of aquifers, as they maintain a saturated source of water that can enter aquifers that are tapped elsewhere by wells.
- Wetlands are like kidneys in that they filter out contaminants, sediment, and otherwise purify water before it enters streams and groundwater.
- Wetlands support biomes with exceedingly high Net Primary Productivity, sequester carbon rapidly, and are unique in that they can accumulate carbon on a continuous basis, creating peatlands, which are ultimately the progenitors of coal beds. As such, they play a key role in the carbon cycle, and in view of anthropogenic carbon emissions from fossil fuel burning, are particularly important in reducing the rate of accumulation of CO_2 in the atmosphere.
- Wetlands serve as nurseries for a great many species, a large fraction of which are endangered, in part due to the loss of wetland area caused by development, agriculture, and other human land use.

Due to these and other important functions of wetlands, the U.S. EPA has developed stringent policies regarding the preservation of remaining wetlands (Peralta et al., 2017), which represent a small fraction of the area of wetlands prior to European colonization of North America. However, regulations take into account only wetland area rather than function, and land owners have been known to be able to destroy a fully functional wetland and replace it with essentially a mud puddle of equal area elsewhere. Unless a newly created wetland performs the functions mentioned above, it is not useful, and in time, a more functional characterization of wetlands may play a role in policy-making.

Limits of Water Supply

The highly heterogeneous distribution of precipitation and runoff has led to a very uneven distribution of water resources around the world. Changes in these water resources have led to the collapse of entire civilizations in the American Southwest, the Middle East, and South America, for example. While water shortage may have led to the collapse of ancient civilizations, modern population growth, industrialization, and evolving diet are currently leading to global stress of the Earth's water supplies. Of the approximately 9,000 cubic km of annual global river runoff that is steadily available for use, humans presently appropriate about half. An increasing fraction of the world's population is being considered water scarce (less than 1,000 cubic m per person per year) or water stressed (less than 1,700 cubic m). As global population grows, and water resources do not, this may be the cause of deprivation and even conflict in the coming decades.

Groundwater can be found in virtually every part of the earth, even (and most important!) under deserts. In many parts of the world, groundwater is the only available source of clean, fresh water. Overuse of groundwater has, however, created some problems that are inevitable anytime a resource is over-utilized. The aquifer can be polluted or intruded by salt water, it can be depleted, and in some cases, the land above it can be caused to subside. In part, these problems are unique to groundwater because it flows very slowly through the pores and interstices in rock formations. This means that if water is withdrawn from wells, it may take a long time to recharge an aquifer. Also, if it is polluted from, for example, leaky underground fuel tanks or surface spills, it is very difficult to clean up.

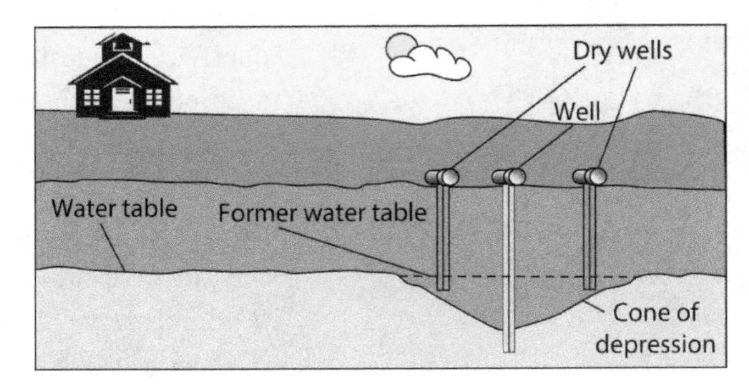

FIGURE 7.12 Multiple wells tap an aquifer
When multiple wells tap an aquifer, their cones of depression interact, and a deeper well may cause shallower wells to run dry.

One can think of a well drilled into an aquifer as a straw inserted part way down into a glass of water. Water is drawn up the straw, and water from the surrounding volume moves in to take the place of what has been sucked out. An aquifer behaves more like a slushy, frozen drink, however. Water cannot move horizontally very quickly toward the straw to replace what is removed, so the water in the overlying volume moves down, depressing the surface of the water relative to the surrounding slushy ice in the case of a drink, and rock in the case of an aquifer. This creates a **"cone of depression"** around a well, causing the water table to locally sink down toward the end of the well (Figure 7.12). The less permeable the rock that makes the aquifer, the deeper and narrower the cone of depression will become, and in extreme cases of rapid water withdrawal, could reach the well end, causing the well to run dry, drawing up air instead of water. In such a case, slowing the rate of withdrawal could give water in the surrounding volume time to move toward the well and keep the water table (base of the zone of aeration) above the well. In many areas, multiple wells are drilled into the same aquifer, and if they are close enough and withdraw water fast enough, the cones of depression could meet, thus causing the entire water table to be depressed, and depending on the rate of recharge, could lead to depletion of the entire aquifer. This is currently occurring in many aquifers, including the famous Ogallala Aquifer in the U.S. High Plains, in central California, in the Sahara

Cone of depression: A part of an aquifer where water withdrawal by a well has lowered the water table locally. Its depth and width depend on the ability of water to move toward the well through the rock and, thus, on permeability.

Desert, and numerous other places where the population relies heavily on groundwater for agriculture and all other needs. When a resource (such as water) is removed and does not recharge at an appreciable rate, it is mining, and groundwater mining is becoming an increasingly troublesome practice as critical water resources are permanently depleted.

> Two people live in adjacent properties and obtain their water from wells drilled in an aquifer to the same depth. One uses more water, and makes a large cone of depression that causes the neighbor's well to run dry periodically. What do YOU think should be done about this?

Another problem that can occur from excessive groundwater withdrawal is land subsidence. Some areas are underlain by a **confined aquifer** in which the water itself helps to support the overlying land surface. It can do this because the aquifer is overlain by an **aquiclude**, or impermeable layer of rock that prevents the water from rising to the surface so that the grains that make up the rock formation (which may or may not be well consolidated into solid rock) can become more compacted. When this water is removed from a confined aquifer through wells, the grains are permitted to become more compacted, and the land above sinks down a bit. This subsidence can be severe in certain places, and one such locality is the San Joaquin Valley in central California, in which agricultural water demand has caused so much water to be removed (permanently—thus mined) from the confined aquifer that the land surface subsided by about thirty feet, and the water table dropped by over four hundred feet. Once a confined aquifer compacts, it loses some of its porosity, or ability to store water, and even if it were allowed to recharge over time, less water would be stored than previously.

Water Pollution

Water is useful for terrestrial ecosystems, agriculture, and especially drinking, only if it is fresh and clean (Figure 7.13) (Zahran et al., 2018; Masten et al., 2017). However, water is an excellent solvent, and thus dissolves many substances that enter the water for any reason. This is critical for biogeochemical cycling, but it can be detrimental when toxic chemicals or other contaminants are added to water resources. Water also carries pathogens and other microorganisms that affect human health, sometimes severely. The chemistry of water can be monitored, and there are limits set for concentrations of various common contaminants, beyond which water is deemed unsafe for various uses, particularly drinking.

FIGURE 7.13 Water pollution

Eutrophication

When the dissolved oxygen content in water falls to critically low levels, animals, such as fish, macroinvertebrates, and others die of asphyxiation. In what seems like a paradoxical mechanism, it is nutrients like nitrogen and phosphorus that lead to **eutrophication** when they enter a body of water in excess quantities (Reddy et al., 2018) (Figure 7.14). When nutrients needed for plant growth, and algal growth in particular, enter a stream, lake, or wetland, a huge bloom of greenery results, mostly algae due to its rapid growth and reproduction rate. The water is essentially fertilized, and the expected bloom occurs. This is fine so far as it goes, for fish, for example. Neither the nutrients nor the algae bother animals. The problem is that all these algae soon die to be replaced by others. When they die, they are decomposed by bacteria in a process that requires oxygen dissolved in the water, so the oxygen content plummets and this is what kills the fish by asphyxiation. (In part, this is a reason you see laundry detergents advertising that they are "phosphate free", so as not to exacerbate eutrophication.)

FIGURE 7.14 Eutrophication

Eutrophication: The reduction of dissolved oxygen in a water body due to the decomposition of large masses of algae that bloom in response to the addition of nutrients (e.g., N, P) into the water from runoff from farms and other fertilized land, as well as sewage treatment plants and other sources of nutrients.

Biochemical Oxygen Demand (BOD)

A measure of the rate at which bacterial decomposition occurs in a water body is the biochemical oxygen demand (BOD) (Zhu et al., 2018). A great deal of decomposition demands a great deal of oxygen, and thus depletes the oxygen available for fish and other animals in the water. Lots of decomposition occurs in the presence of an algal bloom caused by high nutrient flux into a water body, so this is where you see high BOD and in extreme cases, eutrophication. A common place to see such conditions is in the effluent of a municipal waste water treatment plant. Even though the sewer water is treated chemically, is oxygenated, and after treatment, contains low and acceptable levels of bacteria and other contaminants harmful to human and other animal health, it still often contains high levels of nutrients such as nitrogen and phosphorus. These are difficult to remove from waste water and are not harmful to any forms of life. In fact, they are very nutritious for plants (and algae)! Thus, when waste water is poured into a stream (as it must be), even though the water may be clean enough to drink (it seldom is, due to other pollutants already in the stream), it is highly nutritious for algae and they bloom just downstream of the treatment plant, causing high BOD and reducing dissolved oxygen. Farther downstream, the oxygen content recovers, as eventually the decomposition is completed and the water has time to exchange oxygen with the atmosphere (Figure 7.15).

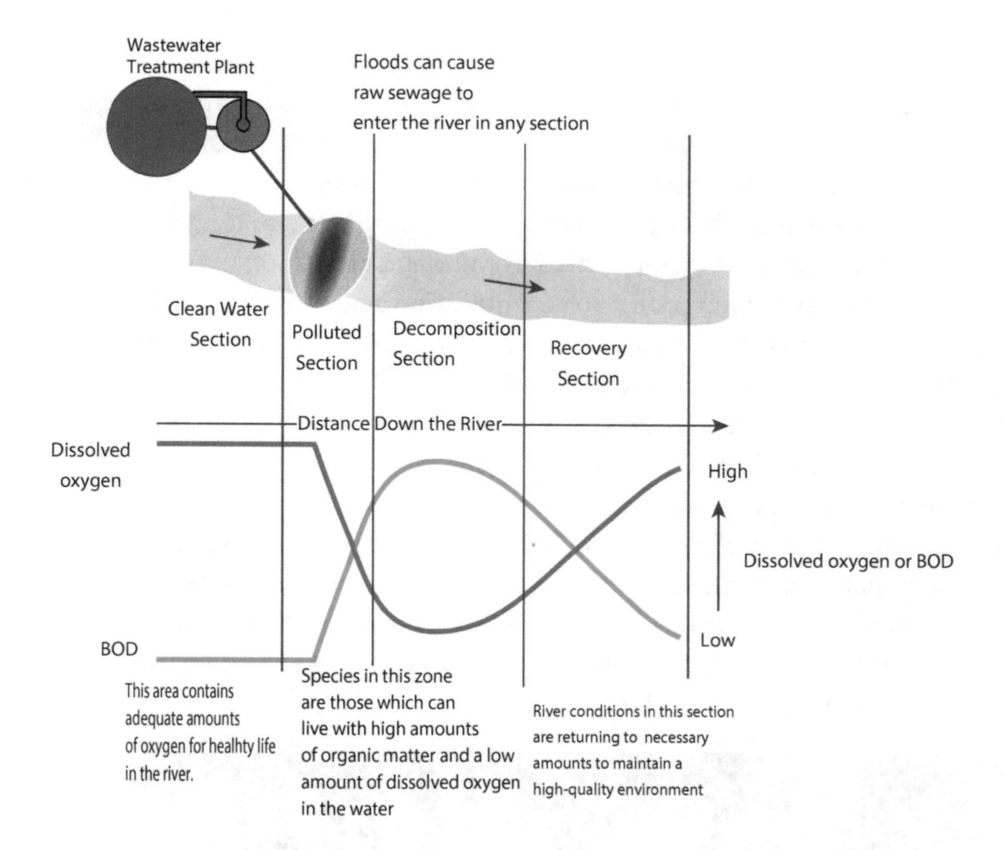

FIGURE 7.15 Biochemical oxygen demand

Biochemical oxygen demand is the amount that decomposing bacteria use up oxygen dissolved in water in order to decompose dead organisms, most notably algae that may bloom in response to introduction of nutrients (e.g., N, P) into surface water.

Fecal Coliform Bacteria

There has been great concern and many news headlines regarding fecal coliform bacteria found at beaches, water supplies, and other water bodies. These bacteria are not actually harmful to humans, and, in fact, are critical symbionts in our digestive tracts. However, they are easy to detect and measure in a sample of water. As such, they are used as a proxy for the presence of fecal matter that may (or may not) contain other bacteria and microorganisms that ARE harmful to human and other animal health. For example, the presence of these bacteria may indicate the presence of viruses such as hepatitis that are much more difficult to identify in water samples. Fecal coliform bacteria are not always used merely as proxies, however, and one form in particular is indeed harmful to human health. *Escherichia coli* (*E. coli*) can cause severe illness and death in humans, and has been the cause of disease outbreaks in a few American cities over the last several decades.

Non-point source pollution: Release of contaminants into water (or soil) from a broad area as a result of rain runoff from the ground surface. This is difficult to attribute to a single person, place or activity. Excess fertilizer from farms dissolved in rain runoff, providing nutrients for eutrophication, is a common type of non-point source pollution, as is oily runoff from city streets.

Surface Water Pollution

It is very easy to pollute surface water. Just dump or spill a contaminant on the ground or into the water and surface water at some scale will quickly become polluted. One can do this in either of two ways. If you concentrate your contaminants and release them from a single place, such as a drain pipe, or tank, it is considered **point source pollution**. This is relatively easy to track to its source. If, however, a contaminant is spread over a large area, such as an agricultural region in which all the farmers apply fertilizers, herbicides, or pesticides, for example, which are carried by rainwater and runoff into local streams, it is considered **non-point source pollution** and is much more difficult to attribute to a particular person, place, or source. There are a number of ways to reduce the rate of surface water pollution, both point and non-point source. The first and most obvious is to reduce the quantity of chemicals used in industry and agriculture so that there are fewer contaminants to worry about. Beyond that, it is possible to treat point-source contaminants at the source through a variety of chemical and biological treatment methods. Non-point source pollution is inherently more difficult to treat, as it enters streams and other water bodies over very broad areas. This is where wetlands and the **riparian zones** (ecosystem adjacent to a stream) bordering all streams are particularly important (Figure 7.16).

Point source pollution: Release of contaminants into water (or soil) from a single place such as a pipe, drain or smokestack.

Non-point source pollution: Release of contaminants into surface or ground water from a broad area rather than a single spot. Examples are runoff from farms, or city streets.

FIGURE 7.16 Riparian zone

In their function as water filters, wetlands can greatly reduce nutrient runoff by taking up nutrients (so terrestrial and wetland plants grow faster, rather than aquatic algae that die and create high BOD and eutrophication), and settling out sediments carrying other contaminants such as oils, pesticides, or herbicides before they reach a stream. From the perspective of cleaning up a polluted stream, the simplest approach is merely to stop polluting it. The residence time of water in a stream is days to weeks, and once pollution sources are halted, the stream will cleanse itself in short order.

Groundwater Pollution

Groundwater is slightly more difficulty to pollute, but it is MUCH more difficult to clean up than surface water (Figure 7.17). If contaminants enter an aquifer through a recharge zone, an entire aquifer can be polluted and all the wells drawing water from it can be rendered useless for human consumption. While it may seem a simple matter of protecting recharge zones, many aquifers have very broad recharge zones, and many shallow aquifers are recharged from the entire overlying area.

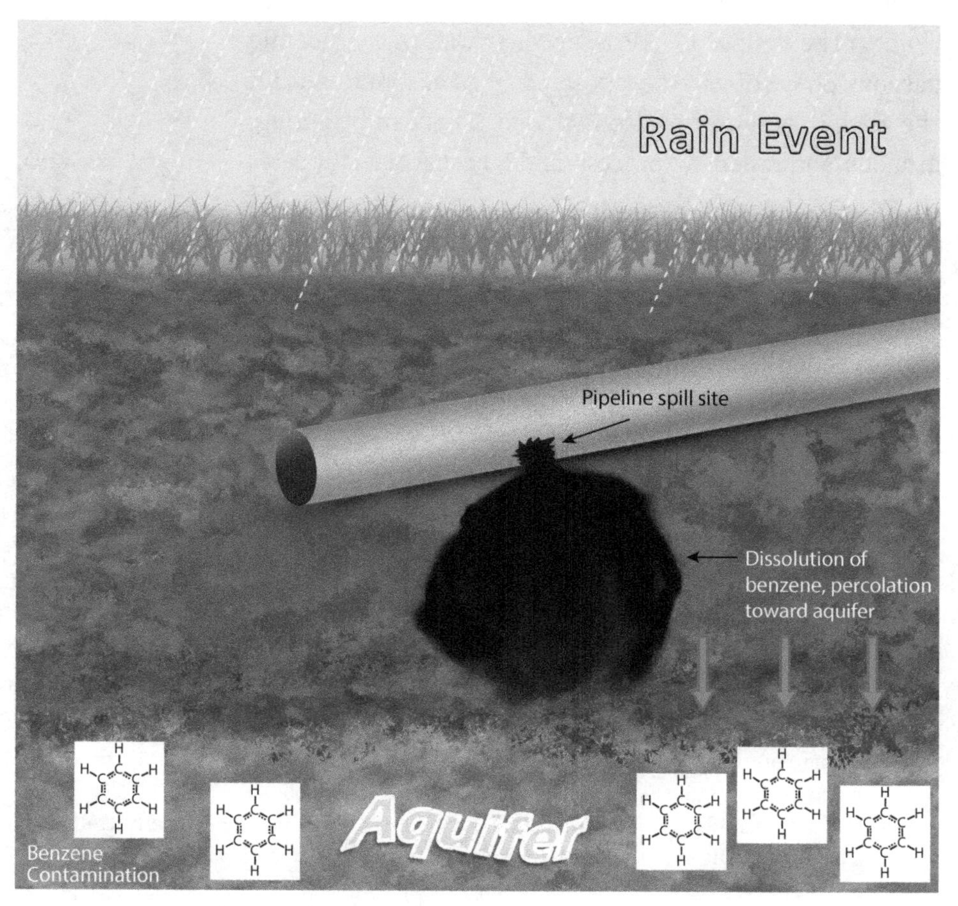

FIGURE 7.17 Groundwater pollution

There are many potential contaminants of groundwater, but there are a couple of particularly troubling classes of chemicals that commonly pollute aquifers. These are **dense non-aqueous phase liquids (DNAPLs)** and **light non-aqueous phase liquids (LNAPLs)**. "Non-aqueous" means that they are difficult to dissolve in water, so they remain at full strength in an aquifer that receives them. DNAPLs are chemicals such as trichloroethylene (TCE) and perchloroethylene (PCE) that sink to the bottom of an aquifer, inevitably passing the level of well ends on the way. These are not chemicals anyone would want in their drinking water. LNAPLs are less dense than water, so float on top of the water table. Common examples are gasoline and many types of oil that can be released from leaking tanks at gas stations, industrial sites, and many other locations.

Cleanup of polluted groundwater is both difficult and expensive. One method is to drill extraction wells at various positions around a contaminant source, accounting for the movement of groundwater downstream in its baseflow (Figure 7.18). These wells draw out the contaminant and adjacent polluted water until the extracted water reaches an acceptable concentration of the contaminant in question. In the case of LNAPLs, another well to extract the gaseous phase of vapor evaporated (e.g., gasoline) within the zone of aeration is also included. This can be a long, arduous, and expensive process, and is best avoided by carefully removing underground contaminant sources such as tanks. Gas stations are increasingly installing above-ground tanks to reduce the huge liability of an underground leak that could go undetected for a long time and be exceedingly expensive to remediate. In many cases, the entire property involved is not worth the cost of the required clean-up, and thus banks are very careful about writing mortgages for such properties.

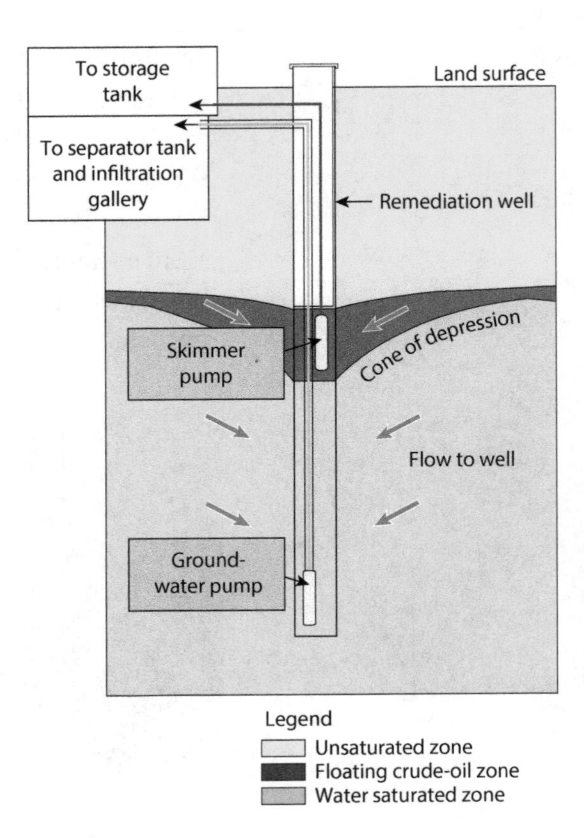

FIGURE 7.18 Aquifer contaimination
When a contaminant enters an aquifer, remediation is difficult and expensive. One approach is to drill a well to pump out the contaminant and the surrounding contaminated water. This prevents the movement of groundwater away from the site by base flow and helps to localize the problem. The remediation system may or may not ultimately remove all of the contaminant.

References

Masten, S. J., Davies, S. H., & McElmurry, S. P. (2017). Flint Water Crisis: What happened and why? *Journal of American Water Works Association*, 108(12), 22–34.

McDermid, S. S., & Winter, J. (2017). Anthropogenic forcings on the climate of the Aral Sea: A regional modeling perspective. *Athropocene*, 20, 48–60.

Peralta, A. L., Muscarella, M. E., & Matthews, J. W. (2017). Wetland management strategies lead to tradeoffs in ecological structure and function. *Elementa-Science of the Anthropocene*, 5(74).

Reddy, V., Fernandes, C., Gasparini, D., & Kurian, M. (2018). A water-energy-food nexus perspective on the challenge of eutrophication. *Water*, 10(2), 101.

Williams, A. S., Kiniry, J. R., Mushet, D., Smith, L. M., McMurry, S., Attebury, K., . . . & Johnson, M. V. (2017). Model parameters for representative wetland plant functional groups. *Ecosphere*, 8(10), e01958.

Zahran, S., McElmurry, S. P., & Sadler, R. C. (2017). Four phases of the Flint Water Crisis: Evidence from blood lead levels in children. *Environmental Research*, 157, 160–172.

Zhu, J., Kang, L., & Anderson, P. R. (2018). Predicting influent biochemical oxygen demand: Balancing energy demand and risk management. *Water Research*, 128, 304–313.

Figure Credits

Fig. 7.1: Copyright © Markus Gayda (CC BY-SA 3.0) at https://commons.wikimedia.org/wiki/File:Wasserl%C3%A4ufer_bei_der_Paarung_crop.jpg.

Fig. 7.2: http://en.wikipedia.org/wiki/File:Water_cycle.png

Fig. 7.3: Copyright © John Kerstholt (CC BY-SA 3.0) at https://commons.wikimedia.org/wiki/File:Rolling-thunder-cloud.jpg.

Fig. 7.7a: https://commons.wikimedia.org/wiki/File:Aral_sea_1985_from_STS.jpg

Fig. 7.7b: https://commons.wikimedia.org/wiki/File:AralSea.A2003283.0705.500m.jpg

Fig. 7.10: https://commons.wikimedia.org/wiki/File:Wetlands_upper_Kanuti_river.jpg

Fig. 7.13: Copyright © Stephen Codrington (CC by 2.5) at https://commons.wikimedia.org/wiki/File:Obvious_water_pollution.jpeg.

Fig. 7.15: Adapted from Peter H. Raven, Linda R. Berg, and David M. Hassenzahl, Environment. Copyright © 2010 by John Wiley & Sons, Inc.

Fig. 7.16: Copyright © Abraham (CC BY-SA 3.0) at https://commons.wikimedia.org/wiki/File:Riparian_zone_of_Klodnica_in_Katowice_Panewniki_2010.jpg.

Fig. 7.17: Copyright © 570ajk (CC BY-SA 3.0) at https://commons.wikimedia.org/wiki/File:Benzene_Transport_to_Groundwater_from_Oil_Spill.pdf.

Fig. 7.18: https://toxics.usgs.gov/highlights/2014-07-03-remediation_challenge.html

CHAPTER 8

Soils, Agriculture, and Food

People, they bred quite a burgeoning brood
Making more hungry mouths needing more food.
Farmers' efforts and toil
Just eroded the soil.
Sustainable farming must now be renewed!

WITH A GLOBAL human population of over 7.9 billion and growing, there is some concern about feeding everyone. As a society, ever since moving on from hunting and gathering, we have been increasingly efficient in our ability to grow our own food. This ability depends heavily on the availability and nature of soils, which, of course, are also essential to natural ecosystems.

Soils: Not Just Dirt

Soil is a thin layer of a mix of mineral grains (45%), organic matter and living organisms (5%), and air and water (50%) at the land–atmosphere interface, and is unique to the earth, being formed by rock, water, and life. It is an active, vibrant place, full of diversity. Yet most people take soil for granted, as it is everywhere: We walk around on it, it gets stuck in our shoes, muddies our children's clothes, and there seems to be an endless amount of it. But there isn't. Soil is a critical global resource, and we are losing it at a rapid rate due to deforestation and agriculture.

Soil is formed by the weathering of rock, so its characteristics are partially determined by the kind of rock in the area (Terra et al., 2018; Hughes et al., 2017). Mineral grains are altered by water and dissolved chemicals, and provide a mineral substrate for soil to form. Plants grow and add organic material to the soil. This decomposes by bacterial and fungal action and makes carbon and various nutrients available to plant roots. The acids produced by root respiration and other organisms help to weather additional rock to continue forming more soil. However, soil formation proceeds at a very slow pace. It takes hundreds of years to form an inch of soil, and it cannot be done artificially by any practical means. (So soil is worth preserving!)

Soil Structure

Natural, undisturbed soils have a well-identifiable structure from top to bottom, consisting of a series of layers, called "horizons," that define a soil profile. At the top, which includes the ground surface, is the **"O" horizon**, consisting mostly of organic matter, leaf litter, and organic waste in the process of decomposing (Figure 8.1).

The next layer is the **"A" horizon**, usually considered topsoil, a mixture of mineral grains and organic matter. This is a critical layer, as much of the biological and root activity in soil takes place here. It is the part that farmers are most concerned about, and the part that they destroy most rapidly. Grasslands have a thick A horizon, and much of grass biomass is in the roots. Rainforests have a very thin A horizon, which is why deforestation is so damaging in these areas, rendering the land veritably useless a short time after deforestation.

Below that is the "B" horizon that accumulates the various nutrients, clays, and dissolved minerals carried downward by infiltrating water from the horizons above. Without much organic matter, it is relatively dense and light colored, and does not retain water as well as the topsoil of the A horizon. Some soil scientists further delineate an **"E" Horizon** that exists in some soils between the A and B Horizons to indicate a zone of heavy leaching (dissolution by water), where metals such as iron, magnesium, and calcium are leached out of the soil, along with clay minerals, to be transported downward into the B Horizon and precipitated there. Leaching is enhanced by the composition of rainwater, which is slightly acidic, as atmospheric CO_2 dissolved in raindrops and forms carbonic acid, giving rain a pH of about 5.6 (7 being neutral).

The lowest layer of anything that could be considered soil is the **"C" horizon,** composed of weathered rock, broken up by physical, chemical and biological weathering processes, and altered chemically in preparation for perhaps, someday, becoming a B horizon. At the base of the entire system is the unaltered rock of the Earth's crust, which serves as parent material for the whole soil. While not part of the soil itself, it could be called a **"D" Horizon**, being necessary to support the soil physically and chemically for the long term.

Life in the Soil

Soils are very much alive, inhabited by a multitude of microorganisms such as bacteria, fungi, algae, protozoa, and microscopic animals. Larger animals also make their homes in the soil, most notably worms, which tunnel throughout the soil, making passages for air and water to enter, and helping to process organic material throughout the soil. Another famous resident is the many species of ants that build their own kingdoms in the soil, complete with complex societies, only some parts of which ever emerge to the ground surface. The part that does moves organic material deeper into the soil, and helps to bring plant seeds to where they can more effectively germinate and grow.

Larger animals such as rabbits, moles, groundhogs, and prairie dogs also live in the soil, constructing large burrow systems that help the soil transmit water, air, and nutrients over large distances underground.

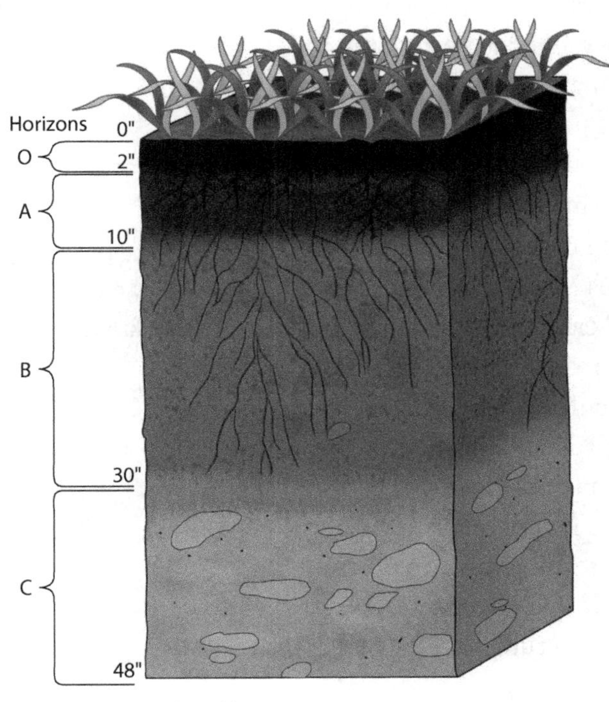

FIGURE 8.1 Soil profile
Depths are generic and can be highly variable between soils.

Fungi play a particularly important role in soils. Plants have a symbiotic relationship with **mycorrhizal fungi** that extend long filaments, called mycelia, from plant roots deep into soils to extract nutrients and water that are then provided to the plant, which, in turn provides the fungi with organic matter as food (Soudzilovsaia et al., 2015; Talbot et al., 2008). While each can live without the other (at least for a while), the relationship between plants and mycorrhizal fungi enable both to grow faster and survive more severe perturbations than they would otherwise.

Food

Although global production of food is at all-time highs, it is unevenly distributed, and some areas (e.g., U.S.) have overabundances of food where many people do not even finish everything on their dinner plate and still face obesity, while people in other regions suffer from undernourishment and malnutrition, with barely enough food to stay alive, much less to lead healthy and productive lives. While this has political, social, economic, and ethical aspects that will remain unexplored in a science textbook, they bear on how regional populations approach agriculture and its impact on soils and the environment.

Mycorrhizal fungi: Organisms in a commensal relationship (typically, and perhaps incorrectly, termed symbiotic) with vascular plants in which the fungi attach to plant roots and extract sugars from the plant to support fungal metabolism (fungi are heterotrophs). While they do this, they provide the plant with nutrients and water. The fungi are able to extract more water and nutrients from the soil, because they have long filaments that permeate the soil and, thus, offer lots of surface area for material exchange.

Agriculture is based primarily on a small number (about 150) of species that have been cultivated and bred over the millennia as food for humanity. The main crops are wheat, rice, maize (corn), and potatoes. Additional crops, such as alfalfa, are planted to feed livestock, which are then eaten by people. Recall the discussion of trophic levels and how 90% of the energy stored in organisms is "lost" when eaten by other organisms. (That 90% was used by the organism for metabolic processes, rather than to build biomass for consumption by another organism.) Therefore, many more people could be fed from the croplands that grow forage crops for animals than can be fed by eating the animals that eat those crops. Animals do not always eat crops that are planted for that purpose. On rangelands, animals graze on grasses that can grow in climates that are not suitable for other crops without irrigation. People cannot digest grass, so animals that can do this can at least provide people with a fraction (about 1%) of the energy stored in grass.

Agriculture: Environmental Impacts of Feeding Humanity

About ten thousand years ago, people figured out that they could get more food if they grew it in place, rather than hunting wild animals and gathering fruits and berries from the countryside. So started agriculture, which began to completely change the face of the planet as ecosystems were coopted to become farms to provide specific food crops for human consumption.

The most significant invention to facilitate agriculture while simultaneously destroying soils and associated ecosystems was the **plow**. By plowing the soil, a number of useful agricultural functions are performed. Plowing crumbles up soil and exposes it to oxygen, kills weeds living on the surface, brings nutrients in the deeper soil to the surface, and creates ridges and furrows that are useful for planting crop seeds. The plow enabled early farmers to increase crop productivity, and thus support more people with a given area of cropland. As such, the plow was part of the modern human population explosion. While this was very useful for early agriculture, it is not sustainable for the long term. Plows mix up the soil and blend the O horizon in with the A horizon and sometimes, in already depleted soils, dig into the B horizon. By exposing a large part of the topsoil to oxygen of the atmosphere, it promotes oxidation of soil carbon, and thus carbon loss, as well as that of other nutrients. So the soil loses fertility, in addition to losing its natural structure, and thus loses the ability to grow crops (or natural ecosystems).

Plow: A device used by farmers to tear up and turn over the upper part of the soil in order to loosen it for the planting of new crop seeds, while killing any weeds (indigenous ecosystem colonizers) that may be starting to grow.

Farmers respond to fertility loss by adding artificial fertilizers composed of nitrogen and phosphorous compounds that are soluble in water and available to plant roots. Nitrogen-based fertilizers are normally produced by fixing nitrogen through the **Haber-Bosch process**, which combines methane (CH_4) and atmospheric nitrogen (N_2) at high temperatures to make ammonia (NH_3) that is useful for plants (Gilbert et al., 2014). In fact, through the Haber-Bosch process, our industrial processes fix more nitrogen annually than all of the bacteria in all the soils of the world. This threshold was passed in the mid-1980s (Figure 8.2).

Haber-Bosch process: An industrial and very energy-intensive means to combine atmospheric nitrogen (N_2) and hydrogen to make ammonia (NH_3). This is done at very high pressures and temperatures and makes ammonia suitable for agricultural fertilizers.

The plow affects soil in other ways as well. The most significant is that it exposes soil to rainwater and wind, both of which can transport soil particles away from a farm where they are exposed by plowing. To date, America, and indeed the world, has lost over a third of its topsoil (O and A horizons) to erosion, most of which has been transported to the Mississippi and other rivers, for deposition behind dams or in the ocean. This is a serious concern for America's "breadbasket" in the Midwest and Great Plains, and across the nation. When an ecosystem is cut down or simply plowed under, it is equivalent to a disturbance that paves the way for the process of ecological succession. Pioneers or colonizers, the early successional plants, are ideally suited to germinating and growing in newly plowed fields. When they do this, however, they are called "weeds" and present a constant problem for farmers who seek to grow a monoculture of the crop of interest without competition for water, nutrients, and sunlight from the early successional plant of the natural ecosystem. This has led to the widespread use of herbicides and the development of genetically modified herbicide-resistant crops that can grow even when general herbicides are applied.

Weathering: The breaking down of bedrock by physical and chemical means. Chemically, water can dissolve some minerals, and acids produced by plants and other organisms can dissolve additional minerals in rocks to help break up solid rock into small pieces. Physically, solid rock can be cracked by tectonic and gravitational stresses, and water in cracks can freeze (and expand), thus further breaking up the rocks, and increasing the surface area for additional chemical weathering.

Erosion: The removal of soil by water or wind. Once soil is being transported, it is considered sediment that will be deposited somewhere downstream or downwind, where currents or winds are reduced. The faster the water or wind moves, the larger the particles that can be moved.

While soil erosion continues throughout all agricultural regions, there have been episodes of enhanced erosion that have had severe consequences on agriculture. The most notable of these was the **Dust Bowl** of the 1930s, when prolonged drought, combined with poor soil management, resulted in wind erosion of vast areas of the Great Plains, with Oklahoma as the poster child (Figure 8.2). In any and all cases, when natural ecosystems are cut down, plowed under, or otherwise destroyed for planting crops, the soil becomes vulnerable to oxidation, erosion, and loss of fertility. Consequently, measures must be taken to minimize the negative impacts of agriculture not only for preservation of the soil, but also for the continued sustainability of agriculture.

FIGURE 8.2 Dust Bowl

Contour Plowing: Plowing of agricultural soil in rows that follow elevation contours so that, rather than creating small gullies for rainwater to flow downhill and carry soil away by enhancing erosion, the gullies form small dams that reduce overland water runoff and help reduce soil erosion to less than "normal" plowing, but still more than that of an indigenous ecosystem

Soil Conservation

Weathering of rocks leads to the formation of soil. The soil can then erode by being carried away by water or wind. It is thus important to distinguish between **weathering** and **erosion**. Plant roots and vegetation in general serves to keep soil in place. However, for farmland, the first thing that is done is removal of all vegetation and roots, in preparation for planting of crops. The bare soil can then erode very quickly. In order to reduce the rate of soil erosion (but not stop it completely), there are several approaches that can be taken by farmers. One key technique is **contour plowing** (Figure 8.3). If one must plow the soil (and one mustn't), one way to reduce the rate of erosion by rain-induced surface flow that carries soil particles downstream is to prevent surface flow. By plowing ridges and furrows across the slope of a hill, rainwater is trapped between the ridges that act like little dams and keep both water and soil from moving downhill. The water eventually soaks into the ground, and the soil stays where it is.

FIGURE 8.3 Contour plowing

This is very effective in preventing washouts, arroyos, and other erosion features in farmlands. It is also easier to manage with a tractor, and is a basic no-brainer for a farmer. It does not, however, prevent wind erosion, or exposure of the soil to oxygen and the consequent loss of carbon and nutrients (Figure 8.4). To reduce prolonged exposure of soil to the air, it is far better to plow in the spring, just before planting, than it is in the fall, after harvesting. Spring tillage enables the stubble and roots of the previous crop to help maintain the soil through the winter. Better still, modern methods of planting do not require tillage at all, and **no-till agriculture** has become an increasingly widespread practice. This requires other measures to reduce pests and weeds, and **integrated pest management** schemes have been developed to accomplish just that (rather than using chemical pesticides, designed to kill animals, of which we ourselves are a variety). The idea behind integrated pest management is to control the overabundance of pests by introducing the appropriate predators that do not harm the crop, and do not eat ALL the pests, but keep them in small enough numbers that they do not significantly affect the crop of interest. Certain types of wasps, praying mantises, ladybugs, and many other species can be used, and if they have an appropriate habitat nearby, can serve to maintain a sustainable, if artificial ecosystem that limits the overabundance of any of the pest species that would otherwise proliferate in the presence of a food crop.

Integrated pest management: The introduction of predator species to eat agricultural pests to keep the pest population low enough to be inconsequential to crop yield.

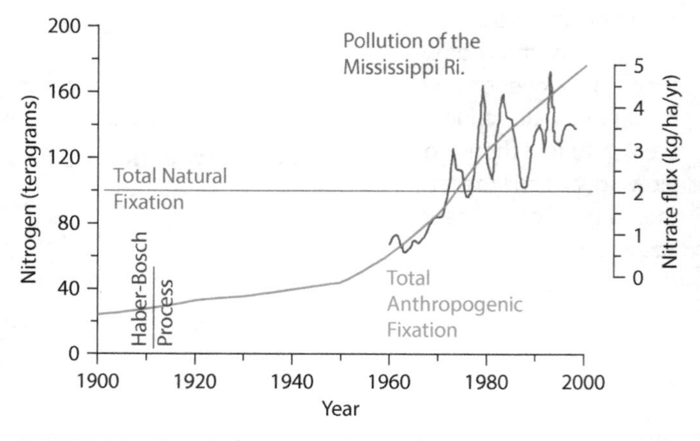

FIGURE 8.4 Twentieth century nitrogen fixation

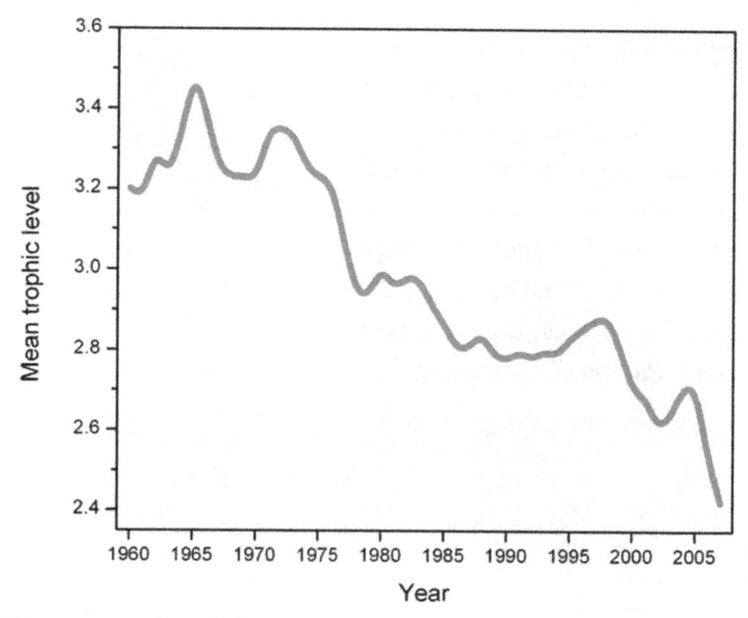

FIGURE 8.5 Fish trophic levels
People like to eat big fish such as salmon, swordfish, and other fish at the top of the food chain. Recall that there are many trophic levels in the ocean (more than on land). As we catch and eliminate most of the big fish, it becomes economically more feasible to catch fish at lower trophic levels. So we are fishing down the food web.

Aquaculture and Mariculture

On land, we abandoned the primary practice of hunting and gathering in favor of agriculture about ten thousand years ago, but in the ocean, we are still mostly hunters and gatherers. We go out in boats, find fish, catch them, and eat them. We have learned to become very effective in this to the point that Daniel Pauly, former Minister of Fisheries of Canada, declared that "We are at war with fish ... And we are winning!" In fact, the fish don't stand a chance, as we have advanced technology that enables us to track them, capture them, and bring them to market most effectively. This has had the effect of wiping out the fish stocks in many of the prime fisheries of the world; most notably the Grand Banks, once teeming with fish, and now essentially barren. With the favored fish at the highest trophic levels depleted, fishermen catch lower trophic levels for market, and as such, are **fishing down the food web** in the ocean (Figure 8.5).

Fishing down the food web: The practice of catching fish from lower trophic levels, because fishermen have already caught so many higher trophic level fish as to deplete their populations below what is practical for commercial fishing.

Aquaculture: Raising fish in confinement for human consumption in fresh water.

Mariculture: Fish (or shellfish) farming in the ocean.

As a result of the sharp decline of global fisheries, some fishermen have taken agriculture into the water, and started farming fish and shellfish both in the ocean and in freshwaters (Figuer 8.6) (Fentry et al., 2017). In freshwater, along streams, lakes, and ponds, fish are raised in enclosures that in many cases obtain food and nutrients from the flow of water in the stream, and in other cases from municipal and other food wastes that maintain a food supply for a high density of fish of the desirable species. In the U.S., this kind of **aquaculture** is most commonly done with catfish in the southern part of the nation. In the ocean, there is much more area to work with, but it can be less convenient and manageable, yet **mariculture** is an increasingly common practice. This is most commonly done with mussels and oysters, which can grow on artificial structures and reefs in great numbers. Another shellfish being farmed is lobsters in New England, where the high density of lobster traps on the ocean floor are a major source of food for a large lobster population. As it turns out, the traps do not actually fool the lobsters, which, as was found when marine biologists placed underwater

FIGURE 8.6 Aquaculture

cameras near the traps, enter and leave the traps at will, enjoying a smorgasbord of bait as food, grazing from one trap to another. Lobstermen catch the lobsters that are "at the table" at any given time. Another major mariculture species is salmon, which is kept in large enclosures (nets) in the ocean and fed fish (killed and processed) that are caught elsewhere in the ocean (Figure 8.7). (No living fish that could be eaten by a salmon would wander into a huge net full of hungry salmon!) However, humans could just as well eat the mackerel, herring, sardines, and other fish fed to the salmon. Recalling the loss of energy between trophic levels, we get only a small fraction of the energy if we

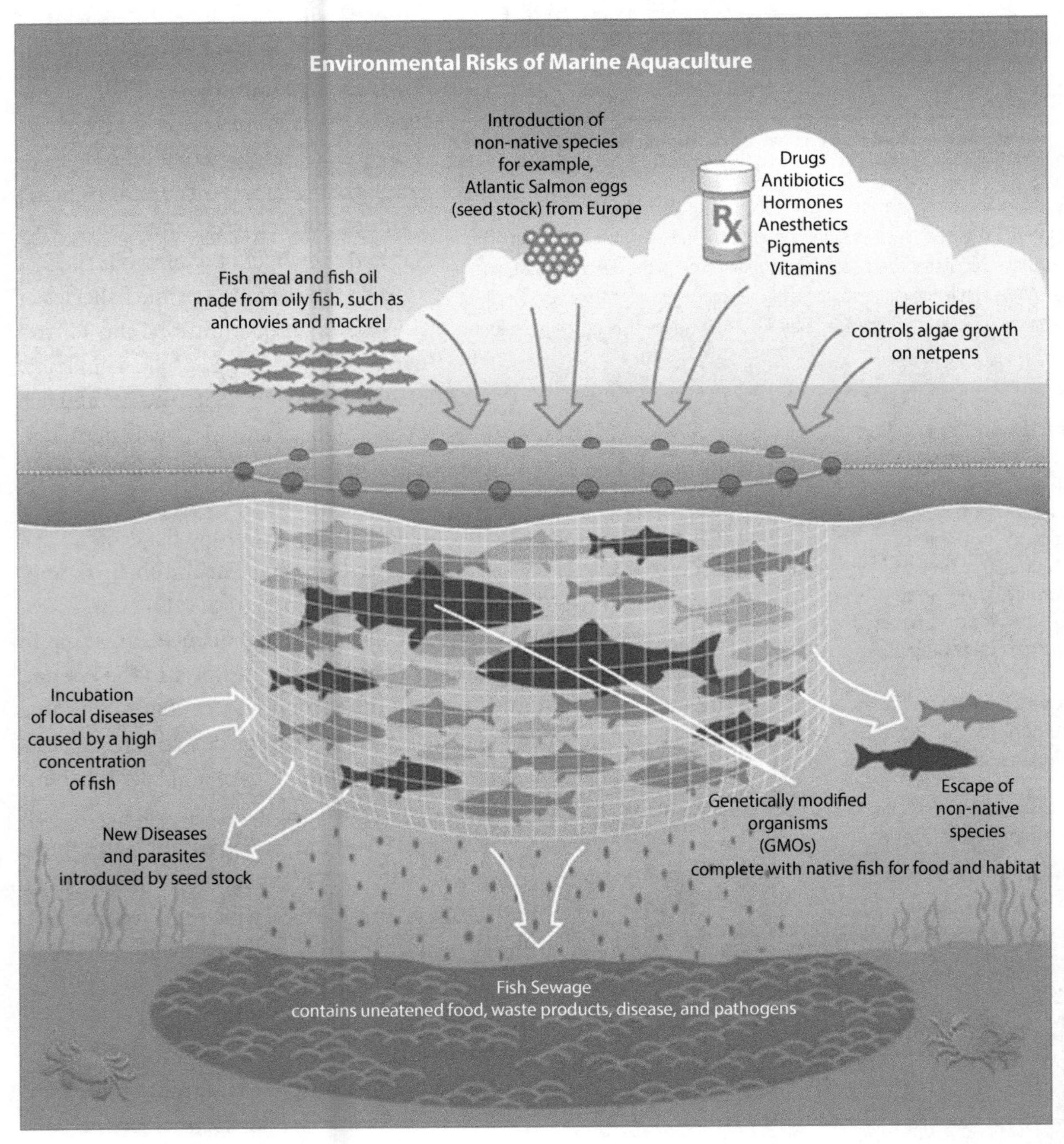

FIGURE 8.7 Risks of aquaculture

Farming up the food web: The feeding of lower trophic fish to commercially preferable higher trophic level fish in the course of fish farming (aquaculture and mariculture). Because about 90% of the energy is lost between trophic levels, this greatly accelerates the rate of overfishing the ocean, because we could have eaten the fish that are fed to the higher trophic level fish. So, farming up the food web actually accelerates the rate of fishing down the food web.

eat the salmon that eat the fish, rather than eating the fish themselves. As such, we are **farming up the food web** and more rapidly reducing global fish stocks than we would if we reverted to hunters and gatherers and ate the fish we are feeding to the salmon.

Genetically Modified Organisms (GMOs)

For millennia, farmers have battled drought, pests, weeds, and plant diseases in their attempts to grow food. The recent capability to alter plants' DNA has made it possible to more effectively avoid crop damage from all of these hazards, and has also produced crops that grow faster, can mature better after harvesting, are frost resistant, and have longer shelf life. This has been a boon to industrialized American agribusiness and to the supply of food for market. It has also raised some questions regarding environmental impacts and human health. There are concerns that consumption of GMOs may harm human health in ways that we do not understand (Toft and Hoyer, 2012; Kuiper, 2008). These concerns are particularly prevalent in Europe, where there are strict regulations regarding GMOs, their use, and the labeling of GMO products in the marketplace. In the U.S., GMOs are much more prevalent, although there are still some concerns. Most U.S. corn, soybeans, and cotton have been genetically modified and most Americans are unaware that they are eating GMOs every day.

It takes ten pounds of sardines to feed and thus produce one pound of salmon in mariculture. If the average person eats one pound of fish per month, how fast will wild sardine populations be depleted if they are fed to salmon, relative to how fast they would be depleted if the sardines are fed to people?

Answer: The sardines would be depleted ten times faster if they are used in mariculture. This is farming up the food web.

The precautionary principle suggests that nothing new be tried until it has been thoroughly tested and found to be safe. To date there has been no discernible harm evident from eating GMOs, so it is not clear how long tests would need to be conducted to determine that they are safe. A decade? Two? A century? Concerns extend beyond human health and into the broader environment. Risks regarding gene-escape from seed or pollen dispersal into the natural environment and the consequent alteration of the genetic composition of ecosystems have caused many to reexamine the value of GMOs. Even in agricultural systems, if herbicide-resistant GMOs cross-breed with wild relatives that are already considered weeds in croplands, and they transfer their herbicide resistance to the weeds, farmers will have an even greater problem with weeds than they have now. Likewise, GMOs designed to include toxins that make them pest-resistant could reduce insect populations to the point that birds and others that depend on them will be negatively impacted. Concerns, like the overuse of antibiotics, center on the emergence of "super-bugs" that are pesticide resistant and become environmentally very damaging. There has been discussion of a "terminator gene" that renders the seeds from GMOs sterile. This would help to prevent cross-breeding and gene-escape into the wild. However, it also contributes to the privatization of food production, as farmers would not be able to use seeds from a previous crop to plant the next crop, as they have done for millennia. Instead, they would have to buy new seeds every year from the corporate producer of the GMO seeds, and there is concern regarding the price of food, and especially the availability in the developing world, which can hardly afford to buy new seeds for every crop.

The advantages and disadvantages of GMOs are numerous and must be weighed carefully before deciding (to the extent that the population has a choice) how food should be produced. Proponents argue that the unknown risks of GMOs should be considered relative to the known risks of NOT using GMOs. If GMOs enable the same amount of crop yield from a smaller area of farmland, much more natural ecosystem can be preserved to provide environmental goods and services lost by conversion to agriculture. Further, if GMOs make it possible to dispense with the use of chemical pesticides (Figure 8.8) (Crowder et al., 2015), food could be safer to eat than that covered in chemical poisons. These are only a few of the issues and considerations regarding the development and use of GMOs for food production. Many more will emerge in the coming years.

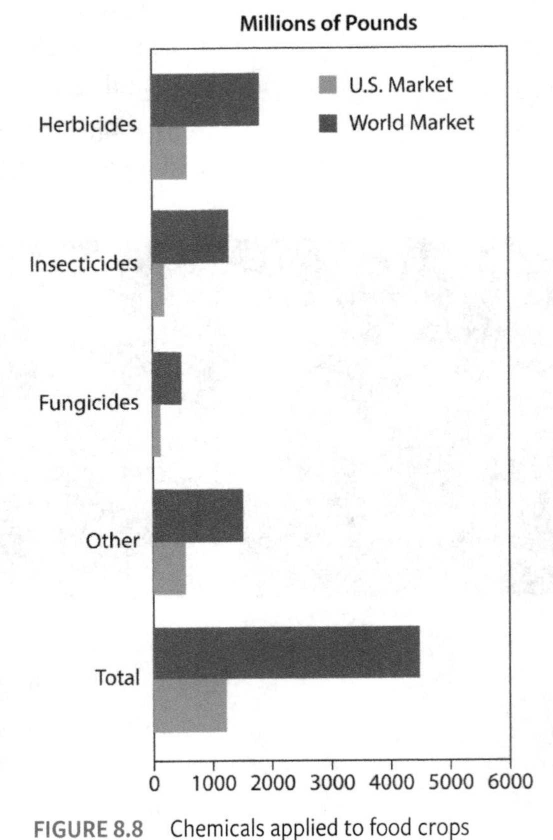

FIGURE 8.8 Chemicals applied to food crops

Organic Agriculture

For millennia, crops were produced and animals raised using simple methods, without genetic modification (beyond breeding and selection), fertilizers (beyond manure), or chemical pesticides. With the advent of chemical pesticides, the Haber-Bosch process for fixing nitrogen to make fertilizer, and even the ability to alter the genetic composition of crops for specific purposes, food production became much more efficient, but concerns arose for the potential contamination of food (Figure 8.8). In 1990, the Federal Organic Food Production Act was passed, providing a set of standards by which organic foods could be identified (Fouilleux and Locanto, 2017). According to the USDA National Organic Standards Board,

> Organic agriculture is an ecological production management system that promotes and enhances biodiversity, biological cycles and soil biological activity. It is based on minimal use of off-farm inputs and on management practices that restore, maintain and enhance ecological harmony.

The standards include the processes of production, some of which may not be measurable in the final product, specific methods for crop production that prohibit a list of substances (both organic and inorganic, including pesticides and herbicides), inorganic fertilizers (e.g., Haber-Bosch-derived N), and, of course, GMOs.

Further, standards for livestock require organic feed (for at least part of animals' lives), prohibit growth hormones and antibiotics (but vaccines are allowed), and require that animals have access to the outdoors for at least part of their lives. Interestingly, farmers are required to provide medical treatment to sick animals, but once doing so, if any of a list of specific medications was required to save the animal, it can no longer be marketed as organic. After a lifetime of medical treatment and medicines provided to people, this raises the question, "How organic are humans?"

Many of the methods involved in organic agriculture serve to preserve soil and to sequester carbon in soils. This has a positive value regarding the climate impacts of agriculture, which typically involve a loss of carbon to the atmosphere, if deforestation or plowing of grasslands is involved. The value judgments regarding the advantages and disadvantages of organic agriculture are not in the realm of science, but the ecological, biogeochemical, and climate implications are at the center of the science that is required to properly inform decisions and policies regarding organic, and indeed, all agricultural practices.

References

Crowder, D. W., & Reganold, J. P. (2015). Financial competitiveness of organic agriculture on a global scale. *Proceedings of the National Academy of Sciences of the United States of America*, 112(24), 7611–7616.

Fouilleux, E., & Loconto, A. (2017). Voluntary standards, certification, and accreditation in the global organic agriculture field: a tripartite model of techno-politics. *Agriculture and Human Values*, 34(1), 1–14.

Gentry, R. R., Froehlich, H. E., Grimm, D., Kareiva, P., Parke, M., Rust, M., . . . & Halpern, B. S. (2017). Mapping the global potential for marine aquaculture. *Nature Ecology & Evolution*, 1(9), 1317–1324.

Gilbert, P. M., Maranger, R., Sobota, D. J., Bouwman, L. (2014). The Haber Bosch-harmful algal bloom (HB-HAB) link. *Environmental Research Letters*, 9(10), 105001.

Hughes, P., McBratney, A. B., Huang, J., Budiman, M., Micheli, E., & Hempel, J. (2017). Comparisons between USDA Soil Taxonomy and the Australian Soil Classification System I: Data harmonization, calculation of taxonomic distance and inter-taxa variation. *Geoderma*, 307, 198–209.

Kuiper, H. A. (2008). Risk assessment strategies for GMOs in a global perspective. *Journal of Biotechnology*, 136(S), S713.

Soudzilovskaia, N. A., Douma, J. C., Akhmetzhanova, A. A., van Bodegam, P. M., Cornwell, W. K., Moens, E. J., . . . & Cornelissen, J. H. C. (2015). Global patterns of plant root colonization intensity by mycorrhizal fungi explained by climate and soil chemistry. *Global Ecology and Biogeography*, 24(3), 371–382.

Talbot, J. M., Allison, S. D., Treseder, K. K. (2008). Decomposers in disguise: Mycorrhizal fungi as regulators of soil C dynamics in ecosystems underglobal change. *Functional Ecology*, 22(6), 955–963.

Terra, F. S., Damatte, J. A. M., Viscarra Rossel, R. A. (2018). Proximal spectral sensing in pedological assessments: Vis-NIR spectra for soil classification based on weathering and pedogenesis. *Geoderma*, 318, 123–136.

Toft, K. H. (2012). GMOs and global justice: Applying global justice theory to the case of genetically modified crops and food. *Journal of Agricultural & Environmental Ethics*, 25(2), 223–237.

Figure Credits

Fig. 8.1: https://commons.wikimedia.org/wiki/File:Soil_profile.png

Fig. 8.2: https://commons.wikimedia.org/wiki/Category:Dust_Bowl#/media/File:Dust-storm-Texas-1935.png

Fig. 8.3: https://commons.wikimedia.org/wiki/File:TerracesBuffers.JPG

Fig. 8.4: https://www.geo.arizona.edu/Antevs/nats104/n_fixation.html

Fig. 8.6: Copyright © Narek75 (CC BY-SA 4.0) at https://commons.wikimedia.org/wiki/File:INTER_AKVA_Fish_Farm_in_Sipanik,_Ararat_marz,_Armenia.JPG.

Fig. 8.7: George Pararas-Carayannis, https://commons.wikimedia.org/wiki/Category:Aquaculture#/media/File:Risks_aquaculture_550.jpg. Copyright © 2000 by George Pararas-Carayannis. Reprinted with permission.

CHAPTER 9

The Oceans

The planet called Earth—it really is blue.
Just look down from space—you'll find that it's true.
We scurry on land
So don't understand
That we need to adopt an ocean world view.

IN THIS CHAPTER, we explore over 70% of the Earth's surface, yet an area that is poorly documented, only marginally regulated, and increasingly impacted by physical, chemical, and thermal perturbations. This book, as well as most human activity and concern, is focused primarily on the other 30%—the land. However, in order to understand the function of the Earth System and global environment, it is essential to at least touch on a few key environmental topics regarding the oceans.

Ocean Circulation

One key aspect of the ocean is its role in the climate system. The cooling and heating of different parts of the ocean surface have set up a global current that circulates through all the oceans in a system that serves to move heat around the planet and place strong controls on global climate. This system is called the "ocean conveyor belt" and is driven by variations in ocean water density, with a key element being sinking of cold, dense water in the North Atlantic (Figure 9.1) (de Carbalho Fereira and Kerr, 2017).

The circulation of the ocean is not only determined by prevailing winds, but also by variations in density of the water determined by both temperature and salinity. In the polar Atlantic, cold, salty water is very dense (for water) and sinks to the bottom of the ocean, providing one of the main drivers of the global ocean circulation system. This dense bottom water flows southward, around the southern tip of Africa, and ultimately into the Pacific, where it surfaces, heats up, and returns as a warm surface current. As it enters the North Atlantic again, it trends along the east coast of North America as the Gulf Stream, finally sinking again in the polar Atlantic near Iceland.

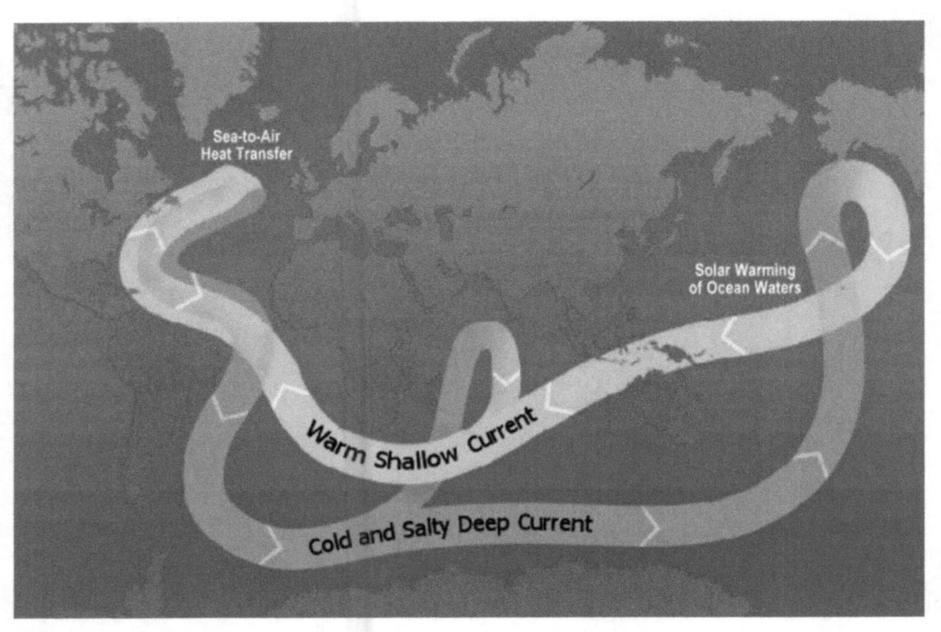

FIGURE 9.1 Ocean Circulatory conveyor belt
The global ocean conveyor belt is driven by the formation of cold, dense water that sinks to the ocean bottom in the North Atlantic (and also around the Antarctic margin). From the North Atlantic, it flows south and east into the Indian and Pacific oceans, mixing with the water in the largest ocean in the world, and rising to the surface where it is heated further to become a warm surface current. This warm surface current moves between Australia and Indonesia, westward across the Indian Ocean, and northward up the Atlantic, where it eventually becomes the very warm Gulf Stream, providing warmth to parts of the east coast of North America and ultimately to northern Europe. By the time it gets to Europe, it begins to cool from evaporation, becomes saltier and denser, and sinks to the bottom again, continuing the cycle of the conveyor belt. As climate changes due to anthropogenic greenhouse emissions, it is likely that the rate of North Atlantic Deep Water (NADW) formation will be reduced due to ice melting and increased European river fresh water runoff, and this will have profound impacts on global climate, as the ocean is the primary source of heat and heat transport globally.

The sinking surface water in the North Atlantic takes with it dissolved nutrients, gases (such as carbon dioxide and oxygen), salt, and any pollutants in the surface waters of the region. By moving CO_2 into the deep ocean, **North Atlantic bottom water formation** serves as a conduit for removing CO_2 from the atmosphere for long-term (thousands of years) storage in the deep ocean. (This is a large discount from global warming due to anthropogenic emissions.) However, reductions in the rate of bottom water formation are already being observed. Warming of the planet (due to excess greenhouse gases from fossil fuel burning) is not evenly distributed, so that the poles warm more than the tropics. This reduces the differential temperature and thus the strength of bottom water formation enabled by temperature-driven density contrast. In addition, melting of land-based glaciers and floating Arctic ice provide fresh water to the surface of the North Atlantic, reducing the salinity-driven density contrast. If the drivers of bottom water formation are impaired in these ways, the effectiveness of the ocean circulation system in helping to absorb anthropogenic CO_2 emissions is

North Atlantic bottom water formation: The sinking of cold, salty, dense water in the northernmost Atlantic Ocean. It is a primary driver of the global thermohaline circulation system (ocean conveyor belt).

reduced, and more of our emissions remain in the atmosphere, leading to more warming and ice melting, thus accelerating the decline in ocean circulation in a positive feedback (vicious cycle). It is not clear at what point a critical threshold will be crossed to trigger this process irreversibly.

RUBBER DUCKIES

A container ship suffered damage in a storm in 1992 and spilled about 30,000 little rubber duckies (the tubby toys) into the North Pacific Ocean (Figure 9.2). (Actually, along with the ducks were beavers, frogs, and turtles.) The little tub toys floated around the ocean, following winds and currents until they washed ashore at various points all over the earth, enabling oceanographers to study the current that carried them around the world's oceans (Hohn, 2011). They first showed up in Alaska, then Hawaii a few years later, and a decade later, made it all the way through the Arctic Ocean to Scotland and even Maine. Presumably some of these toys are still floating around the world's oceans today, but they would no longer be the brightly colored, friendly tub toys they started out as, but rather an encrusted haven for a variety of marine organisms that would render them unrecognizable blobs of floating sea life.

FIGURE 9.2 A rubber ducky

The directions these floating toys traveled were determined by the large-scale ocean gyre of the North Pacific, which, in addition to transporting rubber duckies, also moves heat, nutrients, larvae, and pollutants throughout the ocean. The North Pacific gyre is a large circular current, moving clockwise, that takes about three years to complete a rotation. (An adjacent subtropical Pacific gyre rotates counterclockwise, and in the southern ocean, the subtropical gyres rotate clockwise and the sub-polar gyres rotate counterclockwise, as dictated by the Earth's rotation and the **Coriolis effect** that control atmospheric and ocean circulation.) In the North Pacific, the current within the gyre can move rubber duckies about ten km per day. These gyres spiral water along the surface toward the center, where it slowly sinks to the deeper ocean. Consequently, anything floating in the region of the gyre moves toward the center, and gets concentrated there. This has led to the "Great Pacific Garbage Patch" with floating plastic and debris at high concentrations in the middle of the ocean (Lebreton et al., 2018).

Dansgaard-Oeschger events: 25 episodes of rapid warming during the most recent glacial period, each followed by a gradual cooling back to "normal" glacial conditions.

Heinrich events: Extreme cold events between some of the Dansgaard-Oeschger events, during which large armadas of icebergs melted in the North Atlantic, dropping ice-rafted sediment farther south than it normally extends and adding lots of fresh water to the surface ocean, thus reducing the strength of the ocean conveyor and making regional climate even colder.

Water has a very high value for heat capacity. That is, it takes a lot of energy to heat water, and once it is warm, it stays warm for a long time. Because of this and the sheer mass of the world ocean, ocean thermal structure strongly controls global climate. Any changes in either ocean temperature or the changes in the rate of movement of water in the global thermohaline circulation system would have an immediate and profound impact on climate. This has happened in the past, when the melting of Arctic ice freshened (thus lightened) polar ocean surface water, and thus reduced the strength of bottom water formation and ocean circulation. Over the last one hundred thousand years, rapid climate change events, called **Dansgaard-Oeschger events** and **Heinrich events**, were triggered by alterations in fresh water input into the North Atlantic. This will be explored further in the Chapter 13.

El Niño Southern Oscillation (ENSO)

A clear example of how the distribution of ocean heat controls climate is the **El Niño Southern Oscillation** (ENSO) (Guilyardi et al., 2016). When a pool of warm tropical surface water remains in the western equatorial Pacific, it drives warming and rising of the air above it, and allows upwelling of cold ocean bottom water in the eastern Pacific along the coast of South America (Figure 9.3) (Chen et al., 2017, An et al., 2018). This upwelling brings with it nutrients that feed the blooms of phytoplankton and all the trophic levels that depend on them, including the fish that the local fishermen in their boats are seeking. The warm water in the west and cold water in the east sets up an atmospheric circulation system over the Pacific with air rising in the west and sinking in the east. Cold water cools the air, making it more dense so that it sinks. As it does so, it compresses, warms, and reduces humidity, leading to no rain. Warm water heat the overlying air, causing it to rise and cool, reduce its relative humidity, and cause rain that could lead to flooding and landslides in nearby coastal regions. Sometimes, about every three to seven years, the warm pool of water moves eastward, off the coast of South America. This inhibits upwelling of nutrient-rich bottom water, and the loss of productivity of phytoplankton and all higher trophic levels leads to bad fishing. Typically, the warm pool arrives in December, so the religious local fishermen named the event "**El Niño**" after the Christ child. The warm water off the coast of South America also alters atmospheric circulation, with warm air rising in the eastern Pacific, and sinking in the west. This alteration is transmitted throughout the world, and places can be warmer and wetter, or cooler and drier, during El Niño times on continents as far away as Africa and Europe. During the opposite phase the warm pool moves to the west, and this phase is called **La Niña**.

El Niño: Oceanographic and atmospheric conditions caused by a shift of a warm pool of equatorial Pacific surface water from the western side of the Pacific to the eastern side, closer to South America. This occurs every three to seven years and suppresses upwelling of nutrient-rich deep waters off the coast of South America. Without nutrients, phytoplankton do not bloom as profusely, thus affecting the higher trophic levels and reducing the utility of the fishery off South America.

La Niña: Movement of the equatorial Pacific warm pool to the Western Pacific. This enhances upwelling and nutrient availability adjacent to South America.

Sea Level

Rising sea level has been a concern to scientists for decades, and has recently come to the attention of the general public, mostly through its impact of exacerbating the impact of storms and storm surges, like the ones associated with Hurricane Sandy that devastated the mid-Atlantic and Northeastern regions of the U.S. in October of 2012 (Solecki et al. 2017). Sea level has been an elusive concept for scientists. After all, we measure the height of the ocean relative to the shoreline, where we place tide gauges, yet at the same time, we measure the elevation of the land surface relative to mean sea level. A better way is needed to avoid obvious circularity.

There are two ways to consider sea level. The first is **relative sea level**, which pertains to the relative elevation of the ocean surface and the land surface at a particular spot on the shoreline. This can be easily measured with tide gauges in harbors along the world's coasts. However, tectonic uplift and subsidence, along with other vertical motions of the land surface in response

Relative sea level: The height of the sea surface compared to the height of the shoreline at any particular point. It is affected both by raising or lowering of the sea surface (eustasy) and raising or lowering of the land surface (tectonics or epeirogeny).

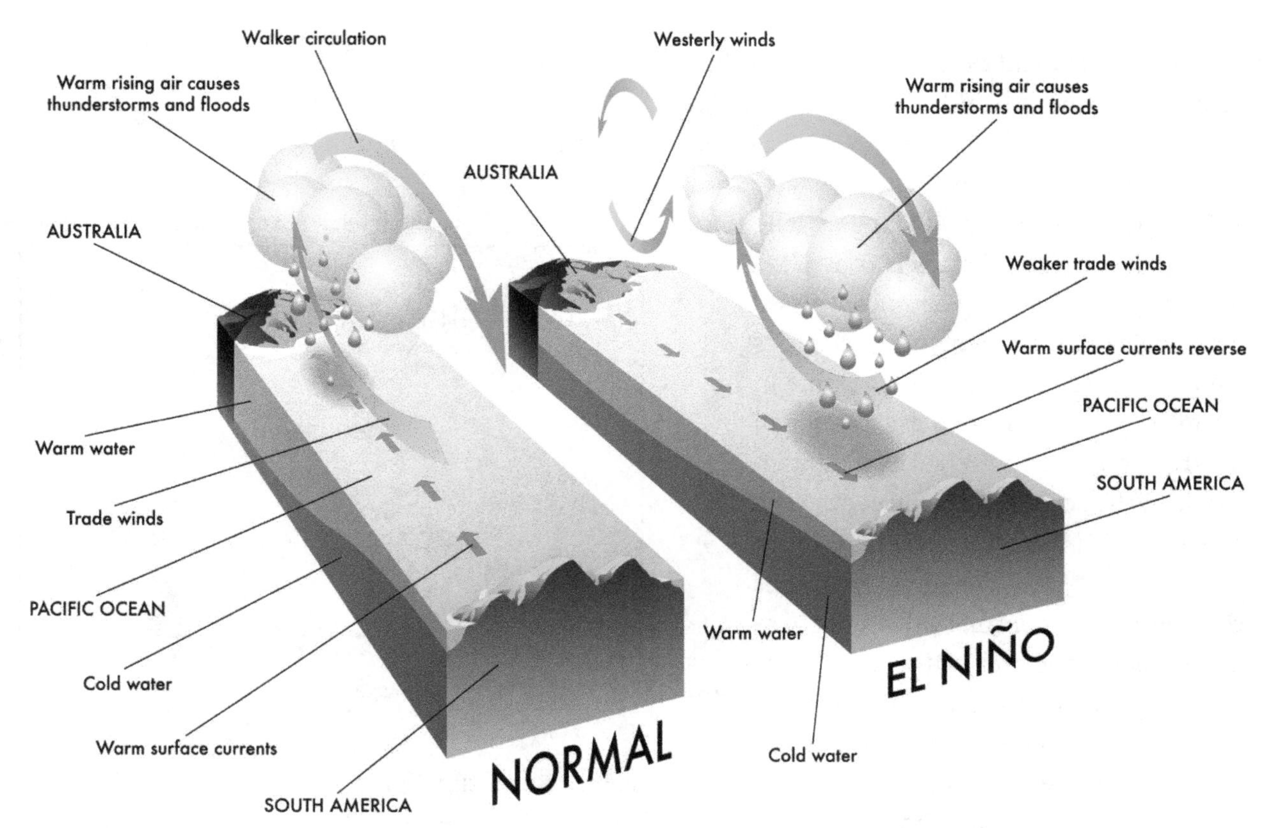

FIGURE 9.3 El Niño

The El Niño Southern Oscillation (ENSO) is a natural cycle in which a warm pool of tropical surface water moves from the western Pacific eastward, and back again every three to seven years or so. When the warm pool is in the west ("normal" or La Niña times), cold, nutrient-rich bottom water rises along the coast of South America, "fertilizing" phytoplankton and all the overlying trophic levels, including the fish sought by South American fishermen. During El Niño times, the eastward migration of the warm pool suppresses upwelling along South America, and the fishery is greatly reduced. Note that atmospheric circulation is also affected throughout the Pacific basin, and globally.

to deglaciation, sediment loading or compaction, sub-surface water and oil removal, and other factors lead to different measurements of relative sea level in different parts of the world. Consequently, relative sea level tells us little about mechanisms for sea level change and provides no information about changes in ocean water.

Eustatic sea level: The height of the sea surface above the center of the earth. It reflects the relative volumes of the world's ocean water and the global ocean basin. It is very difficult to measure for the past, but if tectonics can be quantified, it can emerge from a correction of relative sea level measurements, such as tide gauges. Recently it can be measured with satellite altimeters so modern measurements are much more accurate than previously.

A seemingly simpler, yet more problematic concept is that of **eustatic sea level**, which pertains to the whole world's ocean water. At the most fundamental level, eustatic sea level is the quantitative relation between the volume of the world's ocean water and the volume of the world's ocean basins. We have no way to measure either of these directly, but sea level curves at various timescales have been constructed by scientists on the basis of many proxies of both eustatic and relative sea levels. Even though we cannot measure eustasy directly, we know what causes changes in these volumes, so if we could obtain a record of eustatic sea level for the past, it would provide insights regarding the evolution of these mechanisms. On the basis of tide gauges scattered around the world, and corrected for local land movements by a number of models by several investigators, it appears that sea level rose in the twentieth century (Nerme et al. 2018) at a rate of about 1.5 to 2 mm/yr. In the twenty-first century this has already accelerated to greater than 3 mm/yr (Figure 9.4) (Hu and Bates, 2018).

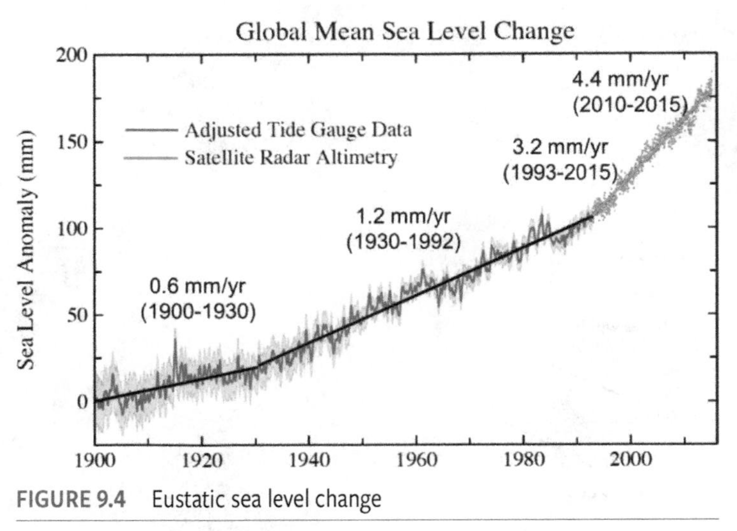

FIGURE 9.4 Eustatic sea level change

Ocean Basin Volume Changes: The volume of the ocean basins is controlled by much slower and long-acting processes of tectonics and sedimentation. Scientists usually consider ocean basin volume in terms of ocean area times average ocean depth. The area is controlled by the extent to which the continents collide with each other and crumple up their edges into mountain belts (like the Himalayas), thus thickening the continental lithosphere and reducing the area of the continents to enlarge the oceans. The depth of the ocean is controlled strictly by ocean age. The ocean deepens at a rate proportional to the square-root of time for the first 70 million years, and inverse exponentially after that. The oldest significant area of ocean lithosphere that remains today is of Jurassic age in the western Pacific. Old, cold, deep ocean lithosphere lies at between four and five km depth, and the average depth of the ocean lithosphere is about four km. When volcanoes and ocean plateaus grow, they displace water and raise sea level. In addition, variations in the rate of sediment input to the ocean from continental interiors can affect the rate at which seawater is displaced. After glaciation, large areas of the continents were stripped of their vegetation cover so that soils and rock were ground into fine sediment that readily was transported by water (mostly) and wind into the oceans. This led to a slight reduction in the volume of the ocean basins. A prime example of this is the extremely thick Pleistocene sediments in the Gulf of Mexico.

Ocean Water Volume Changes: On timescales of human activities that affect sea level, we take the volume of the ocean basins to be invariant, and concern ourselves with the mechanisms that alter ocean water volume. The volume of ocean water is controlled in part by the mass of water, which is in turn affected by the amount of water that is held on the continents as (in decreasing importance) glacial ice, groundwater, and surface water. Ice volume changes in response to climate change, groundwater is changing due to removal (mining) of water from aquifers that recharge more slowly than wells remove the water (mostly for crop irrigation), and surface water are affected by many human activities such as deforestation, wetland drainage, and dam-building. Accounting for just how much water has been added or subtracted from the ocean due to these human activities has been very rudimentary to date.

Ocean water volume is also affected by its temperature, because water, like anything else, expands when heated (aside from the fortunate aberration in water behavior between 0 and 4 degrees Celsius at one atmosphere pressure). As climate changes, warmer oceans make higher sea level, and vice versa.

Steric Expansion of Seawater: Because water temperature affects volume and thus sea level, it would appear important to know both the average and variability of ocean temperatures over time. Unfortunately, the thermal structure of the ocean is poorly understood, even in the present day, and long-term temperature change is not monitored. Thus scientists must rely on ocean models with sparse observational data to construct a picture of ocean temperature distribution in terms of latitude, longitude, and depth. Based on these models, it appears that the mixed layer of the ocean (from the surface to about 25 to 200 m depth, depending on season and position) has been affected by atmospheric warming the most since pre-industrial times, but that warming and expansion is extending to great depths in many places as well. At the present time it is estimated that about half of the inferred sea level rise for the twentieth century has been due to steric expansion (from warming) of the total ocean. As physical oceanographers continue to monitor and understand the intricacies of ocean circulation and thermal structure, more accurate estimates of the impact of ocean warming on sea level will emerge.

Melting Mountain Glaciers: The addition of water to the ocean from the melting of mountain glaciers is somewhat simpler to observe, and has been for centuries in places like the Alps. Global estimates of mountain glacier melt water that has entered the ocean and contributed to sea level rise are about half the observed total rate over the twentieth century of 1.5–2 mm/year, or about .75–1 mm/year from glacial melting. However, this is primarily from small mountain glaciers, and the ice mass balance of the large glaciers in Greenland, and especially Antarctica, has been less well constrained. Recent analysis of Greenland suggests that near the end of the twentieth century, the rate of ice melting significantly increased, and that it is now a major component to sea level rise. In the last decade or two, for the first time in history, seasonal melt water on top of the glacier has been observed to melt its way all the way down to the base, traveling down moulins to flow out of the glacier and into the sea. Careful microwave elevation measurements have also indicated that the glacial surface is dropping to lower altitude. The addition of water to the oceans in this manner appears to be accelerating, and its magnitude is overtaking that of the melting of mountain glaciers.

There is currently about 30 million km^3 of ice in Antarctica, about 3 million km^3 in Greenland, and only about 21,000 km^3 in small alpine glaciers. If this ice were all to melt, how much would it raise sea level, given a global ocean area of 3.6×10^8 km^2?
Answer: Total ice volume is 33×10^6 km^3, as the small alpine glaciers are trivial in comparison. If this were to be spread over the 3.6×10^8 km^2 of ocean (assuming vertical shoreline walls, which is not really correct ...), just divide volume by area to get distance (height) of sea level rise. Here is a big hint for this sort of thing- See what you need to do to the units of measure to get the answer you need- $km^3/km^2=km$. So, 33×10^6 km^3 / 3.6×10^8 km^2 = 0.092 km, or 92 meters. Actually, it is not quite as simple as that. The added load of the water over the ocean crust will isostatically depress the oceans, making them deeper, and to conserve global earth mass, will pop the continents up a bit by redistributing the underlying mantle. Mantle is about three times the density of water, so the oceans will get 1/3 of the added water depth deeper, reducing sea level rise by about 30 m, for a total sea level rise due to melting of about 62 meters.

Antarctic Ice: Meanwhile, the situation in Antarctica is not nearly as clear. While accelerations in the decay of ice shelves has been observed, along with the concomitant flow of ground glacial ice to take the place of the ice that breaks off and floats away, the warming of the surface water of the Southern Ocean could be causing greater evaporation into the warming air, and thus enhanced precipitation (still as snow) over the Antarctic continental interior, but feeding the glaciers that are flowing into the sea, in what may be merely an accelerated hydrologic cycle rather than net loss of mass to the ocean. A great deal of additional research must be conducted before a definitive answer can be obtained regarding the rate of net loss (or gain) of Antarctic ice and its projected changes in the twenty-first century and beyond. In sum, as the mountain glaciers primarily responsible for the ice mass contribution to sea level rise in the twentieth century shrink away, and the large continental glaciers begin to feel the effect of climate change, we might expect that the water mass contribution for sea level change in the twenty-first century will shift as well from mountain glaciers to continental ice sheet loss. While all low-latitude mountain glaciers could melt away within centuries, it would take millennia for Antarctic ice to completely melt, according to current model projections.

Anthropogenic Impacts on Sea Level: Both steric expansion due to ocean water warming and addition of water mass to the ocean from ice melting are considered to represent natural processes of sea level rise. Indirectly, these processes are likely driven at least in part by human activities such as fossil fuel burning that leads to climate changes that drive ocean warming and ice melting, but this is a different matter. Besides, there is a DIRECT influence of human activities on sea level (Sahagian, 2000). The burgeoning human population influences not only ecosystems, soils, and biodiversity, but in the last century, has played an important role in global water balance as well. Several human activities directly transfer water from the continents to the ocean. One of the most important of these is the withdrawal of well water from aquifers in arid and semi-arid regions for the purpose of irrigation of crops for human consumption (and perhaps increasingly, for energy). When water is withdrawn from wells faster than it is recharged by rainfall and infiltration back into the aquifer, it is considered water mining, and in many cases, very old groundwater, considered fossil water, is withdrawn. When fossil water is mined and used for agriculture, it may run off into rivers, and be transported to the ocean to contribute to sea level rise. Alternatively, it may evaporate and be transported in the atmosphere to a region that is more humid, where it may rain out and be transported to the ocean. In neither case does it recharge the aquifer from which it was withdrawn, and thus the net effect is to transfer water from the continental interiors to the ocean, thus raising sea level. The rate at which this occurred in the twentieth century has been estimated to contribute to sea level rise at a rate of about 0.2 mm/yr, based on a partial list of mined aquifers.

An additional direct effect of human activities on sea level comes from deforestation. A great deal of water is stored in a forest, both in vascular water and in the biomass water-equivalent (in addition to water in and around roots below ground). When a forest is cleared, this water enters rivers or the atmosphere and finds its way to the ocean, and raises sea level at a twentieth century rate of about 0.15 mm/yr.

Diversion of surface water that feeds internally draining lakes such as the Aral Sea is another source of anthropogenic sea level rise. While irrigating the Karakum Desert to grow cash crops such as rice and cotton, enough water was evaporated from this internally draining basin and added to other basins that drain to the ocean, to raise sea level at a rate of about 0.1 mm/yr during the latter half of the twentieth century. Additional activities such as wetland draining (usually for suburban development in places like south Florida), overgrazing of marginal lands and consequent desertification and loss of soil water, and other human activities that serve to dry the continents and add water to the ocean, bring a conservative total for twentieth-century sea level rise rate caused by humans to about 0.54 mm/yr.

However, there is one thing humans do that works the other way, and stores water on land that would have entered the ocean, and that is dam-building. A trick that we learned from our friends, the busy beavers, is that when an impoundment is constructed in a river, a lake that holds a large volume of water can be created. We have elevated this practice to the level of the Aswan, Hoover, and Three Gorges dams, to name a few. Over the twentieth century, dams were built that impounded water on land (that would have flowed to the ocean) at a sea level equivalent rate of .52 mm/yr, or an almost exact counterbalance of all the human activities that served to move water off the continents and into the ocean. However, our compilation of dams does not include the millions of small dams built for rice paddies and farm ponds, which could impound as much as the fewer big dams. Further, a dam impounds at least as much ground water as it does surface water, so we may have been impounding the equivalent of 2 mm/yr of sea level during the twentieth century. Keep in mind that because we are discussing RATES, it is the rate of filling of newly constructed dammed reservoirs that matters. As soon as a dam is completed, water begins to accumulate behind it while reducing the river flow downstream. When the reservoir fills, its contribution to reducing the rate of sea level rise is finished because after that point, the flow of water into the reservoir is balanced by the rate of water flow out of the reservoir and downstream. At that point, only another new dam counts toward reducing the rate of sea level rise.

Now consider this: If dam-building in the twentieth century served to lower sea level by somewhere near 2 mm/y, and the 0.5 mm/yr of anthropogenic additions of water to the oceans is subtracted from that (to yield net human influence of -1.5 mm/yr of sea level lowering), then the actual contribution of the natural causes of ice melting and ocean thermal expansion (which are poorly measured a priori) must have been 1.5 mm/yr greater than we thought, but simply counterbalanced by direct human activities so that only 1.5–2 mm/yr were observed over the twentieth century. Here is where things get a bit tricky: The large-scale construction of new dams has led to numerous environmental, political, economic, and social concerns. As a result, there is an apparent global moratorium on dam-building (Three Gorges appears to be the last major dam, and it raised many concerns, with a debate still raging) (Figure 9.5). So what would happen once we stopped building dams to impound new water at the rate that we did in the twentieth century? We should expect the full effect of ice melting and steric expansion to be observed as an increase in the rate of sea level rise of about 1.5 to 2.0 mm/yr of the twentieth century to about 3 to 3.5 mm/yr in the twenty-first century. Indeed, modern measurements (using satellite altimetry) of twenty-first-century sea level rise show just that, so the rate of water warming and ice melting is actually greater than that estimated previously based on an artificially reduced rate of sea level rise. See the IPCC Fourth and Fifth Assessment Reports for further discussion of modern sea level measurements.

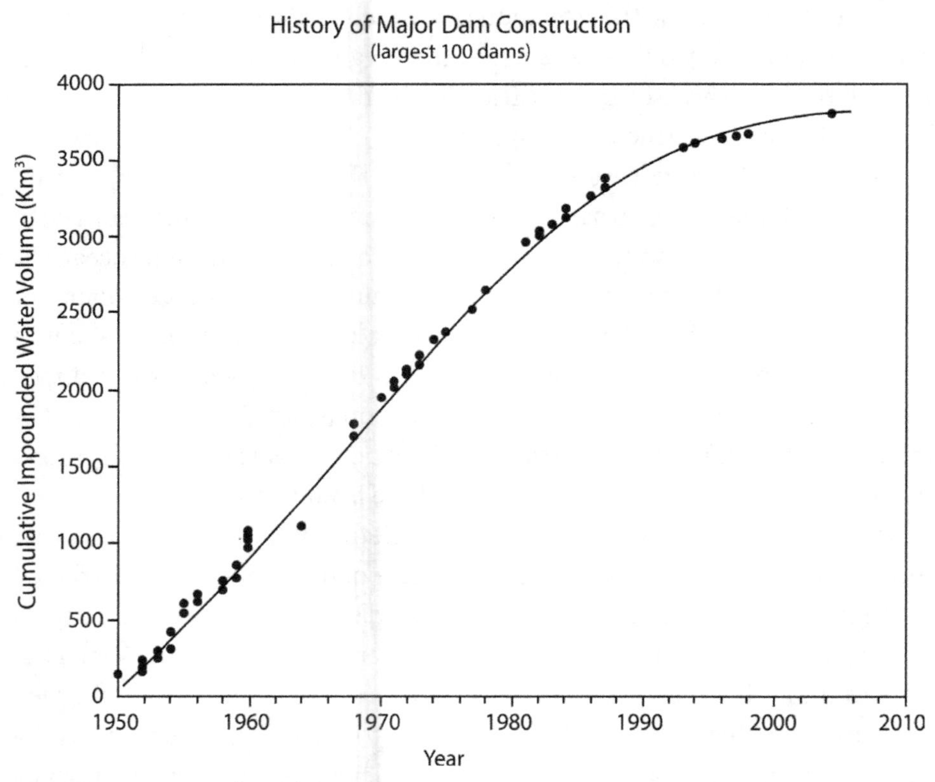

FIGURE 9.5 History of major dam construction
The second half of the twentieth century was a time of unprecedented dam building throughout the world. The rate at which water was impounded in reservoirs behind dams was sufficient to reduce the rate of sea level rise by about 1.5 mm/yr, which was about equal to the observed rate of rise over that time interval. We are no longer building dams like we were in the last century, so we should expect the rate of sea level rise to double in the twenty-first century. This has already been observed. However, the rate of glacial melting and thermal expansion is also expected to increase in the twenty-first century, so sea level may rise faster than currently predicted, with serious consequences for coastal communities such as Miami and New York.

Ocean Chemistry

Water is an excellent solvent, and thus dissolves a whole host of chemicals, most of which find their way to the global ocean. While the main dissolved chemical is salt (sodium chloride, NaCl) with a concentration of about 35 parts salt per thousand parts water, there are many other things dissolved in sea water, many of which are critical nutrients for marine life.

The oceans play a pivotal role in the global carbon cycle, as water can dissolve a great deal of carbon dioxide (CO_2), as anyone drinking a carbonated beverage knows firsthand. The concentration of CO_2 in sea water is in equilibrium with the concentration of CO_2 in the atmosphere. As far as CO_2 is concerned, the ocean is a buffered solution in which CO_2 reacts with the water and forms carbonate (CO_3^{2-}) and bicarbonate (HCO_3^-), the relative amounts of which depend on available hydrogen, or the pH of the water. The more CO_2 dissolved in the ocean, the more carbonate converts to bicarbonate. This **ocean acidification** leaves less carbonate available for marine organisms to make their shells, and it is already becoming more difficult for many marine organisms that serve as the base of the marine food web to exist. The pH of the global ocean was 8.2 in pre-industrial times. Due to anthropogenic emissions from fossil fuel burning, the additional CO_2 dissolved in the ocean has reduced the pH to 8.1 in the present day. This may not sound like much, but because pH is a log scale, this is about a 30% increase in acidity. The ecological impact of this has not yet been severe, but as marine pH approaches 8.0 (another 30% increase in acidity), many marine organisms will face severe difficulties precipitating their shells out of sea water. This could happen only a few decades from now, thus greatly reducing the biological pump that transfers atmospheric CO_2 into the deep ocean for long-term storage. There have been times in the geologic past with high atmospheric CO_2 concentrations, but these were times of warmer ocean water temperatures, such that not as much CO_2 dissolved into seawater to acidify the ocean (recall that cold water can dissolve more CO_2 than warm water).

Ocean acidification: The reduction of pH of ocean water by solution of excess atmospheric CO_2 through the solubility pump. Historical pH was 8.2, but is declining. If it falls below 8.0, marine organisms will have difficulty making calcium carbonate shells, thus threatening the biological pump, as well as the marine ecosystem. With a cold ocean, as present, and with sudden high atmospheric CO_2 concentrations (as we are tending toward due to fossil fuel burning), more CO_2 can be dissolved in the ocean than would naturally be possible, leading to concerns regarding the marine ecosystem.

Coral Reefs

One of the most diverse and important environments in the ocean is the coral reef (Figure 9.6). Corals are actually animals that have a symbiotic relationship with algae that provide food for the corals, which in turn provide CO_2 and nutrients for the algae, as well as a substrate to anchor and live. However, changes in ocean chemistry and temperature throughout the major coral reefs of the world have caused the corals to expel the algae (coral bleaching) and without their symbionts, they die. The die-off of **corals** can have a devastating impact on the rest of the marine ecosystem, as corals provide food, protection, and habitat for many species. As the ocean becomes more acidic, corals, like other shell-forming organisms, have more difficulty precipitating carbonate out of sea water, thus weakening reefs over and above the impacts of bleaching from warm-water events.

The Coast

The world's coasts are varied, including sandy beaches, rocky promontories, mangrove swamps, and many other landforms that depend on the balance of sediment supply, wave energy, winds, and tides. More than half the human population lives in the coastal zone, as does a complex set of diverse ecosystems. This is where urban development and highly productive coastal ecosystem come into conflict. While disruption of ecosystems by drainage of wetlands, filling of estuaries, and destruction of mangrove swamps in favor of shrimp farms leads to loss of the ecosystem functions and services provided by these systems, changes in global climate are also affecting the cities built along the coasts. The impact of Hurricanes Katrina and Rita on New Orleans was devastating, in part due to the loss of protective wetland ecosystems to the south. Rising sea levels and warming ocean temperatures were contributing

FIGURE 9.6 Coral Reefs

FIGURE 9.7 Hurricane Sandy images
Overwash processes involve movement of sediment (e.g., sand) from the shoreface to the bay side of a barrier island. Mantoloking, NJ experienced overwash during Hurricane Sandy in 2012. As sea level rises, overwash enables the barrier island to migrate inland and thus maintain its integrity and continued protection of the mainland.

Barrier island: A long linear island along a coast separating the ocean from a quiet bay. The island is created and maintained by the equilibrium between sea level, waves, sediment supply, wind, and currents.

Overwash processes: The action of storm waves, high tides, and storm surges pushing ocean water up and over a barrier island, to deliver sediment (usually sand) to the bay side, while raising the average elevation of the island. This is how barrier islands keep pace with sea level rise.

factors in the devastation caused by Hurricane Sandy to New York and New Jersey (Figure 9.7), and frequent hurricanes in Florida. These storms hit coastal regions that are largely dominated by **barrier islands** that would protect many parts of the mainland.

Along the East and Gulf coasts of the U.S., there is chain of barrier islands that consist of sand held in place in equilibrium between wave action, wind, longshore currents, and sediment supply from rivers. The barrier islands act like a shield for the mainland, protecting it from storm waves and associated erosion, and create a system of calm bays that are the home of and nursery for a multitude of marine species. Because their positions and shapes are controlled by waves and currents, the barrier islands can absorb storm energy by altering their positions and the distribution of sand. Dune grasses help to stabilize the sand, helping it to accumulate by wind deposition in large piles that maintain an elevation above sea level. In the case of rising sea levels, the islands respond by rolling back toward the mainland through **overwash processes** that wash sand from the shoreface to the bay side, slowly moving the island toward the mainland, and raising it up to keep up with sea level rise. If the island did not do this, it would be flooded and lost to the sea, and the full power of the ocean would impinge on the mainland. Because the barrier islands are where we find some of the world's best beaches, they have become highly developed with roads, buildings, and many groins and jetties that affect the movement of sand. The people who live or vacation on a barrier island are not normally concerned about the preservation of the island through overwash processes, because it is these very processes that involve flooding and washing away of their roads and homes. These processes showed themselves to New Jersey during Hurricane Sandy (2012), when a new inlet formed in the middle of a residential area, and large quantities of sand were moved from the beach to the center of the island and the bay side. "Restoration" efforts are being made to put the sand back where it was in order to preserve the value of private property, paying little attention to the fact that in order to be preserved, the island needs to move toward the mainland in response to sea level rise.

Ocean Pollution

The ocean is the final resting place for much of the trash and pollutants we create on land. Rivers transport materials and dissolved contaminants directly to the ocean, and additional dumping and pollution is done on the ocean itself by ships, oil platforms, and other activities. Much of the pollution is concentrated near the coastlines, and this is also the area of greatest nutrient supply, shallow water, and marine productivity and diversity. As such, there is great concern regarding the health of the coastal waters due to pollution. Shellfish beds are commonly closed for fishing, as the contaminants, both chemical and biological, have exceeded human health standards in many locations. In some areas, oxygen depletion from algal blooms caused by excess nutrient runoff (mostly from farm fertilizer) has led to massive die-offs of all animal marine life, including fish and shellfish. Of particular concern is the very shallow ocean (upper few mm) in which many of the phytoplankton and other primary producers for the ocean reside. Pollutants are much more concentrated in this upper thin layer, and thus may be adversely affecting the basis of the marine food chain.

THE GULF OF MEXICO "DEEPWATER HORIZON" DISASTER OF 2010

On April 20, 2010, the deepest offshore oil well ever drilled failed and uncontrollably released oil and natural gas from its broken wellhead (Konkel, 2018). The erupting plume caught fire at the ocean surface, led to large explosions, and destroyed and sank the floating drill rig, killing eleven workers (out of 126 on board). The fire was later extinguished, but oil continued to gush from the wellhead at an estimated rate of 56,000 gallons per day from April 20 until July 15, when a temporary cap was successfully installed (after months of failed attempts). Relief wells were completed and the problem well was finally considered sealed in September, 2010.

The amount of oil released from the well into the ocean has been estimated to be between 4.4 and 4.9 million barrels. At 42 gallons per barrel, this means about 184 to 205 million gallons were leaked into the Gulf of Mexico in the summer of 2010. The fate of this oil is still under scrutiny, but more was dispersed and decomposed by microorganisms than had been expected. A massive cleanup effort along the Gulf Coast attempted to minimize ecological damage in the coastal zone, and studies are still underway to assess any long-term environmental impacts of the spill (Ainsworth et al., 2018; Rabalais et al., 2018).

Plastic Pollution

It is illegal to dump plastic of any kind into the ocean anywhere. Yet several million tons of plastic are dumped into the ocean every year (Figure 9.8). These are in the form of fishing nets and lines, six-pack holders, plastic bags, and many other plastic products. Because plastic is chemically very stable, it does not decompose in the ocean, so it floats around for many years, giving it ample time to interact with sea life. Many plastic items look like food to fish, birds, and marine mammals, who eat them but cannot digest them.

FIGURE 9.8 Plastic pollution.

They then fill animal stomachs, leaving no room for actual food, and the animals literally starve to death with full stomachs. Animals also get trapped in abandoned fishing nets; entrapped fish cannot eat, and mammals drown very quickly if they cannot reach the ocean surface to breathe. One attempted solution is to use "biodegradable plastic," but this is an oxymoron. Plastic is not actually biodegradable, but gets mixed with biodegradable substances, which, when they degrade, release the plastic in tiny particles that remain in the ocean. Hopefully, they are small enough to pass through animals' digestive tracts, and certainly to not lead to entrapment, but there is growing concern regarding "microplastics" in the ocean (de Sá et al., 2018).

"Red Tide"

FIGURE 9.9 Red Tide

Many beachgoers on the U.S. East Coast have heard of "Red Tide" and the occasional closing of beaches as a result (Figure 9.9). The discoloration of the water is caused by algal blooms that feed on excess nutrients, such as nitrogen and phosphorus, which are delivered to the ocean by rivers that drain fertilized farmlands and even suburban lawns and gardens (Kim et al., 2018). Most algae are not toxic to humans, and the ocean is large enough not to become anoxic as a result of the biological oxygen demand (BOD) from the decomposing algae. However, some algae are highly toxic. One toxic dinoflagellate, Pfiesteria, leads to major fishkills in North Carolina, Virginia, and Delaware, and can affect humans as well with skin sores, and difficulty breathing and concentrating.

Oil Pollution

As anyone making a salad dressing knows—"Oil and water don't mix." Oils are immiscible in water—they don't dissolve, so remain in their pure form to impact organisms throughout the ocean. When oil is spilled into the ocean from, for example, a tanker ship, about a quarter of it evaporates (causing some air pollution, but the impact of that is less than the part remaining in the water). Some of it sinks to the bottom, coating **benthos** (bottom-dwelling animals like clams), and disturbing the water–sediment interface. The rest floats on the ocean surface (Figure 9.10), sealing off the ocean from the atmosphere, causing the upper ocean to become depleted in oxygen (which it normally gets from maintaining chemical equilibrium with the air above). This impacts fish and other marine animals. Those animals that break the surface, such as marine mammals and birds, become coated in oil, which disrupts skin functions in marine mammals and ruins the aerodynamic and insulating qualities of bird feathers. Oil-covered animals do not live long, even after rescue workers help by removing as much oil as they can. Oil also washes onto beaches and impacts coastal ecosystems as well as recreational beaches. However, with a ban on single-hulled ships, that source of oil spills is greatly reduced. The main threat now is blowouts from offshore drilling, such as occurred in the Deepwater Horizon well in the Gulf of Mexico in 2010, which "spilled" far more oil than any tanker ship ever contained. Even that paled, however, to the purposeful release of oil in 1991 during the Gulf War into the Persian Gulf in the Middle East.

Bad behavior and accidental oil spills are not the only sources of oil into the ocean: About 363 million gallons per year comes from runoff from the land, be it from automobile oil changes, leaking lubricants, or other oils that go down the drain and ultimately reach the ocean. Ships clean their bilges at sea, and are another source of oil, as are incompletely burned hydrocarbons from gas and diesel engines. Natural seeps from oil reservoirs are also a significant source of oil into the ocean. All of these sources are larger than oil tanker spills, on average. However, the monumental spills of the Deepwater Horizon (200 million gallons) and Gulf War (500 million gallons) surpass any normal operating spills.

Coastal harbors are polluted by a great many toxic materials discarded by industrial, transportation, and service industries. Many of these attach to sediments and settle to the bottom of estuaries where they remain until disturbed by processes such as storms or dredging. This brings them back into the environment where they can cause harm to marine life as well as humans. Polychlorinated Biphenyls (PCBs) were especially troubling in places like New York Harbor, where a century of dumping led to highly contaminated bottom sediments. The dredging necessary to enable increasingly large ships to enter the harbor brought these PCBs back into the human environment through fish and shellfish, causing concern regarding their health impacts.

FIGURE 9.10 Oil spill

References

Ainsworth, C. H., Paris, C. B, Perlin, N., Dornberger, L. N., Patterson, W. F., III, Chancellor, E., . . . & Perryman, H. (2018). Impacts of the Deepwater Horizon oil spill evaluated using an end-to-end ecosystem model. *PLoS One*, 13(1), e0190840.

Chen, C. & Cane, M. A. (2017). ENSO in the CMIP5 simulations: Life cycles, diversity, and responses to climate change. *Journal of Climate*, 30(2), 775–801.

de Carvalho, F. Luiza, M., & Kerr, R. (2017). Source water distribution and quantification of North Atlantic Deep Water and Antarctic Bottom Water in the Atlantic Ocean. *Progress in Oceanography*, 153, 66–83.

Guilyardi, E., Wittenberg, A., Balmaseda, M., Cai, W., Collins, M., McPhaden, M. J., . . . & Yeh, S. (2016). Fourth CLIVAR workshop on the evaluation of enso processes in climate models, "ENSO in a Changing Climate." *Bulletin of the American Meteorological Society*, 97(5), 817–820.

Hansoo, K., Donhyug, K., & Jung, S. W. (2018). Development and application of an acoustic system for harmful algal blooms (HABs, Red Tide) detection using an ultrasonic digital sensor. *Ocean Science Journal*, 53(1), 91–99.

Hohn, D. (2011). Moby-Duck: *The true story of 28,800 bath toys lost at sea and of the beachcombers, oceanographers, environmentalists, and fools, including the author, who went in search of them.* New York, NY: Viking.

Hu, A., & Bates, S. C. (2018). Internal climate variability and projected future regional steric and dynamic sea level rise. *Nature Communications*, 9(1068).

Konkel, L. (2018). Cleanup in the Gulf: Oil spill dispersants and health symptoms in Deepwater Horizon responders. *Environmental Health Perspectives*, 126(2), 024001.

Lebreton, L., Slat, B., Ferrari, F., Sainte-Rose, B., Aitken, J., Marthouse, R., . . . & Reisser, J. (2018). Evidence that the Great Pacific Garbage Patch is rapidly accumulating plastic. *Scientific Reports*, 8(4666).

Nerem, R. S., Beckley, B. D., Fasullo, J. T., Hamlington, B. D., Masters, D., & Mitchum, G. T. (2018). Climate-change-driven accelerated sea-level rise detected in the altimeter era. *Proceedings of the National Academy of Sciences of the United States of America, 115(9)*, 2022–2025.

Rabalais, N. N., Smith, L. M., & Turner, E. R. (2018). The Deepwater Horizon oil spill and Gulf of Mexico shelf hypoxia. *Continental Shelf Research*, 152, 98–107.

Sá, L. C., Oliveira, M., Ribeiro, F., Rocha, T. L., & Futter, M. N. (2018). Studies of the effects of microplastics on aquatic organisms: What do we know and where should we focus our efforts in the future? *Science of The Total Environment*, 645, 1029-1039. doi:10.1016/j.scitotenv.2018.07.207

Sahagian, D. (2000). Global physical effects of anthropogenic hydrological alterations: sea level and water redistribution. Global and Planteary Change, 25(1–2), 39–48.

Solecki, W., Leichenko, R., & Eisenhauer, D. (2017). Extreme climate events, household decision-making and transitions in the immediate aftermath of Hurricane Sandy. *Miscellanea Geographica*, 21(4), 139–150.

Soon-Il, A., Im, S., & Jun, S. (2018). Changes in ENSO activity during the last 6,000 years modulated by background climate state. *Geophysical Research Letters*, 45(5), 2467–2475.

Figure Credits

Fig. 9.1: Source: https://commons.wikimedia.org/wiki/File:Ocean_circulation_conveyor_belt.jpg

Fig. 9.3: Copyright © 2017 Depositphotos/rob3000.

Fig. 9.4: Source: http://climateadaptation.hawaii.gov/sea-level-rise/

Fig. 9.6: Copyright © Richard Ling (CC BY-SA 3.0) at https://en.wikipedia.org/wiki/Coral_reef#/media/File:Blue_Linckia_Starfish.JPG.

Fig. 9.7: Source: http://geology.com/usgs/hurricane-sandy/

Fig. 9.8: Copyright © epSos.de (CC by 2.0) at https://commons.wikimedia.org/wiki/File:Water_Pollution_with_Trash_Disposal_of_Waste_at_the_Garbage_Beach.jpg.

Fig. 9.9: Copyright © Marufish (CC BY-SA 2.0) at https://commons.wikimedia.org/wiki/File:Algal_bloom(akasio)_by_Noctiluca_in_Nagasaki.jpg.

Fig. 9.10: Source: https://commons.wikimedia.org/wiki/File:Defense.gov_photo_essay_100506-N-6436W-023.jpg

CHAPTER 10

Human Health and the Environment

Our health is important, as mom would insist
Yet toxins we see would make a long list.
Mercury and lead
Can screw with your head
Exposure to toxins we all should resist.

"ENVIRONMENTAL HEALTH" CAN have two meanings; the first, which is the subject of much of this book, is the health of natural ecosystems in terms of functions and sustainability, usually in the context of human perturbations. The second, which is the subject of this chapter, is the impact of environmental degradation of all scales on human health. These scales range from our individual bodies (such as from smoking cigarettes), to our homes (such as from radon), to our cities (such as from water and air pollution), to the entire world (such as from impacts of climate change).

Sources of Environmental Health Hazards

Physical Hazards

Most physical environmental hazards are naturally occurring, with little that humans could do to exacerbate or prevent them. However, some, even at the global scale, have been created by human activities.

Earthquakes occur from the sudden release of elastic strain along a fault in the earth's crust. This strain is caused by differential motion of the tectonic plates, which cause different parts of the crust to move relative to each other, but bend when they get stuck on either side of a fault. When they break free suddenly, it makes an earthquake. People do not create them, nor can we prevent them. We can't even predict when they will occur along the various well-documented faults around the world. The hazard from earthquakes is collapse of buildings and roads leading to immediate injury or death—not the earthquake itself.

Tsunamis are caused by earthquakes that occur at undersea tectonic plate margins (Figure 10.1). The sudden movement makes a long-period wave in the ocean (like the mechanical generation of waves in a wave pool at an amusement park). The wave travels very fast (around 500 mph, or the speed of a commercial jet) and when it reaches shallow water, like any other wave, its wavelength reduces due to interaction with the bottom, and its amplitude increases.

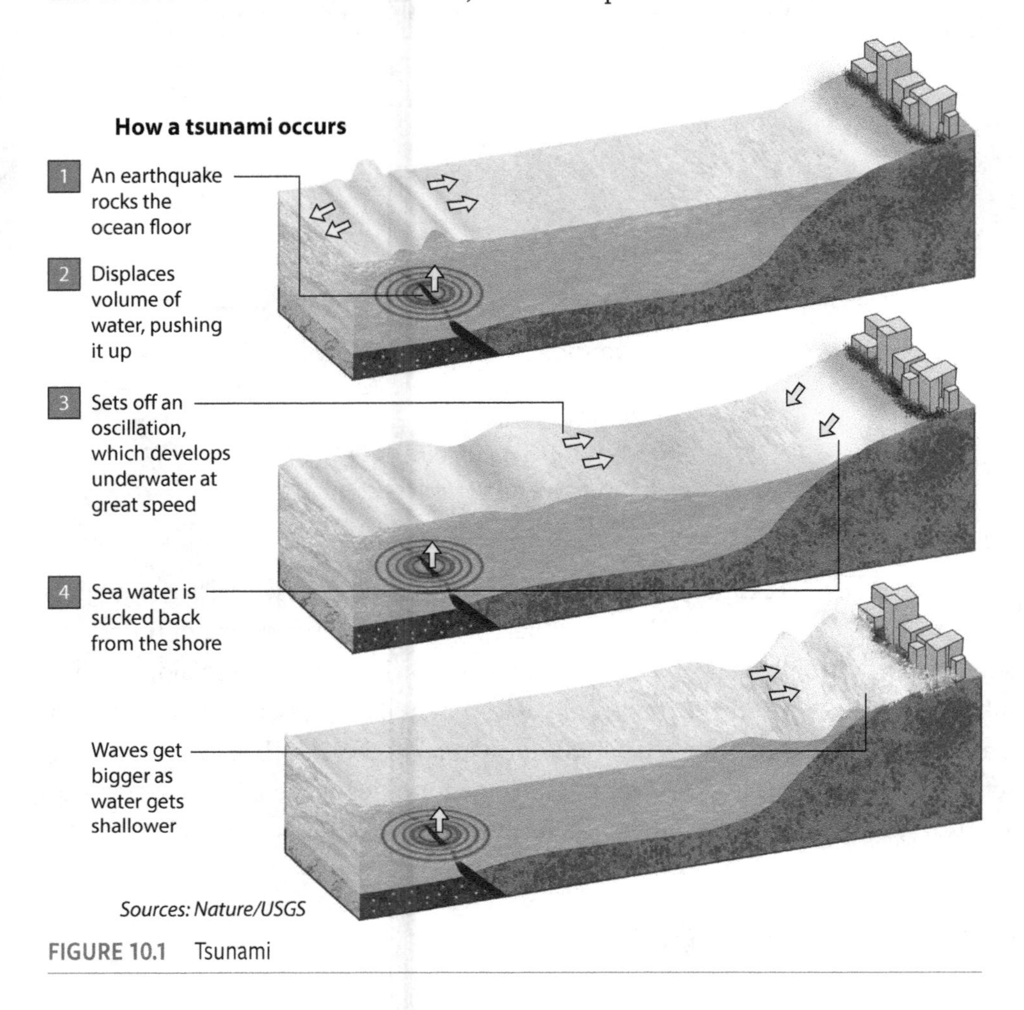

FIGURE 10.1 Tsunami

When it washes onto the shore, the sudden flood washes away buildings, roads, boats, and anything else in its path. When a tsunami in the Indian Ocean struck Indonesia and other coastlines in late 2004, hundreds of thousands of people were killed and millions displaced from their homes. A tsunami in 2011 led to a particularly troublesome problem when it washed into the Fukushima Daiichi nuclear power plant in Japan, leading to a meltdown of the reactors, in addition to killing almost 16,000 people directly.

Storms and other extreme weather events all over the world lead to injury and death. Hurricane winds, storm surges, and associated rain are obvious health hazards, even when they do not disable water and sanitation systems, access to medical facilities, heat, electricity, and other critical infrastructure. Whereas storms have occurred since long before human perturbations of the climate system, one manifestation of global change is the intensification of some types of storms, such as hurricanes, in response to warming of sea surface temperatures due to anthropogenic greenhouse gas emissions.

Floods, sometimes associated with storms, but often merely from extreme rainfall events (without excessive winds) in distant regions upstream, are a health hazard from the obvious drowning, but also from transmission of **disease vectors** such as from animal waste from farm feedlots, contamination of domestic water supplies, loss of medical facilities, and other critical infrastructure. Floods are exacerbated by human land use in the upstream regions of a watershed because paved surfaces, and even agricultural land, allow more rapid runoff of rain water into streams, thus raising rivers downstream more rapidly than they would raise if upstream ecosystems were intact.

Volcanic eruptions present a variety of health hazards, ranging from incineration by lava flows at 1200°C or by pyroclastic density currents, to respiratory effects of ash fall, to burial by volcanic mudflows or suffocation from volcanic gases (Cronin et al., 2014). Volcanic ash drawn into modern high-temperature jet engines melts internally, coating turbine surfaces and rendering engines ineffective, which is why flights were grounded and airports closed throughout England and some of northern Europe during the Icelandic eruption of Eyjafjallajökull in 2010. Ash can leach potentially toxic elements into agricultural fields and contaminate water supplies, thus leading to additional health hazards. Ash can also affect the functionality of electrical infrastructure because it is conductive when wet, posing serious implications for the management of healthcare facilities and general public services. There is no human influence on the processes that cause volcanoes to erupt, but increasing population and settlement density in the vicinity of volcanoes increases vulnerability of people and infrastructure to volcanic hazards.

CFCs: (chlorofluorocarbons) were used for decades as refrigerants and propellants for spray cans until it was discovered that they destroy the critical ozone layer in the stratosphere. With the loss of stratospheric ozone, much more ultraviolet light reaches the ground, leading to skin cancer and damage to plant tissue. See chapter 12 for further details on stratospheric ozone.

In an unusual volcanic hazard, 1,700 people were found dead, along with their cattle and all other animals, in the vicinity of Lake Nyos, Cameroon, in 1986. Plants were unaffected, yet every animal died. The cause was determined to be the outgassing of CO_2 from volcanic activity beneath the lake. The CO_2 accumulated in bottom water, then when oversaturated, bubbled up and came out all at once into the air. The CO_2 made a blanket of inert gas that asphyxiated all animal life (including human). After this tragedy, a pump was installed to make a fountain bringing bottom water to the surface and spraying into the air to release CO_2 gradually to avoid sudden release in the future.

UVB: Ultraviolet radiation from the sun between 280 and 315 nanometers (nm). Most solar ultraviolet is in the longer wavelength UVA range (315–400 nm), but it is the UVB that leads to cellular disruption in humans and other organisms.

Ultraviolet Radiation may be the most common physical environmental hazard, as it affects everyone all the time, as well as all ecosystems, both terrestrial and aquatic. With the reduction in stratospheric ozone concentration (Figure 10.2) due to the addition of chlorine from industrially produced **CFCs**, **UVB** reaches the ground at levels not seen since the evolution of land plants and animals (hundreds of millions of years ago). The immediate impact of exposure to UVB on human skin is sunburn, to the point that it is now wise to use sunblock when spending significant amounts of time in direct summer sunlight. This was not the case only a few decades ago, when there was more ozone in the stratosphere to absorb the UVB. This issue will be discussed further in Chapter 12 in the context of the atmosphere. Extended exposure to UVB and repeated sunburn cases have been shown to lead to elevated incidence of cataracts in the eye, as well as skin cancer due to disruption of skin cell DNA.

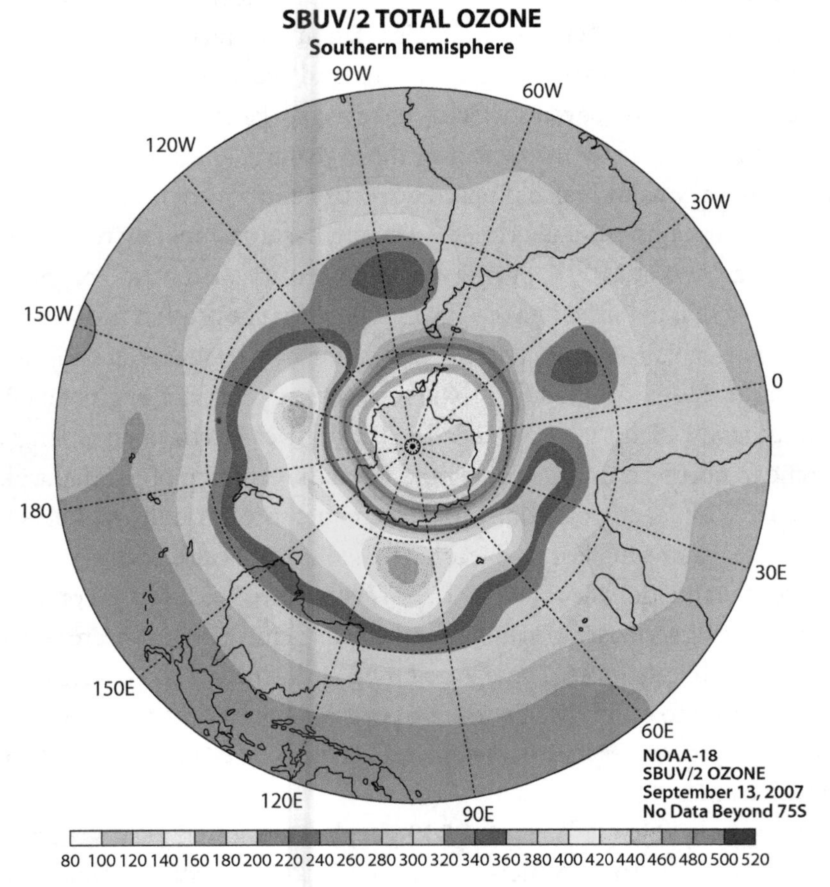

FIGURE 10.2 Ozone Hole

Chemical Hazards

There is a long list of chemicals released into the environment that adversely affect human health. A few of these are discussed here, but there are many more, some of which may not yet be recognized.

Hormonally Active Agents (HAAs) are a broad class of chemicals that trigger hormonal responses in people (and other animals) (Anwer et al., 2016). These chemicals, when ingested, mimic the role of various hormones in the body, and serve as **endocrine disruptors**. Of considerable concern are those that act like female sex hormones, and are termed **xenoestrogens**. These chemicals are suspected to increase breast cancer rates in women as well as reduce sperm counts in men, in addition to various deformities of wildlife. There are many common chemicals that act as endocrine disruptors, such as bisphenol A (BPA) found in plastic linings of metal cans and in plastic water bottles,

Endocrine disruptors: Chemicals that interfere with the endocrine system in humans and other animals. The endocrine system consists of the glands that produce and release hormones that control the function of various body organs and functions, including reproduction, growth, cellular metabolism, heart rate, and many other critical functions.

Xenoestrogens: Endocrine disruptors that behave like female sex hormones.

Polychlorinated Biphenyls (PCBs): A class of numerous (209) chlorinated artificial compounds based on attachment of chlorine atoms to a double benzene ring (biphenyl) composed of carbon atoms. The structure is similar to dioxin, another highly toxic substance. PCBs were made between 1929 and 1979 (when they were banned in the U.S.) for many industrial uses, because they were not flammable, were a good electrical insulator, and could be made with a range of consistencies from fluid to almost solid. When ingested, they serve as an endocrine disruptor, impair the reproductive system, and cause cancer.

polychlorinated biphenyls (**PCBs**) used in electric transformers and other equipment, polybrominated diphenyl ethers (PBDEs) used as fire retardants, and dichlorodiphenyltrichloroethane (DDT) a controversial insecticide famous for reducing the incidence of malaria in Africa.

Methyl Mercury is a compound of mercury $(CH_3Hg)^+$ that interferes with brain function, leading to developmental abnormalities in humans, somewhat like cerebral palsy. It accumulates in animals because it attaches itself to fatty tissue, which is then consumed by other animals, concentrating its way up the food chain. Some high trophic-level fish have elevated levels of methyl mercury (Martinez-Salcido et al., 2018). Elemental mercury is not nearly as toxic as methyl mercury. In order for elemental mercury to convert to methyl mercury, it must first be oxidized, and then bonded to C and H (methylation). The behavioral disorders caused by poisoning by mercuric nitrate used in felt-making for hats in the nineteenth century rendered people "mad as a hatter."

Dioxin may be the most carcinogenic chemical ever known (Figure 10.3). It forms from the oxidation of dichlorobenzene, making 2,3,7,8-tetrachlorodibenzo-p-dioxin (dioxin for short). It was a component of "agent orange" that was used in Vietnam in the 1960s as a defoliant, causing a wide variety of serious health issues for American soldiers and Vietnamese alike. It is also formed inadvertently from the burning of a variety of organic compounds that contain benzene rings. Like methyl mercury, it accumulates in fatty tissue of animals, concentrating its way up the food chain.

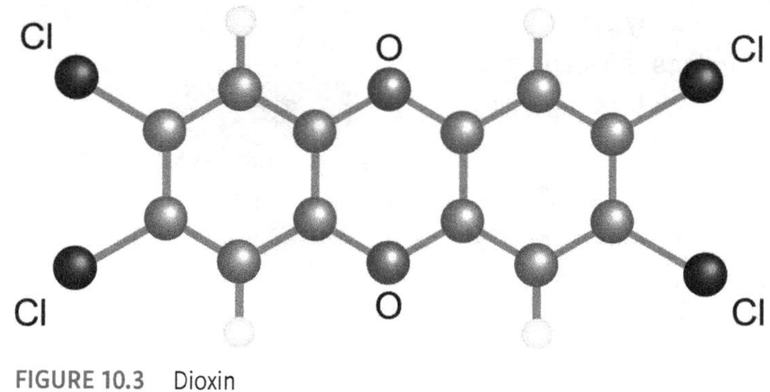

FIGURE 10.3 Dioxin

Hexavalent Chromium (Cr^{+6}) is a carcinogen that is used in the production of paints, stainless steel, metal coatings, and many other applications (McLean et al., 2012). Some industries allowed spills of Hexavalent Chromium to enter surface and groundwater, leading to greatly elevated rates of cancer in local residents. A famous case that occurred in Hinckley, CA, was publicized in the movie "Erin Brockovich."

Ozone (O_3) in the troposphere, at ground level is a lung and eye irritant, being a powerful oxidant. Because the third O atom is loosely bound to the O_2 molecule, it readily combines with other substances, oxidizing (i.e., burning) lung tissues, causing epithelial cells to become disrupted and leak enzymes into the airways. Ozone affects plant leaves in the same manner, leading to the collapse of cells on leaf surfaces (Lefohn et al., 2018). Ozone is formed in chemical reactions with nitrogen oxides, volatile organic compounds, and sunlight, typically in traffic-laden cities in mid-afternoon. More will be discussed about ozone in Chapter 12.

Lead (Pb) is a simple metal element found in minerals and useful for many applications due to its high density and malleability, and low melting point. It had numerous industrial applications in glass, batteries, and paint until 1978; and until 1995 in the U.S., gasoline, in which it served to reduce engine "knock" by enabling smoother burning, and to prevent the microwelding of valves on their seats. By now it is banned in paint and gasoline in most countries due to the neurological impacts of ingesting lead. When ingested (eating, drinking, or breathing), lead preferentially takes the place of other metals, such as iron, that are critical to metabolic function (Cabral et al., 2015). This leads to gene malfunction by the loss of controlling proteins that also affect blood pressure, and most importantly, brain function, especially in developing children, who may have additional exposure by eating lead-containing paint chips from old dwellings. (Crudely put, mercury may make you crazy, but lead makes you "less intelligent.") A recently publicized incident was the case of lead in the public water supply of Flint, Michigan (Figure 10.4) (Zahran et al., 2017).

FIGURE 10.4 Contaminated water protest, Flint, MI

In this case, aging infrastructure involving water pipes made of lead caused high levels of dissolved lead in drinking water throughout a city that was already beleaguered with economic distress since the bursting of the bubble of American car manufacturing. A switch in water source from the Detroit River to the Flint River, followed by management led to public scandal, yet the basic problem remained that an unhealthy amount of lead remained in the public water supply due to supply lines made of lead pipe. In part, this might have been prevented by a standard water treatment technique by orthophosphate, which creates a coating on pipes that prevents the lead from dissolving into the water, but this was not done. The lead pipe will always remain a health threat until they are replaced.

Arsenic is a naturally occurring element that is found in high concentrations in groundwater in some places. It is useful for making semiconductors, especially for lasers. It can be ingested by drinking/cooking with groundwater that contains arsenic and by breathing air with high concentrations of arsenic emitted by fossil fuel burning. Health impacts include skin lesions, stomach pain, nausea, diarrhea, loss of feeling in extremities, paralysis, blindness, and cancer (Bolt, 2012). It is found in particularly high concentrations in the groundwater of Bangladesh, and simple technologies have been engineered to help remove arsenic from the water at well sites.

Other Hazards

It is impossible to even mention all the human health hazards found in the environment, but a few more will be presented here, although this is still an incomplete list.

Particulates serve as irritants to lungs and eyes, and some are particularly damaging in fibrous form. Particulate matter in the 2–10 micron size range

Alveoli: Tiny "balloons" in the lungs that enable oxygen exchange into the blood (and CO_2 out).

Asbestos: A group of silicate minerals whose crystals form long fibers that have been found to be useful in the production of thermal insulation materials. However, the long thin fibers also irritate and lead to scarring of lung tissue (e.g., alveoli) and reduce lung efficiency (e.g., pulmonary fibrosis) and can lead to uncontrolled cell growth and reproduction (cancer).

Allergens: Anything that is breathed, touched, or ingested that triggers an excessively strong response in the immune system that is "fooled" into perceiving a threat to the body, thus producing lots of immunoglobulin E antibodies, which then produce histamines that make the allergic reaction.

Immunoglobulin E: An antibody produced by our immune systems with the perceived threat of an infection, leading to tissue inflammation, often in response to allergens.

Smoking cigarettes: JUST DON'T DO IT!

Alpha radiation: Helium nuclei emitted during nuclear decay of radioactive elements.

Beta radiation: Electrons emitted during nuclear decay of radioactive elements.

Radon: A radioactive, invisible, odorless, tasteless gas in the Uranium-238 decay chain that seeps into basements where it decays to Polonium. The Polonium is inhaled into lungs where it quickly decays by alpha decay that

(PM_2 to PM_{10}) remain suspended for long periods in air (Ding et al., 2005), and are thus available for inhalation (larger particles get stuck in the nose and mouth, so they don't enter the lungs). In the lungs, they lodge in the **alveoli** (tiny air sacs), and are difficult to remove by coughing or other means. They irritate the linings and reduce air-exchange efficiency, leading to asthma and other respiratory conditions as well as cancer. Sources of particulates include burning of all fossil fuels as well as biofuels, and especially smoking cigarettes, which do not limit their effect to the smokers, but to all around them. Some particulates are long and fibrous, the most famous being **asbestos**, widely used as a thermal insulator. The long, sharp shards enter the lungs and irritate the tissues, leading to asbestosis involving scarring of lung tissue, rendering it ineffective at exchanging oxygen for breathing. A special class of particles is **allergens**, usually dust, plant pollen, or pet dander that leads some people to generate **immunoglobulin E** as a defense mechanism against what the body thinks is a parasitic infection. The greatest environmental health hazard involving particulates is the simple act of **smoking cigarettes**. Many consider this a hazard class of its own, because it is optional, expensive, addictive, and deadly.

Noise pollution does more than disturb the peace. In many people, continued exposure to elevated levels of noise leads to stress-related illnesses, poor sleep, and in extreme cases, loss of hearing.

Electromagnetic fields have been blamed for numerous health problems, including loss of hearing, neurological disorders such as autism, and of course, cancer (Kjellkvist et al., 2016). After years of testing under controlled conditions, studies within various agencies have concluded that there is insufficient evidence for any harmful effects of normal electromagnetic fields. However, research and observation continues. Common fields are created by consumer electronics such as computers, TVs, and cell phones, as well as infrastructure such as cell towers and power transmission lines. However, in extreme cases of powerful fields in microwave frequencies, people exposed can be literally cooked as if they were in a microwave oven. This is theoretically possible, but has never happened as far as anyone knows.

High-Energy Radiation. The highest frequency (energy) forms of electromagnetic radiation, X-rays and Gamma rays, carry sufficient energy to strip electrons off atoms inside the human body (Figure 10.5). When the ionized atoms are in DNA molecules (thus breaking up the DNA) the most serious health impacts are felt, as the cells no longer can replicate properly, leading to cancer, reproductive problems, and other disorders. Other forms of radiation, not in the electromagnetic spectrum, include **alpha** and **beta** radiation, which are merely helium nuclei and electrons, respectively, and they, too, can disrupt DNA and other organic molecules in cells. Radioactivity stems from many natural processes as well as nuclear (fission) power generation and associated waste. Management of these materials is a critical (and controversial) aspect of nuclear power, explored in Chapter 11. One natural source of radioactivity is **radon** that accumulates in basements. The alpha radiation that is emitted into lungs when inhaled disrupts DNA and thus cellular function, leading to cancer and lung disease. The details of radon are explored in Chapter 12.

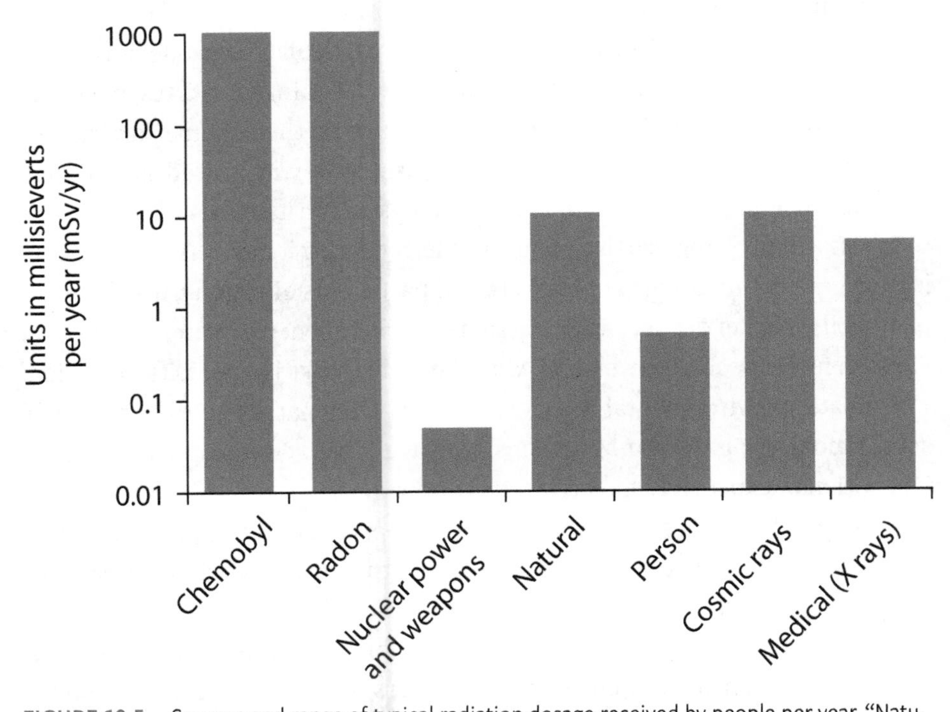

FIGURE 10.5 Sources and range of typical radiation dosage received by people per year. "Natural" includes rocks and soils.

Infectious diseases are becoming increasingly problematic as human population density increases and greater local and global transportation becomes common. The most common environmental source of disease is surface water that can carry a multitude of pathogens. In parts of the world with little or no sanitation facilities, waterways essentially serve as open sewers, and diseases such as **cholera, dysentery, typhoid fever, polio, schistosomiasis, cryptosporidiosis**, and many others can be readily transmitted to the human population.

Other **disease vectors** include biting insects such as mosquitos that can transmit **malaria** in many tropical regions. Dichlorodiphenyltrichloroethane (DDT) is an insecticide found to be very effective against mosquitos, and reduced the incidence of malaria in Africa for many years, before being banned for its own adverse health effects. The relative harm of malaria vs. DDT is still being discussed, and some African countries are using DDT again to reduce malaria. In the eastern U.S. (and northern California) **Lyme disease**

FIGURE 10.6 Deer ticks that carry Lyme disease

shoots a high-energy helium nucleus through tissue, thus disrupting cellular DNA (and leading to cancer). Inhaled radon is commonly exhaled before it decays (half-life of a few days). However, Polonium in the air from Radon decay that is inhaled is more likely to decay in the lungs (half-life of just a few minutes) and lead to cancer.

Cholera: Bacterial infection of the small intestine leading to diarrhea and vomiting.

Dysentery: Inflammation of the colon by infection by protozoa, bacteria, or parasites, leading to severe diarrhea and blood loss.

Typhoid fever: Debilitating and often deadly bacterial infection caused by ingesting contaminated food or water, leading to high fever, delirium, intestinal hemorrhaging, encephalitis, dehydration, and other symptoms.

Polio: Viral infection of throat and intestines that leads to paralysis but has been essentially eliminated in most of the world by the polio vaccine.

Schistosomiasis: A parasitic tropical disease caused by a group of parasites that live in freshwater snails, travel through the water to enter human skin, and lay eggs that irritate the gastrointestinal and other systems, causing diarrhea, abdominal pain, coughing, fever, and other symptoms.

Cryptosporidiosis: Infection by a protozoan parasite in water that is resistant to usual treatments, such as chlorination, so must be carefully filtered from drinking water. Infection causes diarrhea, stomach pain, fever, vomiting, dehydration and other symptoms, but can be overcome by the human immune system.

Malaria: Infection by a protozoan parasite carried by tropical mosquitos. It infects the blood and liver and leads to high fever, vomiting, muscle pains, and headaches. Control has met limited success by elimination of host mosquitos using DDT, and there is controversy over which is worse for people, DDT or malaria.

Lyme disease: Bacterial infection from tick bites that often makes a bull's-eye rash initially, then leads to fatigue, muscle and joint pain, fever, and swollen lymph nodes. It is easily treated if diagnosed early.

Disease vectors: Pathways for the pathogens that cause disease to reach humans.

Threshold dose: The amount of a toxin (or medicine) below which there is no response in anyone.

Effective dose (ED-50): The amount of a medicine that has a positive effect on 50% of people that take it.

Toxic dose (TD-50): The amount of a toxin that causes a negative reaction in 50% of people exposed.

Lethal dose (LD-50): The amount of a toxin that kills 50% of people exposed.

is a bacterial infection carried by ticks who transfer the disease from mice and deer to humans (Figure 10.6). With intensifying encroachment of human settlements on natural ecosystems, infected ticks come into contact with humans more frequently, and the incidence of Lyme disease has been increasing.

The 2020 Coronavirus (COVID-19) pandemic may be related to climate in two directions. The first is the marked reduction in greenhouse gas emissions caused by the closure of many of the world's industrial and transportation sectors. This demonstrated that it is indeed possible to halt climate change (which could have been done without "social distancing" as well). The second is that land use and consequent reduction in animal habitats may have brought wildlife into closer contact with human settlements, more readily transferring viruses between animals and human populations. At the time of writing, details and outcome remain unknown, and remain a fertile area of future research.

Dose-Response Functions

When something is ingested (be it a toxin or medicine), it is considered a dose, and each individual person has a different response (Figure 10.7). In many cases, there is a **threshold dose**, below which there is no effect, but some toxins, such as dioxin, have a zero-threshold dose. Even a little bit is toxic. Many substances are necessary for the human body, and small doses are needed for metabolic function. An **effective dose** (ED) has a beneficial impact. (TD-50 for 50% of the people) (Figure 10.7). A **toxic dose** (TD) causes health problems for 50% of the people exposed. Because of people's different responses, some remain unaffected, while others are severely impacted. A **lethal dose** (LD) kills 50% of the people.

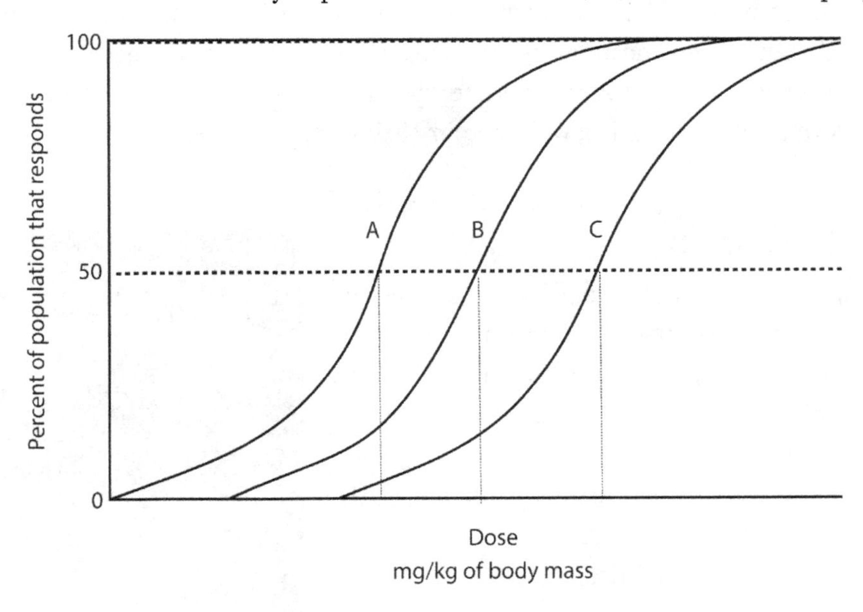

FIGURE 10.7 Dose response-antagonist
Dose Statistical response of a population to a hypothetical toxin. Note that the curves overlap such that what is lethal to one person may not even be toxic to another. A is the dosage that has an impact on 50% of the population. B is the dosage that is toxic to 50%, and C is the dosage that is lethal to 50%.

Some elements and compounds are necessary for life, yet toxic in overdoses. Fluorine is one such element (found as fluoride, F_2), which is needed for strong teeth and bones, yet in sufficient doses, can cause abdominal pain, convulsions, vomiting, and even heart attack. A corresponding dose response curve shows that insufficient amounts lead to poor health, there is an intermediate range of maximum benefit, and in excessive amounts, can lead to health problems and even death (Figure 10.8).

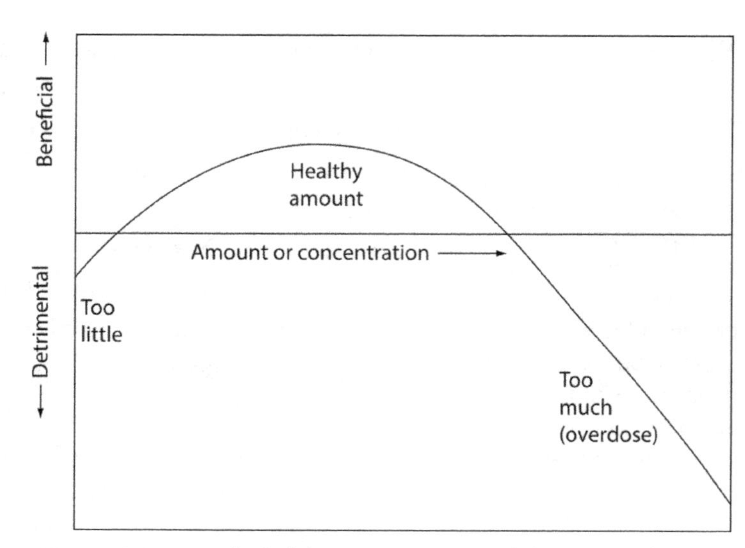

FIGURE 10.8 Hypothetical dose-response curve. If not enough of the substance is ingested, poor health results. An intermediate range provides maximum benefit, while overdose is harmful.

Biomagnification

When toxins are ingested by animals from food (or water, or air), they can be either expelled or absorbed. Many toxins are hydrophobic, meaning that they do not dissolve in or get attracted to water. These toxins bind to fatty tissue in animals, and include substances like methyl mercury, dioxin, and various endocrine disruptors. As such, when animals are eaten by other animals, the toxins are not expelled or metabolized, and remain in the predator (Zenker 2014; Zenker et al. 2014). Recall that only 10% of the energy is retained by the predator, but most of the toxins are transferred to the predator. Thus, the higher up the food chain you go, the more concentrated toxins become in animals. Because the ocean has so many trophic levels, top predators like salmon and swordfish have elevated levels of toxins, with mercury being of special concern.

Risk Assessment

An important aspect of human health is assessing the risk of adverse effects of exposure to various hazards, be they physical, chemical, biological, or otherwise (Ferguson et a., 2017). Risk (R) is normally quantified as the product of probability of occurrence (P) times the severity of impact (I).

$$R = P \times I$$

> In Figure 10.7, Curve "C" represents the response of the population to a "Lethal dose" that kills 50% of the people. At the bottom of the curve shows the dose that kills the first and most sensitive person. What percentage of the population does not even notice this same dosage? Answer: drawing a vertical line up from the start of Curve C, it intersects Curve A at about 35%, so 35% of the population does not even notice the dose that kills the most sensitive people. People respond very differently to toxins, allergens, and other substances.

The usual steps to determine risk are to

1. Identify the hazard. What is the source of concern? Can it be isolated from other hazards?

2. Dose response. What is the dose-response curve of the population exposed?
3. Assessment of exposure. What is the dosage to be received, and by what population?
4. Characterize the risk. With the trivial calculation above, a value, even relative, can be assigned and the nature and severity of the risk can be determined.

Some hazards are very unlikely, yet devastating. The most extreme example of that would be a major asteroid impact on the earth, leading to mass extinction, as apparently occurred at the end of the Cretaceous, about million years ago. It is very unlikely that another of that magnitude will strike in our lifetimes, but if it does, it will be bad. Other hazards are more frequent, but not quite so catastrophic. Tsunamis occur from time to time, volcanoes erupt, and severe droughts and floods occur every few years. Some hazards are very frequent, even constant, with only minor individual impacts. The health hazard of breathing in particulates from diesel engine emissions is ever-present, and most people who live in or near cites are exposed constantly, yet few are particularly afraid of the threat to health it poses. Each of these examples has roughly the same total risk with varying relative contributions of P and I. Public policy and health management seeks to minimize total risk, and when plotting probability against impact (Figure 10.9), one aims to stay as near to the origin (lower left) as possible, where there is little chance of anything happening, but if it does, it will not be so bad anyway. The various areas in Figure 10.9 can be likened to characters in Greek mythology, whose stories are based on various characterizations of risk.

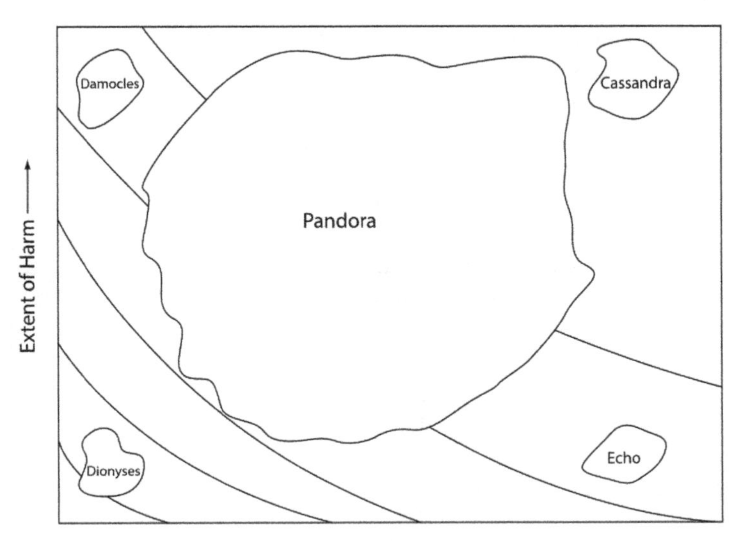

FIGURE 10.9 Characterization of risk in terms of probability versus severity of impact. Lines of "isorisk" have the same risk but in different proportions of probability and impact. One seeks to remain in the realm of Dionysus, the god of poetry and wine, but sometimes one finds one's self with Damocles, who ruled with a large sword dangling over his head. It was unlikely to fall, but if it did, his days were done. Echo was constantly annoying, but posed little immediate threat. The place one never wants to find one's self is with Cassandra, who had foreknowledge that doom was approaching the city, but was cursed by the gods so that no one would ever believe her. Thus destruction was certain, and the impact was severe. In the case of Pandora (and her box), you just don't know what you are going to get, but it will probably be bad.

References

Anwer, F., Chaurasia, S., & Khan, A. A. (2016). Hormonally active agents in the environment: a state-of-the-art review. *Reviews on Environmental Health*, 31(4), 415–433.

Bolt, H. M. (2012). Arsenic: an ancient toxicant of continuous public health impact, from Iceman Otzi until now. Archives of Toxicology, 86(6), 825–830.

Cabral, M., TOure, A., Garcon, G., Diop, C., Bouhsina, S., Dewaele, D., . . . & Verdin, A. (2015). Effects of environmental cadmium and lead exposure on adults neighboring a discharge: Evidences of adverse health effects. *Environmental Pollution*, 206, 247–255.

Cronin, S. J., Stewart, C., Zernack, A. V., Brenna, M., Procter, J. N., Pardo, N., . . . & Irwin, M.(2014). Volcanic ash leachate compositions and assessment of health and agricultural hazards from 2012 hydrothermal eruptions, Tongariro, New Zealand. *Journal of Volcanology and Geothermal Research*, 286, 233–247.

Ding, G. A., Chan, C. Y., Gao, Z. Q., Miao, Q. J., Li, Y. S., Cheng, X., . . . Miao, Q. J. (2005). Vertical structures of PM10 and PM (2.5) and their dynamical character in low atmosphere in Beijingurban areas. *Science in China Series D-Earth Sciences*, 48(2), 38–54.

Ferguson, A., Penney, R., & Solo-Gabriele, H. (2017). A review of the field on children's exposure to environmental contaminants: A risk assessment approach. *International Journal of Environmental Research and Public Health*, 14(3), 265.

Kjellqvist, A., Palmquist, E., & Nordin, S. (2016). Psychological symptoms and health-related quality of life in idiopathic environmental intolerance attributed to electromagnetic fields. *Journal of Psychosomatic Research*, 84, 8–12.

Lefohn, A. S., Malley, C. S., Smith, L., Wells, B., Hazucha, M., Simon, H., . . . & Gerosa, G. (2018). Tropospheric ozone assessment report: Global ozone metrics for climate change, human health, and crop/ecosystem research. *Elementa-Science of the Antropocene*, 6(28).

Martinez-Salcido, A. I., Ruelas-Inzunza, J., Gil-Manrique, B., Natares-Ramirez, O., & Amezcua, F. (2018). Mercury levels in fish for human consumption from the Southeast Gulf of California: Tissue distribution and health risk assessment. *Archives of Environmental Contamination and Toxicology*, 74(2), 273–283.

McLean, J. E., McNeill, L. S., Edwards, M. & Parks, J. L. (2012). Hexavalent chromium review, part 1: Health effects, regulations, and analysis. Journal of American Water Works Association, 104(6), 35–36.

Scientific Committee in Emerging Newly Identified Health Risks. (2015). Opinion on potential health effects of exposure to electromagnetic fields. *Bioelectromagnetics*, 36(6), 480–484.

Tuyet-Hanh, T. T., Minh, N. H., Vu-Anh, L., Dunne, M., Toms, L. M. Tenkate, T., . . . & Harden, F. (2015). Environmental health risk assessment of dioxin in foods at the two most severe dioxin hot spots in Vietnam. *International Journal of Hygiene and Environmental Health*, 218(5), 471–478.

Zenker, A., Cicero, M. R., Prestinaci, F., Bottoni, P., & Carere, M. (2014). Bioaccumulation and biomagnification potential of pharmaceuticals with a focus to the aquatic environment. *Journal of Environmental Management*, 133, 378–387.

Figure Credits

Fig. 10.1: Copyright © Sam1353 (CC BY-SA 4.0) at https://commons.wikimedia.org/wiki/Category:Tsunami#/media/File:Tsunami_formation_.png

Fig. 10.2: Source: https://commons.wikimedia.org/wiki/Category:Ozone_layer#/media/File:-Satellite_Image_of_Ozone_Hole_in_2007.jpg

Fig. 10.3: Copyright © 2011 Depositphotos/natalia2484.

Fig. 10.4: Copyright © Edward Kimmel (CC BY-SA 2.0) at https://commons.wikimedia.org/wiki/File:Climate_March_1085_(34368550705).jpg.

Fig. 10.6: Copyright © 2016 Depositphotos/Goldfinch4ever.

Fig. 10.7: Copyright © Dylan2106 (CC BY-SA 3.0) at: https://commons.wikimedia.org/wiki/File:Dose_response_antagonist.jpg.

CHAPTER 11

Energy

They searched high and low leaving no stone unturned
They dug wide and deep, gladly found stuff that burned
They powered their act
But learned that in fact
Sustainable power use had to be learned.

At its most fundamental level, energy is the ability to do work. To place this in context in a simple way, from high school physics, one recalls that work = force x distance (W = Fd) and force = mass x acceleration (F = ma) so when it comes down to it, work is mad. Because energy provides us with the ability to do work (of all kinds), it has become a primary driver of modern society, and our dependence on energy for modern technology has led to the burning of fossil fuels and the resulting suite of environmental impacts. Overall, energy cannot be created or destroyed, but it can be converted from one form to another. (Actually, mass can be converted to energy and back again, so that mass-energy is always conserved, but we will not delve into special or general relativity for the purposes of this book.)

Kinetic energy: Energy of motion E = ½mv2 where m=mass and v=velocity. Heat is the combined, total kinetic energy of all the molecules in a substance. Temperature is the average kinetic energy of the molecules in a substance.

Potential energy: Energy stored in various fundamental "forces of nature," including gravity, electricity (electromagnetic), and nuclear forces.

There are two basic types of energy—**kinetic** (energy of motion) and **potential** (stored energy). Energy can be stored in numerous ways, but, as discussed in Chapter 5, there are only three fundamental forms linked to the three fundamental forces of nature. These are gravitational, electrical, and nuclear. (There is actually a fourth, the "weak" force involved in nuclear fusion and radioactive decay, and as critical as it is for the universe, we will ignore that for now.) The weakest fundamental force is **gravity**, yet gravitational potential energy is evident in our everyday lives, and drives many processes throughout the environment (e.g., hydrologic cycle, wind, and the very existence of a concentration of mass called the Earth, or the Sun). When a mass is lifted against gravity, energy is stored by that mass against the gravity field such that it can be allowed to sink again while doing work (performing a function). Gravitational attraction (never repulsion) decreases as the square of the distance between two masses.

The strongest known force in the universe is aptly named the "strong force," or **nuclear** force. This acts only at a very short distance (also true of the weak force), rather than decaying as a square of distance. It binds protons and neutrons together in an atomic nucleus, and can store great quantities of energy that can be released to serve as the basis for modern nuclear energy facilities.

Electrical energy is based on an intermediate-strength force, between gravity and nuclear. It is the force between charged particles (like protons and electrons) and electric (and electromagnetic) fields, and thus performs a great many functions. This causes electrons to seek proximity to protons in every atom in the universe. They are prevented from getting too close by limited quantized energy levels that they can occupy, and this leads to the various specific arrangements of electrons around a nucleus and gives us all the chemistry (including life) that we know. Electric (electromagnetic) energy can be transported as photons when electrons shift energy levels (thus creating or absorbing a photon to conserve energy), some of which shine on us as visible light from the sun and serve as the basis for photosynthesis and thus most life on Earth. The motion of electric charges in matter makes the electric currents that drive organisms as well as modern technology. Both gravity and electric fields weaken as a square of the distance between objects.

Sometimes, the different aspects of electrical potential energy are given their own names, such as "chemical energy," which involves the interactions between atoms. When wood or coal is burned, for example, energy is released when atoms are recombined into electrically more strongly bound molecules (e.g., hydrocarbons converted to carbon dioxide and water). "Radiant energy" is the energy carried by electromagnetic waves such as visible light, x-rays, and radio waves. These waves can be absorbed by electrons in atoms, and move the electrons into higher energy levels. This is how the ozone layer of the atmosphere shields the Earth's surface from harmful ultraviolet rays coming from the sun, for example.

"Thermal energy" is actually a form of kinetic energy at the molecular level. When molecules move or oscillate, they carry kinetic energy, which, when aggregated over a material, is called heat. When two substances of different temperatures are put in contact with each other, heat can flow from the higher-temperature substance to the lower, causing them to equilibrate at an intermediate temperature. This is called heat **conduction** (Figure 11.1). Hot matter emits infrared radiation as electrons change energy levels and emit photons, and this can also transfer energy and warm up nearby (or far away) matter when its electrons absorb the photons. This is heat **radiation**, and old-style room radiators used this mechanism (Figure 11.2). Yet a third way to transfer heat is by moving something hot into a place that is not as hot. When this is done horizontally, it is called advection, and when it is done vertically as a result of less dense hot material expanding and buoyantly rising to take the place of cooler material, it is called **convection**. In the environment, conduction, convection/advection, and radiation are all important mechanisms of heat transfer.

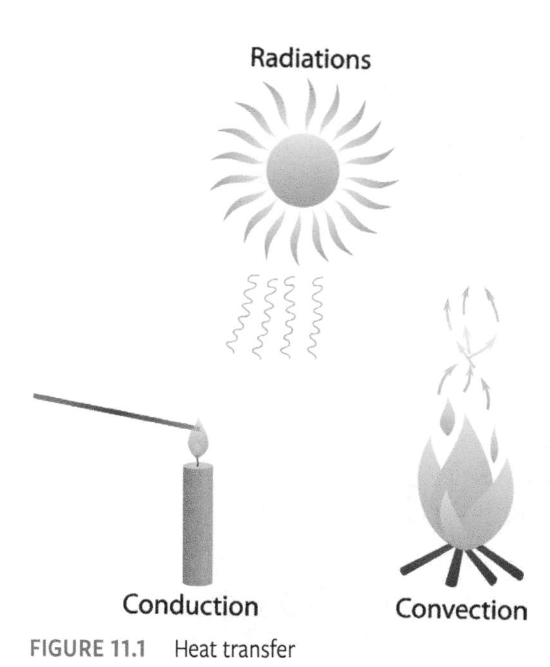

FIGURE 11.1 Heat transfer

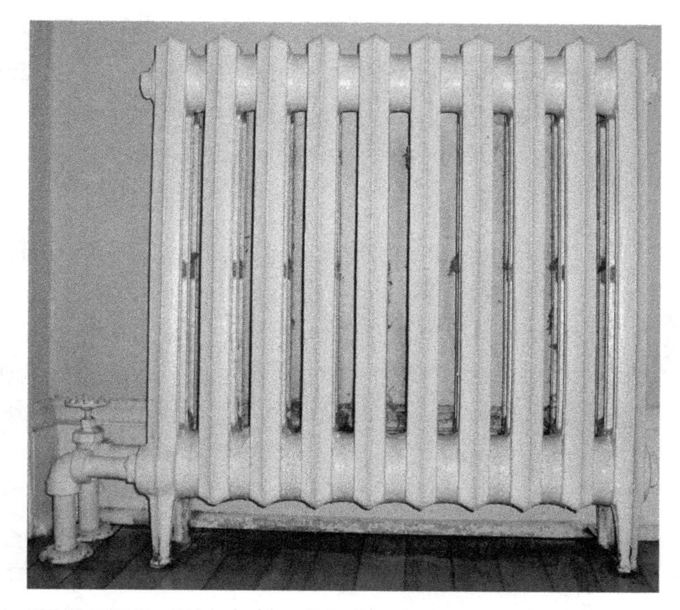

FIGURE 11.2 Household radiator

Each of the three main forces serves as an energy source for modern technology, and means have been devised to extract energy from each. Although the most common is electrical (basis for all chemistry, including all burning, all life, and sunlight) gravity and nuclear are also important players in our energy portfolio.

The Basic Laws of Thermodynamics

There are a few fundamental rules that govern the conversion of energy form one form to another. Some of these seem intuitively obvious, while others are a bit more abstruse. The laws are numbered 0 through 3 rather than 1 through 4 because the most fundamental law was defined last, thus added to the front of the line, as the 0th law. They are paraphrased in simple terms as follows:

> 0th Law: If two things are in thermal equilibrium with a third, they are in thermal equilibrium with each other (all the same temperature ...).
>
> 1st Law: Energy can be transferred from one form to another, but the total is conserved (cannot be created or destroyed: You can't get something for nothing—at best you can break even.)
>
> 2nd Law: Energy flows only from hotter to colder bodies or from those with more energy to those with less (energy can only become more disordered, and diffused, thus increase entropy: You can't break even—you lose on any energy exchange, thus making some unusable heat).

3rd Law: As temperature approaches absolute zero, entropy approaches a constant, so absolute zero is unattainable (you can't take away energy from something that has less energy than that with which it interacts, so you can't cheat on the 2nd law ... Sort of enforcing the "Robin Hood of energy" aspect of the 2nd law).

The 0th law needs no explanation. The 1st Law insists that we can neither create nor destroy energy. The 1st Law DOES allow for conversion of energy from one form to another. It indicates, for example, that we can allow something to fall toward the ground, and convert the gravitational potential energy it has to kinetic energy of motion (downward), and then convert it again to another form of energy, such as electrical. This is how a hydroelectric power plant works, for example. Water falls downward, losing gravitational potential energy to kinetic, and converts its gained kinetic energy to electricity by doing work on a turbine that drives a generator. Alternatively, the wind can transfer its kinetic energy to a windmill blade, thus turning a generator to make electricity, or lifting water from a well against gravity. As such, the 1st Law is a simple matter of converting energy from one form to another.

The 2nd Law is a bit more involved, but most critical. Although the 1st Law allows for conversion of energy, the 2nd Law exacts a "tax" from any conversion, so that you cannot convert ALL of the energy into a new form. There must always be some unusable heat generated, so it is never 100% efficient. This means that the net effect of any energy conversion is to degrade the "quality" of the energy involved, thus increasing what is called the entropy of the system, and indeed of the entire universe. Energy quality can be viewed as the extent to which energy is concentrated in space. Essentially, you cannot use something with unconcentrated energy to further concentrate energy in something that is already more highly concentrated than what you are using to concentrate it. An easy way to visualize this is to consider paint colors—you can make the color of a pastel more intense by adding a more concentrated color. However, you can never make the already more concentrated color more intense by adding some diluted color. If you want to make it less intense, you cannot remove the concentrated color anymore, either- it is already mixed in. Nuclear energy involves lots of energy in a very small place (atomic nucleus), and is thus very high quality. After a nuclear reaction, however, there is less nuclear energy and more electromagnetic energy (light) and heat (random molecular kinetic energy), which are both far less concentrated forms of energy, and thus lower quality. Once you have such a conversion, there is no going back to the original situation because you cannot use less-concentrated energy to make more-concentrated energy (just like the water colors). Because of this one-way rule, the greater the dilution of energy in the conversion, the more is wasted as irretrievable heat, and thus the lower the efficiency of the conversion. Therefore, it would be wasteful to use nuclear energy (most highly concentrated) to heat a house (most diluted energy). A less inefficient use of nuclear energy could be, for example, to produce electricity directly, as this is the next most concentrated form of potential energy, and from this, to organized kinetic energy such as the motion of a machine or vehicle, rather than to the thoroughly disorganized kinetic energy involved in heat.

The 3rd Law is a special case of the 2nd Law, and focuses on the inability to reach absolute zero temperature. This is because in order to reach absolute zero, some very cold bit of matter would have to transfer energy (allowable by 1st Law) to some other not-as-cold piece of matter (not allowed by 2nd Law).

Renewable and Non-Renewable Energy

Places where energy is stored for convenient conversion to other forms that we find useful are called energy sources. These can include petroleum, wind, coal, etc. These sources of useful energy are categorized into two types—the commonly used terms are "renewable" and "non-Renewable." Renewable (sustainable) energy sources produce energy at a rate that can keep up with the rate at which they are used. An example is wind that drives windmills that produce electricity by turning generators. No matter how much of the wind gets used for windmills, it does not deplete the wind. It will not run out. Another example is biomass burning. When a tree or fallen branch is burned for cooking or heating, another grows to take its place. As long as the global rate of cutting and burning does not exceed the rate of regrowth, biomass burning is renewable. However, as soon as the rate of cutting exceeds that of regrowth (which it did already centuries ago), it is no longer considered renewable. Most forms of "renewable" energy are not really renewable, but essentially inexhaustible, being sustainable indefinitely. Most of these energy sources are driven by the sun, including solar (of course), wind, hydropower from dams, and biomass, whether is it burned directly or used to make ethanol or other fuels. Of these, only biomass grows back and is actually renewable. The others are driven by the sun directly, and will be available as long as the sun shines (another several billion years or so) regardless of how much of them we use. Another way of putting this is that these inexhaustible energy sources cannot be exploited at a rate any greater than the rate at which the sun supplies them. In this way, they are in stark contrast with fossil fuels such as coal or oil. Another alternative energy source is the nuclear fission of uranium, which is also non-renewable. Uranium can run out. Nuclear fusion of isotopes of hydrogen to create helium is essentially inexhaustible, if only we could build the technology to create an economically viable fusion-based power plant.

Non-renewable energy sources include all the fossil fuels (oil, natural gas, coal), a well as uranium used in nuclear fission plants. Of course, any source that is used up faster than it is replenished is considered non-renewable. In the case of the fossil fuels, they form slowly over millions of years.

In the case of uranium, it forms in a supernova, and we do not expect one anytime soon in our general vicinity of the galaxy. As such, the amount of non-renewable energy we have stored in these sources is all we will ever get. When we use them up, they will be gone. However, the issue with fossil fuels is not that they will be completely depleted, but the concern is more that the growing demand for them can exceed the rate at which they can be supplied by mining and drilling. At that point, energy shortages will be a more immediate concern, and this is expected soon in the case of oil, and several decades later for coal and natural gas. A more pressing issue is the consequence of fossil fuel burning regarding climate change. There is enough fossil fuel remaining in the ground to change climate far beyond what modern civilization would consider comfortable or even acceptable. This is further explored in Chapter 13.

Fossil Fuels

Fossil fuels are energy resources derived from the altered remains of living organisms that were buried by sediments and exposed to elevated pressures and temperatures for millions of years. Because of the long time it takes to create these materials, they are non-renewable, meaning that more will not be made during anyone's lifetime, or even during modern civilization of humanity on the earth (Table 11.1). There are three basic forms of fossil fuel—oil, natural gas, and coal.

Oil

Oil, otherwise known as petroleum, does not come from dead dinosaurs, despite some perceptions, cartoons, and commercial ads. Oil is the altered remains of marine micro-organisms (plankton) that died, settled to the ocean floor, and were buried by later sediments before they had a chance to decompose (oxidize). Over millions of years, the thick piles of sediments over them buried them so deeply that they were subjected to high pressures and temperatures, driving off the oxygen to mature so that a goo was left behind, consisting only (mostly) of hydrogen and carbon—thus the term "hydrocarbon," (Tissot et al., 1974; Requejo, 1994; Requejo et al. 1992). In order for oil, once formed, to be useful, it must be accessible for drilling from

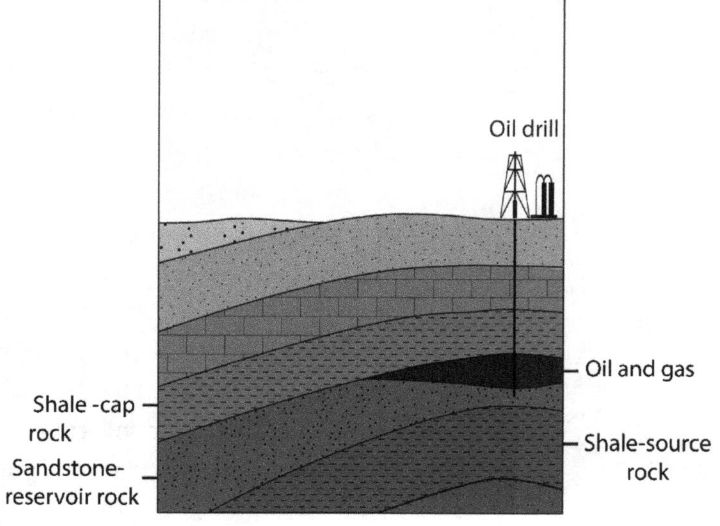

FIGURE 11.3 Structural traps
Structural traps for the concentration of oil and gas. The hydrocarbons migrate slowly from a source rock, upward into a reservoir rock that is confined by a caprock. The trap could be due to a fold in the strata, or a fault that moves an impermeable layer of rock to overlie a dipping (tilted) permeable reservoir rock.

Source rock: The geologic formation that contained organic matter that matured into hydrocarbons over tens of millions of years.

Reservoir rock: The geologic formation overlying a source rock into which hydrocarbons migrate, which is both porous and permeable (like a good aquifer) so that the hydrocarbons can be easily extracted.

Caprock: An impermeable geologic formation immediately over a reservoir rock that prevents hydrocarbons from migrating farther upward, leading to high concentrations of hydrocarbons in the reservoir rock.

oil wells, so it has to move, or migrate from the broad areas where it formed, called **source rock** to become concentrated in a much smaller space, called a **reservoir rock** (Figure 11.3). Because oil is less dense on average than water, and the ground is saturated with water, the oil slowly moves up the slopes of deformed sedimentary layers to get caught in stratigraphic or structural traps confined by a **caprock**. These are the places that oil companies drill to pump out the oil to sell in the market.

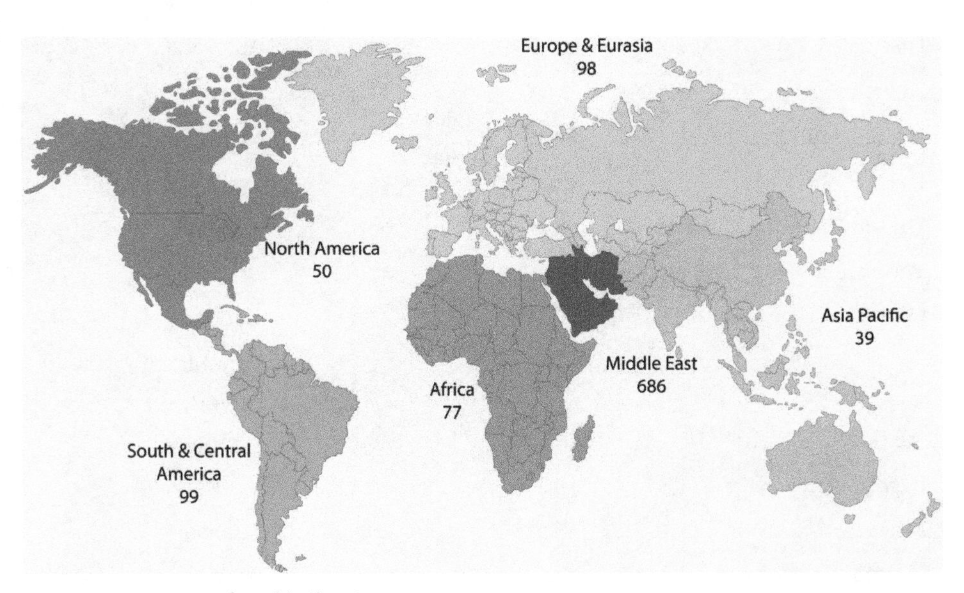

FIGURE 11.4 Map of world oil regions
Oil is unevenly distributed around the world, with much of it residing in the Middle East. The original oil stocks have been unevenly exploited, with the large stock in the U.S. (where the first commercial oil wells were drilled) being mostly depleted. Figures for each region are remaining oil reserves in billions of barrels.

Structural trap: Shape of originally horizontal caprock that has been deformed or faulted to trap hydrocarbons in a specific part of a reservoir rock.

Stratigraphic trap: Concentration of hydrocarbons in a specific part of a reservoir rock due to non-horizontal original deposition of the reservoir and caprocks.

In order for a caprock to effectively concentrate gas or oil, it needs to be deformed so that it has a structure that catches rising fluids that are lighter than water, such as oil and gas. A **structural trap** is an umbrella-shaped dome or other shape that captures rising oil and gas and because the caprock is impermeable, it creates a concentrated reservoir into which a well can be drilled and oil (and gas) can be extracted conveniently. A structural trap can be caused by a fold in sedimentary rocks, or by a fault in tilted (dipping) strata that places an impermeable rock in the way of rising oil or gas.

A **stratigraphic trap**, on the other hand, is a place where the rock type changes, thus causing migrating oil and gas to be funneled into a more highly concentrated location. A stratigraphic trap could be caused by a pinchout of the lithology, or an unconformity in which dipping strata are cut by erosion and then covered by flat-lying strata, or it could be a reef covered by impermeable shale, for instance.

TABLE 11.1 OIL RESERVES IN VARIOUS COUNTRIES.

Country	Estimated Ultimate Recovery (Bb)	Remaining Oil Stock as of 2000 (Bb)	Country	Estimated Ultimate Recovery (Bb)	Remaining Oil Stock as of 2000 (Bb)
Saudi Arabia	273.2	179.6	Egypt	15.5	7.1
Russia	264.6	123.0	Argentina	14.8	6.9
Canada	64.2	37.7	Angola	10.6	6.8
Iran	129.6	77.7	Australia	12.4	6.7
Kuwait	103.5	71.3	Oman	14.7	6.5
United Arab Emirates	85.4	63.9	Malaysia	11.0	6.3
Iraq	109.0	62.4	Syria	8.2	4.9
United States	271.2	62.2	Yemen	6.1	4.8
Venezuela	115.1	60.8	Ecuador	6.9	4.0
China	86.1	40.1	Gabon	5.8	2.9
Norway	42.4	28.5	Republic of Congo	3.6	2.3
Mexico	58.8	28.4	Denmark	3.2	2.2
Nigeria	48.8	28.1	Vietnam	2.7	2.1
Libya	48.2	26.3	Tunisia	2.7	1.5
United Kingdom	44.2	26.0	Brunel	4.6	1.4
Indonesia	38.1	17.7	Peru	3.5	1.3
Algeria	26.5	12.8	Trinidad & Tobago	4.5	1.3
Brazil	18.2	12.4	Romania	6.3	1.1
Qatar	17.4	10.9	Italy	2.0	1.2
Colombia	15.5	10.3	Cameroon	2.0	1.0
India	13.6	7.8	Papua New Guinea	1.0	0.7

In order to burn the oil removed from reservoirs, it must be refined and turned into something like gasoline or diesel fuel. This is done in large refineries. In a refinery, the crude oil is chemically altered (cracked) so that the chains of carbon atoms with hydrogen atoms attached take shape in specific forms that can be burned in predictable ways.

Today oil is found on most of the world's continents, but is distributed very unevenly (Figure 11.4 and Table 11.1). The processes that form oil are very specific, and lots of conditions must be satisfied for a useful oil resource to be exploited, especially the proximity of a source rock to a reservoir rock and a good stratigraphic or structural trap. These conditions are very well satisfied in the Middle East, where much of the world's oil is found.

Natural Gas

When oil matures further under high pressure and temperature deep underground, the carbon atoms no longer bond to each other, becoming surrounded only by hydrogen atoms. This makes natural gas, or methane, which is a gas (not liquid) that rises to the top of reservoirs, and often sits on top of oil beneath it, trapped by the cap rock. Natural gas is the most evolved fossil fuel, and because it has the lowest ratio of carbon to hydrogen, burns the most cleanly, forming only water and carbon dioxide as a result of burning. However, natural gas is difficult to transport (because it is a gas), and does not sell at as high a price as oil. For this reason, for many years, it was simply burned off from the tops of oil reservoirs so that oil companies could get to the more valuable oil beneath. Now, companies go through great pains to extract natural gas through "unconventional" means such as by fracturing shale source rocks.

Natural gas, because of its chemistry, provides the most energy per carbon dioxide molecule emitted from burning. This is a second reason that it is the cleanest fossil fuel. Consider the number of chemical reactions that release energy in order to form more tightly bound molecules that include oxygen (through the process of oxidation, or burning). Coal is simply carbon, so when it is combined with oxygen, every reaction produces a molecule of CO_2. One reaction, one CO_2 emitted. Now consider natural gas, or methane, CH_4. When a molecule of methane is combined with oxygen, the carbon makes a CO_2, but the four hydrogens also combine with oxygen to make two water molecules, so there is a total of three reactions that produce energy for every CO_2 emitted, so we get about twice as much energy per CO_2 released from natural gas than from coal. We do not worry about the water emitted, as it enters and then rains out of the atmosphere in the normal hydrologic cycle, but the CO_2 remains in the atmosphere as a greenhouse gas for millennia.

Marcellus Shale: A Devonian age fine-grained source rock containing large amounts of natural gas (methane) in the Appalachian basin of New York, Pennsylvania, and West Virginia. Exploitation has led to considerable controversy on economic and environmental grounds.

There has been a great deal of interest recently in the **Marcellus Shale** as a source of natural gas in the northeast U.S. This shale is part of the Appalachian Basin, having been deposited in the foreland basin of the rising Appalachian Mountains in the Devonian period, about 385 million years ago. The gas-bearing shale extends from West Virginia to New York State, with a large area in Pennsylvania, the region of greatest economic interest and environmental concern.

In exploiting the Marcellus Formation as a gas source, the common knowledge ideas of tapping reservoir rocks in which fossil fuels have been concentrated is completely ignored. The Marcellus Formation is actually a source rock, from which the natural gas never (yet) migrated into a more permeable reservoir rock, to be trapped by a caprock. This is a clear demonstration that most of the easy sources of fossil fuels have already been exploited, and that we are now desperately seeking fossil fuels in their original source rocks. The problem with doing this is that source rocks (shales) have very low permeability, so gas (and oil) does not move through the rock to enter a well pipe for extraction to the surface.

To circumvent the permeability problem, petroleum engineers have developed a means of creating permeability where there was none in the rock. They do this by fracturing the rock with high-pressure water mixed with lubricating fluids and sands (Figure 11.5). This breaks up the shale so that along a limited set of fractures, gas can seep through the broken rock and into the well. However, it breaks up the rock along only a few planes—it does not give the entire shale the permeability of a sandstone. Consequently, only a very small fraction of the gas can be extracted. To enhance the efficiency of the technique, a single well is drilled, from which multiple horizontally drilled arms or spokes are sent, to fracture or "frack" a greater area than would be possible from a single vertical hole.

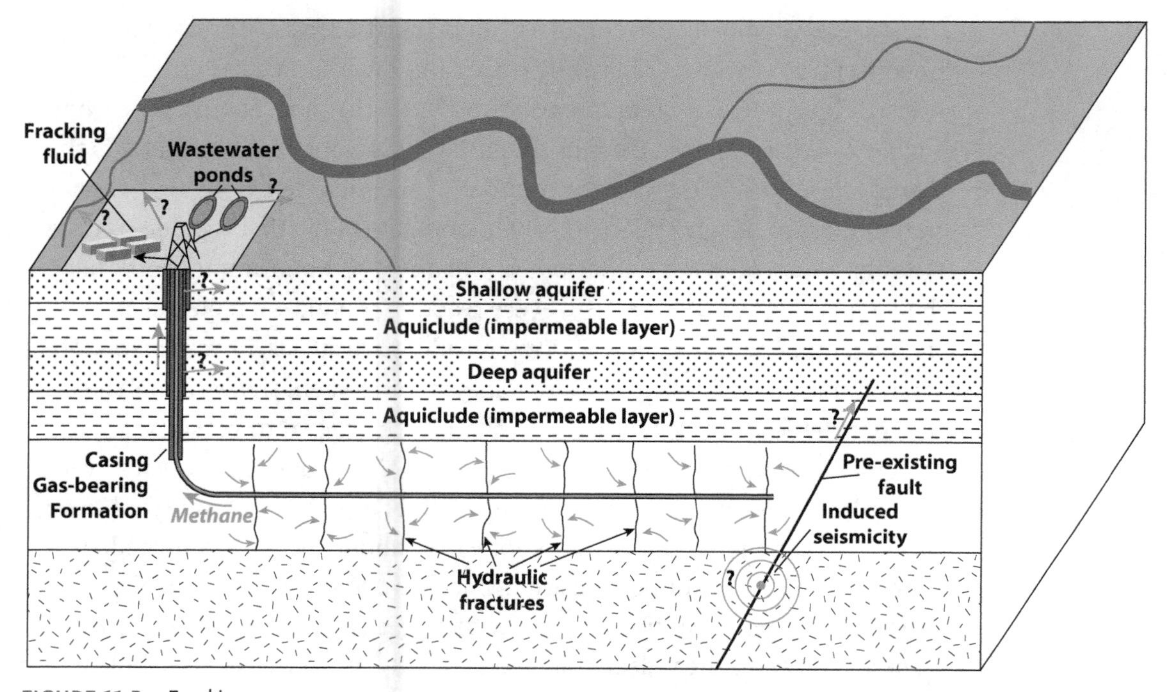

FIGURE 11.5 Fracking
Simple depiction of hydraulic fracturing (fracking). This diagram shows only one of several horizontal arms that extend from the well. Although the rocks that contain gas are much deeper than aquifers used for extracting ground water for domestic use, there is concern about leakage of chemicals used in fracking fluids from the well casing as it penetrates the aquifers, and additional concerns about the fate of these chemicals when they are stored in ponds at the surface.

Exploitation of the Marcellus Shale has raised a number of environmental concerns. Aside from the surface disruptions to the productive ecosystems of the temperate northeast U.S., people have become concerned about drilling fluids and natural gas (methane) entering aquifers and domestic water supplies (Eisenberg, 2015). There are stories of peoples' kitchen sink faucets lighting on fire from the concentration of methane in the water. Further, concerns regarding the fate of drilling fluids and fracking fluids remain. If these enter surface streams, there could be environmental consequences as well (Robinson, 2012; Rozell and Reaven, 2012, Rozell, 2012). Because this technology has been a recent development, there has not been much time for these concerns to be validated or negated through extensive environmental monitoring programs.

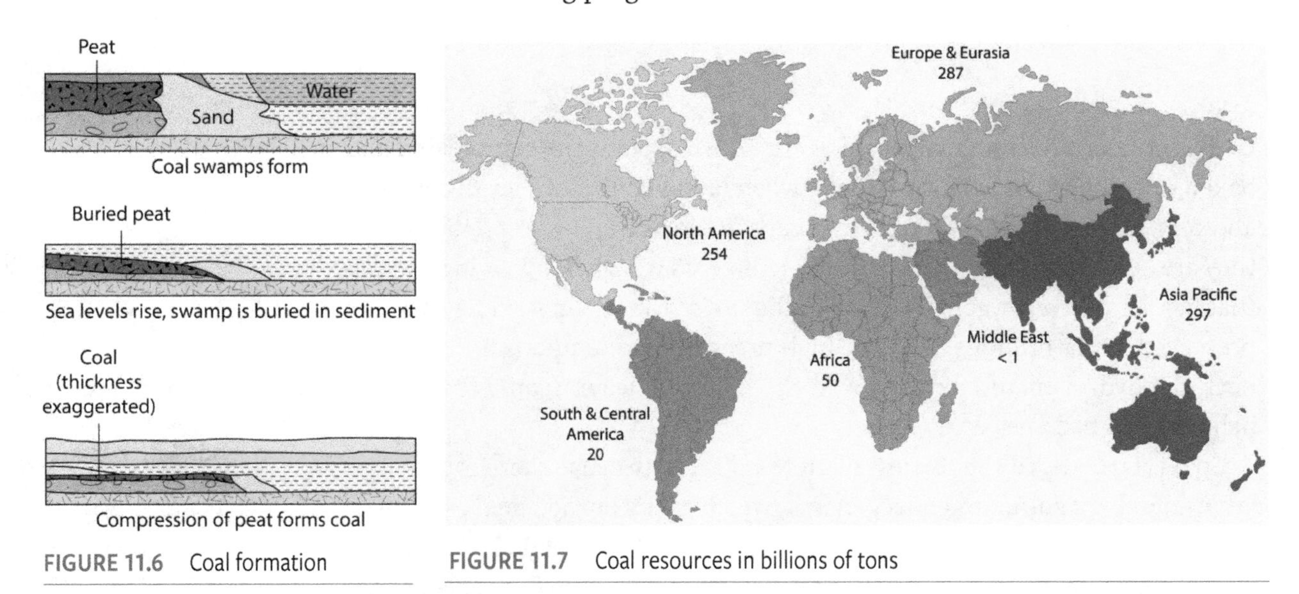

FIGURE 11.6 Coal formation

FIGURE 11.7 Coal resources in billions of tons

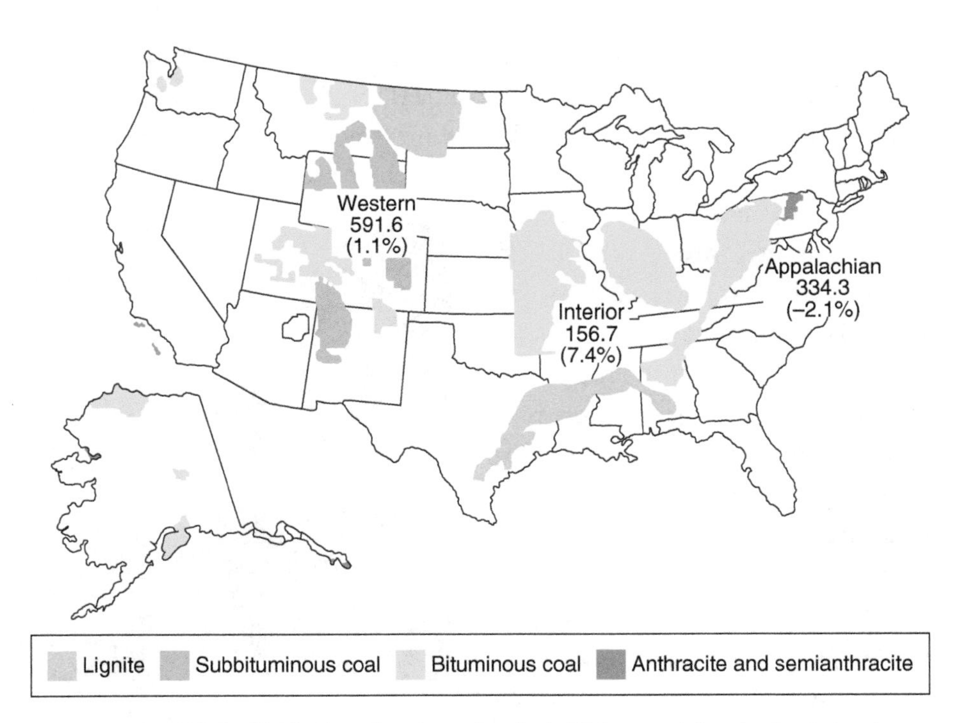

FIGURE 11.8 U.S. Coalfields. Note that the anthracite in PA is essentially mined out already, and lesser quality coals are being burned nationally.

The controversy of the Marcellus Shale as a natural gas source continues, and serves to demonstrate the extremes to which our society will go to obtain more fossil fuels from an increasingly depleted planet to feed an increasingly energy-intensive economy.

Coal

Coal is not derived from marine micro-organisms, but is the compressed and baked remains of land plants. Usually when land plants die, they decompose and oxidize, and ultimately turn into carbon dioxide and water. However, if they are quickly buried by other plants, sediments from a flood, or in a swamp that has very little oxygen dissolved in the water, they do not oxidize and are preserved. Over millions of years, high pressure and temperature drive off both the hydrogen and oxygen from the material, leaving only the carbon behind. This becomes coal (Figure 11.6).

Coal is considered the dirtiest of all fossil fuels, because there is usually a large amount of sulfur and nitrogen in it, which, when burned, makes sulfuric acid and nitric acid, which cause acid rain and other forms of pollution. There is an advertising campaign about so-called clean coal, but the fact remains that of all fossil fuels, coal produces the least energy per carbon dioxide molecule emitted to the atmosphere, thus enhancing the greenhouse effect and leading to global climate change.

Acid mine drainage: The addition of acid-forming compounds (often from pyrite, FeS) in coal mine residual tailing piles into streams by solution in rainwater runoff, rendering the stream water acidic (sulfuric acid) as well as iron-rich. The iron concentration makes a yellow-orange color in the water, and has been so pervasive in these streams that it was given a name "yellow boy."

Coal is found almost everywhere in the world, and remarkably, is almost absent from the Middle East (Figure 11.7). The purest, hardest form of coal, anthracite, was found in Pennsylvania, and has been extracted over the years so that there is very little left (Figure 11.8). However, there are large amounts of lesser types of coal, such as bituminous, and the dirtiest type, lignite, throughout America and all over the world. This has scientists worried that instead of developing new, renewable energy sources, many people in countries such as the U.S. and China will increasingly burn coal and thus greatly increase the rate of global warming by emitting large amounts of carbon dioxide into the atmosphere. Further, the mining of coal has led to denudation of large areas of the countryside in places like West Virginia and Pennsylvania, and the streams that flow past former coal mines become contaminated with acids (Figure 11.9). This **acid mine drainage** makes them uninhabitable for fish and other wildlife as well as impossible for people to drink (Banerjee, 2014). Ultimately, the avoidance of coal-burning is one of the most important reasons to develop renewable energy.

FIGURE 11.9 Acid mine drainage.
Streams flowing through the tailing of abandoned coal mines dissolve iron and sulfur (mostly from the mineral, pyrite) and thus makes the water very iron rich and acidic (with sulfuric acid). Pennsylvania, and then West Virginia, have the most miles of acid mine drained streams in the U.S.

TABLE 11.2 PRODUCTION AND SUSTAINABILITY OF FOSSIL FUELS.

	Coal	Gas	Oil
Global total resources	14 trillion (tons)	20 quadrillion (cubic ft) = 2 × 10**16	6 trillion barrels
Global reserves	1 trillion (tons)	1 quadrillion (cubic ft)	1 trillion barrels
Production rate for last 100 million yrs	140,000 tons/yr	200 million CF/yr	60,000 barrels/yr
U.S. consumption	1 billion tons/yr	30 trillion cf/yr	7 billion barrels/yr
U.S. per capita consumption	3 tons/person/yr	100,000 cf/person/yr	22 barrels/person/yr
Number of people sustainable (production rate divided by per capita consumption)	44,000 people	2,000 people	3,000 people

If everyone in the world consumed fossil-fuel-derived energy like the average American, all fossil fuels would run out in a few decades (coal lasting the longest). Conversely, if we all really wished to consume fossil fuel reserves sustainably, at the rate at which they are produced by geologic processes, the global population could be no more than about 4,200 people, the population of a small town. The figures in this table are approximate and rounded roughly (order of magnitude) so they are easier to remember. They are also changing constantly, but the inevitable conclusion is that if the whole world succeeded in consuming fossil-fuel-derived energy like Americans, we would have to reduce global population to a very small number to avoid running out of fossil fuels very soon.

SUSTAINABLE FOSSIL FUELS?

In round numbers, if the total global oil resources are about six trillion barrels, and they formed over the last 100 million years, the rate of global oil production was about 60,000 barrels per year (Table 11.2). The U.S. consumes about 7 billion barrels of petroleum products per year (EIA, 2013), which, for 315 million people is 22 barrels per person per year. So the number of people that can be sustained by the global production of oil by geologic processes is 60,000/22 = about 3,000 people. Add 14 trillion tons of coal produced over the same 100 million years (140,000 tons/yr), with a U.S. consumption rate of a billion tons/yr (3.2 tons/person/yr), another 44,000 people can be sustained in the world on fossil fuel energy. Add natural gas to this, and 20 quadrillion (2 × 10**16) cubic feet total global reserves formed in 100 m.y., (200 million cf/yr), with a U.S. consumption of 30 trillion cf/yr (3 × 10**13 cf/yr) (100,000 cf/person/yr) supports another 2,000 people in the world. (These numbers do NOT count the fossil fuels used in producing the imported products that Americans buy every year [emergy, short for "embedded energy"], so are extreme underestimates.) In total, if we could economically extract ALL fossil resources economically, fossil fuels could sustainably support a global human population of about 50,000 people who consume energy like Americans. Current global population is considerably greater than this, which is part of why there is a call for transition to sustainable energy sources.

How many Americans could be supported sustainably on fossil fuels if only global reserves were available to use? Answer: For coal, the production rate (tons/yr) would be global reserves (tons) divided by 100 m.y., or 1×10^{12} tons / 1×10^{8} yr = 10^{4} tons/yr = 10,000 tons/yr. To get number of Americans, divide production rate (tons/yr) by consumption rate per person (tons/person-yr) to get people alone in the numerator. So 10,000 tons/yr/3 tons/person/yr = 3,333 people. This is much less than if total resources were available. Do the same for gas and oil.

Coal is most commonly burned in electric power plants to provide electricity to the grid. Because coal is so plentiful, it is relatively cheap, if you do not account for the environmental damage it creates. However, if you build in the price of cleaning up coal mines to prevent acid mine drainage, treating rivers and streams throughout the country to restore their proper acidity, the health costs of particulates from coal-burning, and most importantly, the cost of climate change to the societies of the world, coal would be by far the most expensive energy source, including presently existing renewable sources such as wind, solar, and tidal.

Nuclear Energy

First of all, nuclear is pronounced new-klee-ar. Not new-kew-lar!

Nuclear energy is the strongest kind in the universe (that we know of) and comes from binding energy that holds together protons and neutrons in an atomic nucleus. This does not involve oxidation (burning) of carbon-based fuels, and so does not emit CO_2 into the atmosphere to exacerbate global warming. The binding energy (per nucleon) depends on how large a nucleus is, with the greatest energy per nucleon in iron and nickel. The per-nucleon binding energy decreases as you get to larger nuclei, as well as smaller nuclei (Figure 11.10).

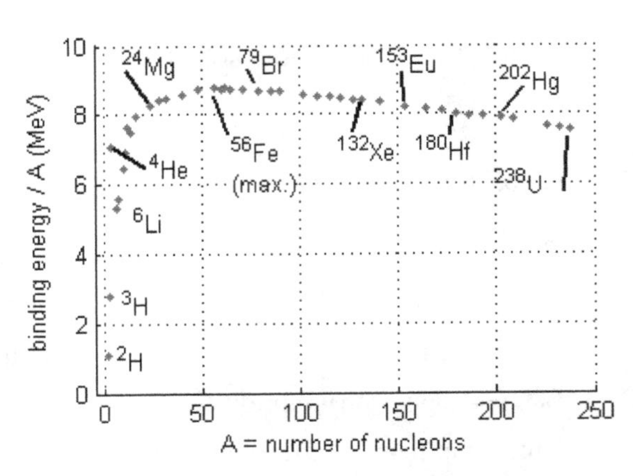

FIGURE 11.10 Curve of Nuclear Binding Energy. The nuclear force holds protons and neutrons together in a nucleus; the strength of the inter-nucleon bond depends on the size of the nucleus. The strongest bonds are within the iron nucleus, gradually tapering to larger nuclei, but steeping dropping off in smaller nuclei. This is why fusion can release much more energy per nucleon than fission. Controlled fusion reactions are difficult to accomplish, and research is currently underway.

FIGURE 11.11 Nuclear plant surrounded by farmland

Fission: The splitting apart of a large atomic nucleus (e.g., uranium), triggered by bombardment by neutrons either from an external source or by other fission reactions in the nuclear fuel.

Fusion: The joining together of small atomic nuclei (e.g., hydrogen) by forcing them together to overcome the electronic repulsion of the protons and allow the nuclear force to bind them together. This powers the sun and all stars.

Radioactive decay: The spontaneous change of an element to another by emission of a helium nucleus (alpha decay) or electron (beta decay). A gamma ray is also emitted in most cases.

Nuclear reactions cause nuclei to split apart (**fission**), join together (**fusion**), or alter their internal structure and emit small particles (**radioactive decay**). Each of these causes a nucleus to move along the binding energy curve to another position closer to iron (the most strongly bound nucleus), releasing energy accordingly. A primary feature of this curve is that it is much steeper to the left of iron than it is to the right, where the heavy elements are located. This means that much more energy can be obtained per nucleon from moving from light elements up toward iron than down from heavy elements.

Fission

We take advantage of this in nuclear power plants (Figure 11.11), which, at present, rely on fission of large nuclei like uranium to release energy. This energy is used to heat water to steam and drive a steam turbine, just like the heat generated from burning coal.

Daughters: The large fragments resulting from fission of a very large nucleus.

Parent: The large nucleus that divides into daughters during fission.

Fission is accomplished by hitting a large nucleus with a neutron to cause it to become unstable and split into two or more fragments (Figure 11.12). These fragments, **daughters**, move away from the original site of the **parent** nucleus, leading to intense heating of the material in the nuclear reactor. Water is piped through the reactor, heating to steam to drive turbines that drive electric generators. According to the 2nd Law of Thermodynamics, using highly concentrated nuclear energy to make the least organized form of energy, heat, is inefficient, leading to rapid increase in entropy, but this is not an issue that has been of great concern to date.

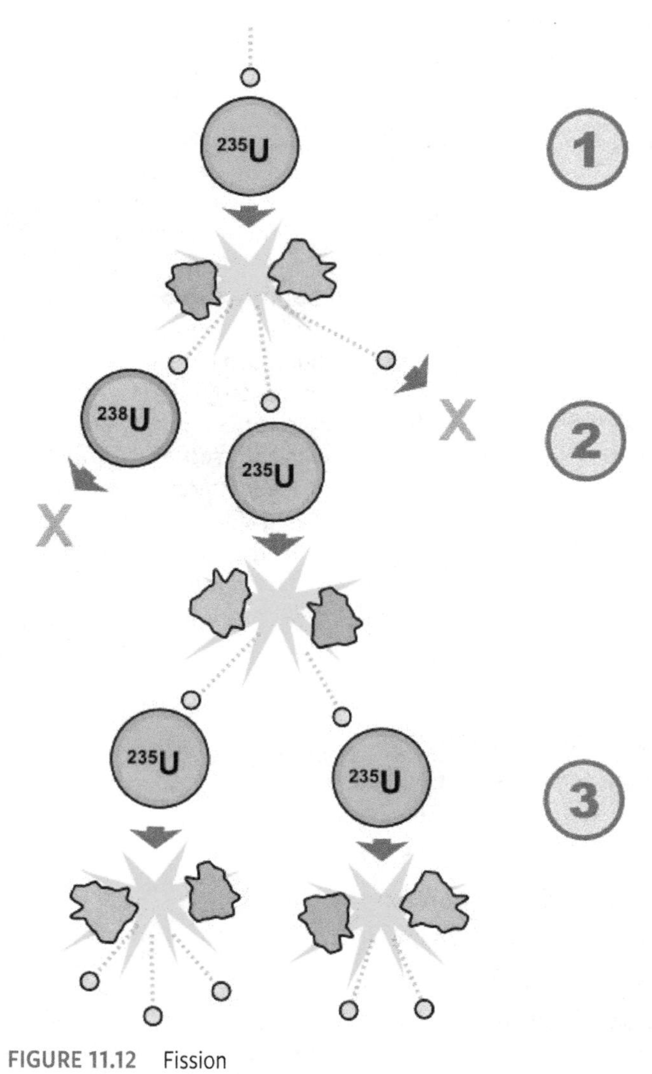

FIGURE 11.12 Fission

The daughters of fission reactions in nuclear power plants are themselves radioactive with a wide spectrum of half-lives, from seconds to millions of years. Consequently, they must be removed and stored for millennia (at least) so that the radioactivity does not enter the environment, ecosystems, food, and humans. This is one of the primary environmental concerns regarding nuclear power as an energy source.

Fusion

Nuclear fusion requires bringing small nuclei so close together that they can feel the attraction of the nuclear force between them. To do this means overcoming the repulsive electric force of positively charged protons and thus exceedingly high temperatures. Fusion occurs within sun (and all stars) due to the intense gravitational pressure that presses hydrogen atoms together at a high temperature to fuse and make helium. As such, our fossil fuels and most renewable energy sources were ultimately derived from fusion within the sun. We can recreate a local environment that leads to fusion by heating deuterium and tritium (isotopes of hydrogen) to tens of millions of degrees (hotter than the center of the sun!) by magnetically containing an ionized gas (it is a **plasma** at that temperature, stripped of its electrons). This is difficult, and while some small progress has been made in experimental reactors (tokamaks), we are still without a practical technology to accomplish self-sustaining fusion from which more energy can be extracted than is expended to create the environment to produce the fusion reaction. Further research is sorely needed to develop fusion to a practical level because unlike fission, which relies on relatively rare and unevenly distributed uranium for fuel, the isotopes of hydrogen needed for fusion are so ubiquitous in the ocean that fusion could satisfy our current rates of energy consumption for many thousands of years.

Plasma: A state of matter (like a gas) that consists solely of nuclei with the electrons stripped away. This gives them all a strongly positive charge that must be overcome by kinetic energy if they are to collide to produce nuclear fusion.

Nuclear Radiation

Nuclear radiation is actually the emission from a nucleus of helium nuclei (**Alpha**), electrons (**Beta**), or high-energy photons (**Gamma rays**). When a nucleus is large with an overabundance of protons in the proton/neutron ratio, it can reach a more stable state by turning a proton into a neutron and emitting an electron and an antineutrino. If there are too many neutrons in a nucleus (for the available energy levels), a neutron can turn into a proton by emitting a positron and a neutrino. The process is controlled by the weak force—that rarely discussed (but very important) fourth force of nature—which can also be related to the electric force in the initial discussions of a unified field theory. When a nucleus is large enough, it can spontaneously emit a helium nucleus (alpha decay), consisting of two protons and two neutrons, bringing it down the binding energy curve a bit toward iron. The third kind of radioactive decay, gamma decay, emits a very high-energy photon, or gamma ray, created by nucleons in the nucleus moving from a higher-energy to lower-energy state. This can occur with or without an accompanying alpha or beta decay. Because nuclear energy states are exceedingly strong, the steps between them are large, and the energy of the photon is great (commonly hundreds of KeV). It can be thought of in a similar way as electrons moving energy levels around orbitals of an atom and emitting photons of visible light with much lower energy than gamma rays (like how any light bulb works).

Alpha radiation: Helium nuclei emitted during a nuclear decay of a radioactive element.

Beta radiation: Electrons emitted during the nuclear decay of a radioactive element.

Gamma radiation: Electromagnetic photons (like light, but much higher frequency and energy) emitted during nuclear reactions when nucleons settle into lower nuclear energy levels within the nucleus. (The nuclear equivalent to visible light being emitted when electrons settle into lower energy levels around a nucleus in an atom.)

Nuclear radiation is a health concern because alpha, beta, and gamma particles all strip away electrons from atoms in an organism (such as your body), and thus lead to cell and DNA damage. The ultimate health effects of this are normally cancer and leukemia, and in extreme cases, loss of hair and bone marrow, hemorrhaging, and other potentially deadly disorders. Commonly organisms such as animals ingest radionuclides (radioactive atoms) that produce radiation within the organism, leading to ionization and cell damage. In most cases, however, low concentrations of such radionuclides do not noticeably damage organisms, and they go through their life cycle never noticing the low level of radioactivity within them. However, when they are preyed upon by other animals, the radionuclides are preferentially transferred to the predator, such that the higher trophic level you get in the food chain, the more bioaccumulation of radionuclides (and other contaminants such as mercury). For this reason, humans are most concerned, but plants less impacted.

Public Perception of Nuclear Energy

Public perception of environmental risks is generally a fickle entity, being driven by events, misperceptions, and often a lack of knowledge of the underlying science (Figure 11.13). Nuclear power became a significant part of our electric generating capacity in the 1970s until the first nuclear accident occurred at Three Mile Island in Pennsylvania in 1979, after which an effective moratorium on new nuclear plants occurred until about 2002. During this time, public opinion dropped as fears grew of nuclear radiation in the vicinity of Three Mile Island, and indeed concerns extended to all nuclear power plants. The disaster at Chernobyl in the Ukraine in 1986 led to further decline in public support for nuclear energy, but remarkably, despite its severity relative to the insignificant Three Mile Island event, no new regulations were enacted in the U.S. because of the perception that the Russian reactor and containment design was flawed and unreliable in any case. Then, as American gasoline prices increased in the last decade, public opinion increasingly favored nuclear fission reactors, despite heated debates regarding nuclear waste, most notably with regard to a proposed waste storage facility in Yucca Mountain, Nevada. Then, in 2011, a tsunami triggered by a magnitude 9.0 earthquake struck Japan, and the Fukushima Daiichi nuclear plant was damaged to the point of releasing radioactive material into

Majority of Americans Now Say They Oppose Nuclear Energy

Overall, do you strongly favor, somewhat favor, somewhat oppose or strongly oppose the use of nuclear energy as one of the ways to provide electricity for the U.S.?

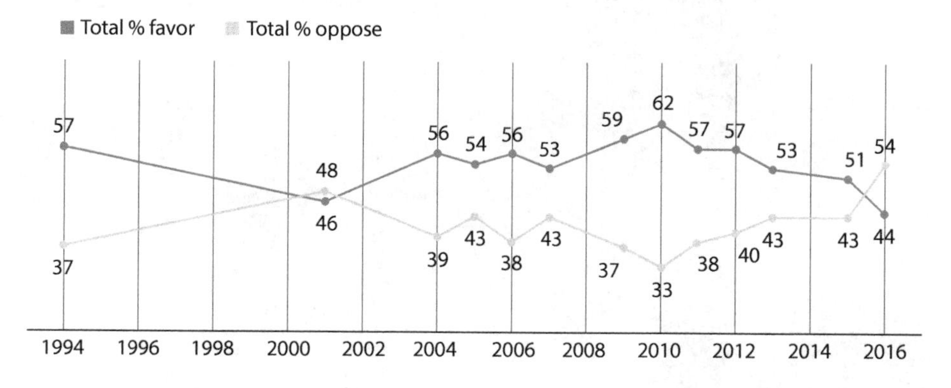

Note: Surveys in 2001–2009 and 2012 asked this question of a half sample

FIGURE 11.13 Gallup poll on nuclear energy
Although public opinion regarding nuclear power (from fission reactors) waxes and wanes with each nuclear accident and the variable price of gasoline, the general trend was to be more favorable in the first decade of the 21st century, and then to become less favorable until the present time.

the local terrestrial and marine environments. Because the Japanese technology was perceived to be top quality, this brought public opinion back to a new low in America and throughout the world (Figure 11.13). Ironically, this negative public opinion may serve to inhibit fusion research, as many people do not understand the distinction between fission and fusion. However, the dip in public opinion was short-lived, and support has been over 50% in the U.S. since 2011.

Sustainable Energy Sources

Fossil fuels have enabled the rapid growth of modern technological society by providing huge amounts of energy that can be exploited in a short time. In the case of oil, gas, and coal, this energy ultimately came from the sun over many millions of years, having been converted to biomass by photosynthesis, and became concentrated chemically to hydrocarbons that could be re-oxidized (burned) with the release of energy. The fact that it took millions of years to accumulate the stored chemical (electrical) energy and that we have the means to extract it as fuel and burn it within decades, makes these fossil fuels that

FIGURE 11.14 Sustainable energy

are not renewable in the lifetime of human civilization, and not replaceable by any means. This has caused concern and even alarm throughout the world, triggering research into alternatives to fossil fuels. The key appears to be to use at least part of the economic and technologic boom enabled by one-time use of fossil fuels to develop new technologies and social structures that will enable future generations to obtain and exploit energy sources and thus fulfill their needs the way we do. This is the basis of sustainability. However, as we will see, access to sufficient energy is not the only concern, as the products of combustion, most notably CO_2, are becoming an increasingly serious problem, as they contribute to global climate change. Alternative energy sources must be developed to not only supply sufficient energy for the world's needs (Chu, 2012; Kammen, 2006), but also to do it in a manner that does not produce any unwanted side effects, such as global warming and its various environmental impacts (Figure 11.14) (Huge et al., 2011; Kaygusuz and Kaygusuz, 2004).

Hydropower

Hydroelectric power generation is the largest source of non-fossil-fuel-derived energy (Figure 11.15). People have been using hydropower since long before electricity came into play. Water wheels were used for mechanical processes such as grinding grain (tedious to do by hand) and for increasingly sophisticated mechanical equipment such as lathes for turning wood dowels, spindles, and bedposts. With the advent of electric lights (and motors, and now electronics ...) the need for electric power arose. It was discovered that one could spin a wheel (turbine) by allowing water to rush through it, and by building a large dam, a large **hydraulic head** (water pressure) could be developed, and water could be shot through a set of turbines connected to electric generators to generate huge amounts of electricity constantly. The taller the dam, the more pressure, P, that can be developed ($P = \rho gH$ where ρ is water density, g is the acceleration of gravity, and H is the height of the dam).

Hydraulic head: The pressure at the base of a column of water due to the weight of the overlying water.

Hydropower may be the most efficient means of electricity production in that the amount of energy required to build a dam and maintain the turbines is small compared to the amount of energy that can be generated over the

FIGURE 11.15 Hydropower and Hoover Dam

lifetime of the dam. This aspect of energy efficiency will be explored in the context of **energy payback ratios** (Gagnon 2008).

Hydropower is not a panacea for energy production with regard to environmental impacts. A dam alters the entire ecosystem upstream from a river or stream along with its riparian zone, floodplain, and associated ecological communities, to create a lake with still water and a very different ecosystem. This alters a number of environmental processes. With respect to the water itself, a lake, having a much larger surface area than a stream, can evaporate much more water into the atmosphere, particularly in arid regions. This reduces the flow downstream. While dams do not stop much of the water (aside from the difference in evaporation), they DO stop all of the sediment that would be transported by a stream. Sediment, be it pebbles, gravel, sand, silt, or clay, can be transported only by moving water, more rapidly moving water carrying larger particles, along with smaller. In the lake behind a dam, however, virtually all sediment being carried by the water settles out. This deprives the downstream stretches of sediments, and erosion can ensue, with the eroded sediments being caught by the next dam downstream. Another impact of dams is biogeochemical. Dissolved nutrients such as nitrogen and phosphorus are carried by a river, as are small amounts of dissolved silica (SiO_2) derived from silicate minerals such as quartz. This silica would normally be delivered to the ocean in the coastal zone, where the relatively quiet water would allow blooms of diatoms, beautiful microorganisms that make their delicate and sometimes intricate shells of glass (silica). However, if a dam creates a quiet lake upstream, the diatoms can bloom in the lake, using up the silica there, leaving the coastal zone depleted of silica. The nutrients are still in the water, however, so algae take advantage of them and in the absence of diatoms, harmful algal blooms (e.g., "red tide") can develop. So while building dams is an effective means of generating electricity, it is by no means environmentally impact free.

Biomass Burning

Humans have been burning biomass since the discovery and control of fire. Burning wood or peat for heating in cold climates and for cooking in all climates is still a common practice. In much of the developing world, biomass burning is the primary energy source. The massive scale of global deforestation (currently in the developing world, and in previous centuries in Europe and America) has caused some to look elsewhere for biomass energy, and sources of biomass other than trees are increasingly being used throughout the world. Many societies produce so much waste that it has become a viable fuel source. In some places, fast-growing crops are being raised specifically for biomass burning. Some crops are being converted from food crops to energy crops. Corn in the U.S. is increasingly being processed to produce ethanol, an alcohol that releases energy when burned (Pimentel, 2013). This practice has come under attack on several fronts, as potential food-producing areas are being converted from food production. Some object to this on ethical grounds of some people going hungry while others drive SUVs. Perhaps even more seriously, the energy required to produce ethanol from corn is considerable and will be discussed later in this chapter with regard to energy payback ratio.

Solar Energy

Solar energy is the ultimate source of almost all of the world's energy sources. Fossil fuels all come from ancient plants (marine or terrestrial) that grew from sunlight. Wind is generated from differential heating of the planet by the sun.

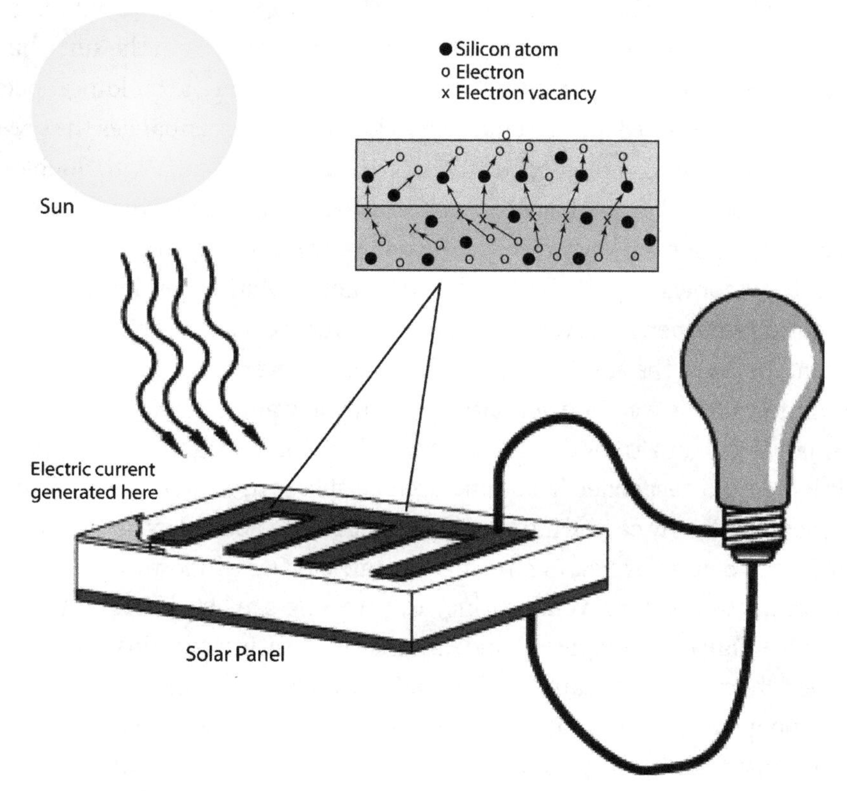

FIGURE 11.16 Basic Photovoltaic cell

In a photovoltaic cell, a silicon lattice (Si has a valence of 4) is "doped" with impurities with 5 (n-type, e.g., P) electrons in their outer valence shells in slabs (wafers) attached to slabs with 3 (p-type, e.g., B) electrons in their outer shells. Although the charge is neutral, there is no place in the lattice structure for the extra electrons of the N-type side to make bonds. They are free to roam. Meanwhile, the P-type side has too few electrons in the outer shell, and the lattice bonds are incomplete. The extra electrons from the P side find a "home" on the N side to make all the bonds with the silicon lattice. Because there are now more electrons than protons on one side, and vice-versa on the other, this makes an electric field (P side positive because it is missing electrons now). When the sun shines onto the N-side of the cell, a specific wavelength knocks the extra electrons out of their adopted "homes" in the bonds within the silicon lattice. Now free, electrons find themselves in an electric field, and thus move in response to the electric force, making a current. The current flows through a wire to a motor, light, battery, etc., to do work for you on the way back to the P side, where the electrons find that all the lattice positions are filled, so they return to the N side to complete the lattice structure, waiting for another photon of light to knock them out again, and the cycle continues.

Rivers used for hydropower are part of the hydrologic cycle that is driven by evaporation and precipitation, all driven by the sun. Biomass to be burned is all grown with sunlight. In fact the only energy sources NOT created by the sun are nuclear (both fission and fusion), geothermal, and tidal power.

Passive solar energy: The use of the sun in building design that optimizes the heating from the sun in winter while preventing excess heat in summer.

Passive solar design of buildings involves taking advantage of solar heating when you want it and avoiding the heat when you want cooling. With windows that face the south (in the mid-latitudes of the northern hemisphere), sunlight can enter the windows and heat interior spaces in the winter when you need heat the most. Overhangs mounted over the windows do not interfere with the incoming solar radiation in winter because the sun is low on the horizon in winter. This allows for the greenhouse effect to trap heat in the building, as glass is transparent to visible light (the primary radiation from the sun), but is partially opaque to the infrared radiation re-emitted from the building. Special window glass designed for low emissivity (low-E) further enhances the greenhouse effect by being more opaque to infrared. This, combined with double- or even triple-pane glass with special non-conducting gases between the panes (such as argon or even krypton), greatly reduces heat loss by infrared radiation out of the windows, especially at night, while not inhibiting daytime heating by sunlight. In summer, however, the overhangs over the windows block incoming sunlight, because the sun is high overhead, so the overhangs cast shadows on the windows. This way, you get heating when you want it in winter, and not in summer. Deciduous trees can add to this effect when planted to the south, as their leaves will cast shade in summer, but be absent in winter, when you want solar heating. In any case, window shades that can be opened by day and closed at night can help solar heating in winter, while closing windows and shades in summer daytime can prevent heating, and opening can help cooling at night.

Active solar energy: The use of energy from solar heating that is intensified by various means. It can boil water, heat other fluids, or even produce electricity directly in a photovoltaic cell.

Photovoltaic cell: A semiconductor device that converts photons of light to electricity (see Fig. 11.18 photo caption).

Active solar technologies include intensified heating of fluids such as oils or other chemicals that can then be used to boil water to steam to drive turbines and generate electricity. The heating is accomplished by large arrays of mirrors that focus onto a point or a pipeline containing the fluid.

Perhaps the most intriguing active solar technology is the **photovoltaic cell**, which converts sunlight directly into electricity (Figure 11.16) (Gong et al., 2015; Ritzen et al., 2017; Lee et al, 1997). One of many advantages of such technology is that it can be used anywhere that the sun shines, without the need for supporting infrastructure. Consequently, it has become very popular in the developing world, where there are few other options for electricity generation. At present, photovoltaic cells are rather inefficient, converting less than 20% of solar energy into electricity. However, research is ongoing, and by capturing a broader spectrum of incident solar energy, efficiencies are being improved, and costs are being reduced.

FIGURE 11.17 Windmill energy

Wind Energy

Wind power has been popular for many centuries for sailing ships and performing work such as grinding grain, even long before the industrial revolution. Wind is created by differential heating of the planet by the sun, so is essentially another form of solar energy, as are most energy sources we use. Modern windmills are aerodynamically more efficient than the ancient windmills consisting of wooden frames supporting cloth blades. Connected directly to electrical generators, the modern, tall, large-blade systems generate hundreds of kilowatts to a few megawatts each, and when grouped into large wind farms (Figure 11.17), produce many megawatts as long as the wind blows more than about 5 m/s (11 mph). Often, such wind farms are built on grazing lands and other agricultural areas that are relatively unaffected by the forest of wind turbines (Davis et al., 2018; Johnson, 2017; Hdidouan and Saffell, 2017).

As with all human activities, windmills have some environmental impacts. In the U.S., there has been a great deal of resistance to the establishment of wind turbines and wind farms. Much of this resistance stems from some people's opinion that large windmills are ugly, and spoil their view of the landscape. While most are quite happy with windmills placed elsewhere, they do not want to see them nearby—a classic case of "Not In My Back Yard" (NIMBY). The main environmental impact (literally) of windmills pertains to birds that fly into the turning blades. Every year, thousands of dead birds are found at the bases of windmills. This appears to be an environmental problem until one puts it into perspective. Housecats kill more birds, yet no one objects (by policy) to housecats. The greatest killer of birds, however, is buildings. Birds apparently cannot see window glass, and collide with windows to the point that a billion birds are killed annually in the U.S. by buildings. Compared to thousands killed by windmills (a million times less), it would seem that this basis for obstructing wind power is weakly founded.

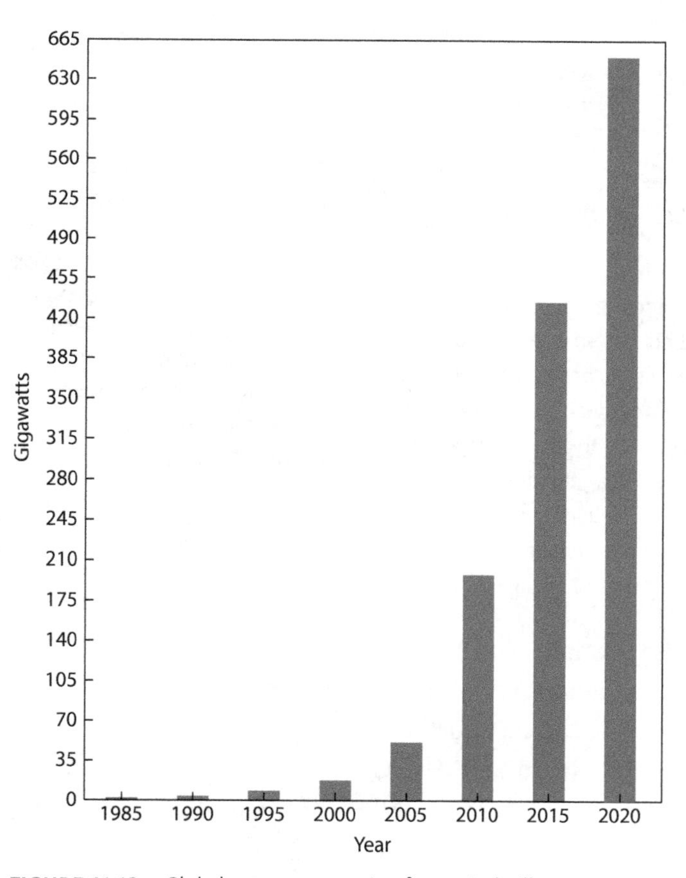

FIGURE 11.18 Global energy generation from windmills

One issue with wind power is that the wind does not always blow at a steady rate. Often there is little or no wind (you need about 5 m/s, 11 mph, to be viable), and sometimes there is far too much, as when a hurricane or tornado comes through. In many locations, there is no guarantee that the wind will blow when you need it most—when electrical demand is greatest, as in with air conditioning during the day in summer, and lighting and heating at night in the winter. If there were more effective means for the storage of electrical energy, the effectiveness of wind power could be enhanced. Nevertheless, wind turbines continue to be produced and installed throughout the world, increasing power output, and gaining a greater fraction of total energy production annually (Figure 11.18).

Geothermal Energy

As the name implies, geothermal energy draws power from the heat of the earth (Mock et al., 1997). At great depth, the rocks that make up the Earth's mantle are so hot that they are partially molten. While the mantle is normally far too deep to draw energy from on a practical basis, there are many locations where its heat is transported to the surface or near surface by volcanic activity. In these regions, magma heats the near-surface groundwater, and in some cases, the water flashes to steam (and make geysers), which, when harnessed, can be used directly to drive turbines in power plants. In other cases, steam is not generated, but the hot water can be used for heating buildings. In some places, wells are drilled to some depth, and water (or other fluids) is pumped to depth until it turns to steam, which is then brought back to a power plant to drive the turbines. In these ways, geothermal energy can be used to produce electricity.

Unfortunately, most parts of the world do not have a volcano located nearby. Still, these areas could also benefit from the use of geothermal energy in the form of **ground source heat pumps** (Figure 11.19). The basic idea is to take advantage of the fact that below the depth that is thermally affected by the seasons (about 3–5 m in most of the U.S., and up to 5–6 m in parts of Alaska), the temperature of the ground maintains a constant temperature year-round (about 58°F in the temperate U.S., and in the 40s in Alaska). If this heat can be brought to the surface in winter, it can heat buildings at least partially, so that other means of heating (e.g., solar, fossil fuel, electric) have a head start to bring interior spaces to a comfortable temperature. In winter in the U.S. northeast, for example, when the temperature is below freezing, starting at 58°F (14 °C), or even 50°F (10°C) accounting for imperfect heat transfer) rather than 20°F (-12°C) is a great bonus in efforts to heat a space to 65°F (18°C) or higher. Then in summer, bringing the interior temperatures down to anywhere near 58°F (14°C) takes care of any air conditioning needs altogether! By running

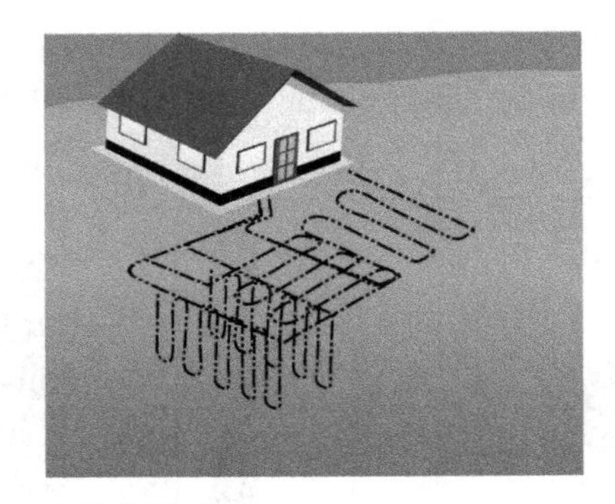

FIGURE 11.19
Ground source heat pump. Geothermal energy is not limited to volcanic areas. For home heating or cooling, tubes of water (or other fluids) circulate underground bringing the fluid to ground temperature of about 58 °F year-round. In the winter, this is a big head start on heating interior spaces, and in the summer, it is cool enough for air-conditioning.

Ground source heat pump: A system that enhances the efficiency of motor-driven heat pumps for home heating and cooling by providing a base temperature for the compressor for heat exchange for cooling in summer and heating in winter.

a fluid (water and antifreeze) through a loop in the ground, heat (or cold) is brought into a building and into a heat exchanger, which preheats (or cools) air entering a heat pump and into a forced air ventilation system that circulates throughout a building. This greatly improves the efficiency of heating and cooling in homes and buildings anywhere, so that anyone can use geothermal energy. Challenges remain, however. Materials for the loops buried underground are either metal or plastic, but metal tends to corrode, especially under certain groundwater conditions, while plastic is a poor heat conductor so is less effective at transferring heat from the ground to the circulating fluid and back. Further, it is critical to size a system properly, or the coils themselves will heat or cool the surrounding ground, thus leading to loss in efficiency. Nevertheless, the improved efficiency provided by ground source heat pumps may become a growing strategy in an overall energy portfolio for the country and the world.

Tidal Energy

The world ocean represents an immense reservoir of energy; its heat controls global climate, and its movements involve great amounts of energy. There have been many schemes for extracting that energy for societal use, and some have proved more practical than others (Lande-Sudall et al., 2018). The most practical appears to be energy associated with the tides in regions that have large tides (megatidal). In areas like the Bay of Fundy (Canada), western France, and southwest England, tides of near 15 m lead to currents into and out of the coastal tributaries that can be used to generate electricity. There are two ways that this can be accomplished. The first (only conceptual) is to place turbines in the currents to generate electricity directly from the currents (like windmills). The second is to build dams that block the incoming (or outgoing) tide until the water level difference on each side of the dam is greatest, and then open the gates to drive the water through turbines, just like in the case of a dammed river for hydropower. The greatest disadvantage of such systems is that power is generated for only a short time each day. Also, damming coastal tributaries interferes with sensitive ecosystems. Finally, there are only a few places on Earth with sufficient tidal range to make this worthwhile at all.

Marine Hydrokinetic Energy

Marine hydrokinetic systems are mechanisms that extract energy from movements of ocean water. This can be done in a number of ways, but has not yet been implemented in nay practical manner. As such, it is all still theoretical. One means is to place large turbines (propellers, like windmills) in major ocean currents such as the Gulf Stream, from which ten to 50 gigawatts of energy could be extracted by strategically located turbines. Another way to obtain energy from ocean water is from waves that cause floating buoys to rise and fall, driving water through vertical hoses containing impellers (turbines) that drive electric generators. Alternatively, "heave-surge" devices are long series of, semi-submerged articulated tubes that are moored at one end so that passing waves make the sections pivot vertically at each of the connections of the sections. At each joint, hydraulic devices resist this bending motion, forcing fluid through a tube and spinning a turbine to generate electricity. In each of these theoretical technologies, the practical matters of transporting electricity to shore, obstruction of ship and boat traffic, and cost, are yet to be addressed, and they all remain conceptual ideas, perhaps someday to be added to our energy portfolio.

Energy Conservation

While we have a portfolio of sustainable energy sources, we would be hard-pressed to produce as much energy from these as we have been obtaining from the concentrated energy in fossil fuels.

The accumulation and concentration of fossil fuels, our current primary energy source, took hundreds of millions of years. This concentration of energy has enabled us to consume energy at a very rapid rate globally, about 17 terawatts (17×10^{18} watts). It would be difficult to produce that much energy from sustainable energy sources in the future, especially if our population increases, poverty declines, and more of the world is in a position to use energy in the thriving global economy we all seek. Thus it will be important to conduct our daily business, achieve our goals, and advance global knowledge, security, and general well-being with much less energy per person. There are many ways to do this, and the first (low-hanging fruit) is energy efficiency. More than half of energy generated in the world is wasted. By streamlining our processes in industry, agriculture, transportation, buildings, and product consumption, we can greatly reduce energy needs without reducing what we do. The energy saved by doing things more efficiently is called **negawatts** in that it subtracts from the total energy needed to function.

Energy reduction strategies can be developed for individual homes with regarding to lighting, heating and air conditioning, and appliances. At the community level, energy consumption can be reduced by rearranging the layout of homes, businesses, and schools in ways that do not require cars to be driven as far (or at all). For example, many towns currently have single-use zoning, in which only one type of land use is allowed, be it single-family homes, multi-family housing, commercial space, retail businesses, etc. It is forbidden to live where you work, shop, or go to the doctor, so you have to drive there. Mixed-use zoning can greatly reduce car travel miles and the associated energy consumption. Another way to organize communities is in clusters, centered on a transportation hub, such as a train station, enabling community functions such as schools and shopping locally, and easy access to broader goods and services such as various employment, cultural events, sports, etc. When new communities are established in such energy-efficient ways, it is called **smart growth.**

Negawatts: The reduction of power usage by enhancing efficiency.

Smart growth: Strategic community planning to reduce transportation needs and, thus, energy consumption, enhance ecosystem function, and stabilize hydrology.

Reduce product consumption: Just don't buy so much stuff. Take care of and continue to use the stuff you already have. Fix it if it breaks—don't just throw it away.

Emergy: The energy embedded in products; i.e., the energy required to produce products.

Another means to reduce energy consumption is to **reduce product consumption.** Great quantities of energy are necessary to produce the goods we use every day. Energy is expended in mining raw materials and transporting them to factories, where more energy is used to produce items. Then more energy is used to transport them to the marketplace, and in many cases, additional energy is involved in the use of the product (e.g., car or TV). The energy that it takes to make and deliver something is called **emergy** for embedded energy. While a great many companies and products label themselves as green, the greenest product is always the one you leave on the store shelf so that another one does not have to be made to replace it.

Energy Payback Ratios

We have many options for energy sources. Throughout human history, people have had to expend energy to get energy. At the most basic level, we eat food to gain the strength to gather fuel wood to burn, and we did this for millennia.

ENERGY CHOICES

As we transition away from fossil fuels for reasons of climate restoration, other environmental impacts, or simply dwindling supply, it will be necessary to develop a new portfolio of sustainable energy sources to satisfy society's needs. The balance between hydropower, solar, wind, tidal, and geothermal will necessarily depend on the local environmental condition—topography with an active hydrologic cycle, Sun vs. clouds, steady winds, volcanoes nearby, macrotidal shoreline, etc. We should each consider where we live/work and think about what would be the ideal suite of sustainable energy sources for the region—it will vary from place to place.

Thus, it cost us some energy (from food) to manually get more energy (from wood). The ratio of energy gained from the wood to energy expended by effort fueled by food was not much greater than one. If the energy gained from an energy source is not much greater than the energy invested to gather it, then the endeavor is hardly worth it. This relation between energy gained and energy invested to get it can be considered an **energy payback ratio** (EPR) (Figure 11.20). If the ratio is greater than two, it may be worth pursuing as a source of energy. Throughout human history, the evolution of energy sources has tended to increase the ratio, from human labor gathering fuel wood, to using (and feeding!) horses, wind, waterwheels, coal, whale oil, petroleum and natural gas, and most recently, uranium (Gagnon, 2008). As technology advanced, the increase in energy payback ratio enabled more energy-intensive activities to be accomplished, culminating in the second half of the twentieth century with the highest composite ratio of all time, between about five and ten in the industrialized world. This can be attributed to easy access to high-grade coal, giant and supergiant oil fields, major hydroelectric plants on large rivers in the most ideal localities, and the development of more efficient wind turbines,

Energy payback ratio: The usable energy obtained from an energy source divided by the energy required to extract and provide that energy.

photovoltaics, and the utilization of waste biomass. However, as fossil resources become increasingly depleted, more energy is required to extract energy resources such as coal, oil, and gas, resulting in a decline in EPR. This naturally drives up the consumer cost of energy, and this is being experienced in the twenty-first century already.

With the discovery of coal to replace wood fuel, concentrated accumulations of energy could be exploited, and despite its being irreplaceable (fossil), was mined and burned very cheaply, increasing EPR for the industrial revolution. Petroleum increased EPR further, reaching about 100 for high-quality Saudi oil, or convenient American oil. (Transportation energy costs must be factored into EPR calculations, so energy sources near markets have an advantage.) The development of electric transmission and distribution systems also helped increase EPR, as it enabled generation of electricity at centralized and efficient locations, and electrification of industry, homes, and some transportation systems. Nuclear power plants further increased EPR, as they produce vast amounts of energy from readily obtained uranium (although the energy involved in decommissioning nuclear plants and long-term storage of radioactive wastes is still under discussion).

Near the end of the twentieth century, the most readily available fossil fuel resources were already found and exploited, and somewhat less convenient sources were explored. This required more energy and thus began to reduce

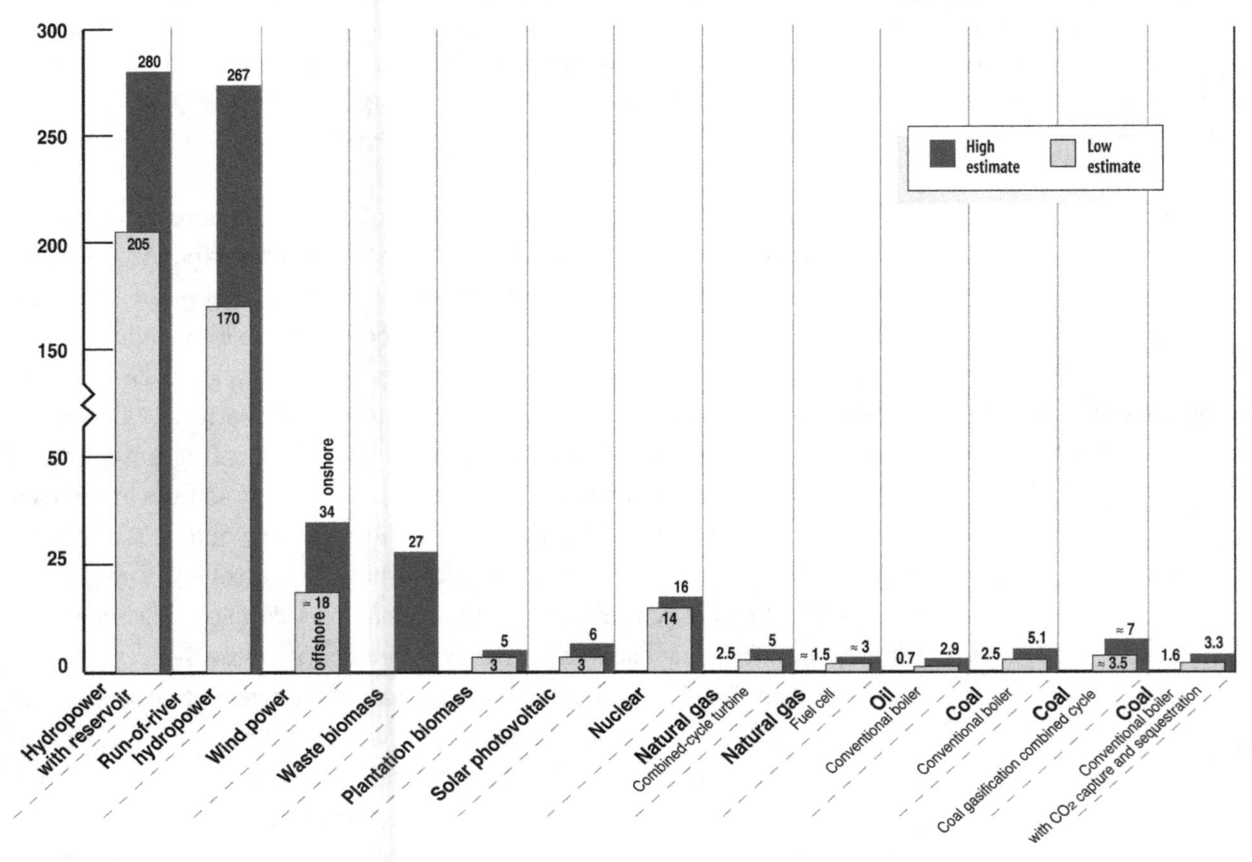

FIGURE 11.20 Energy Payback Ratio

EPR. Drilling for oil offshore requires more energy than drilling on land. Obtaining natural gas from shales (source rock rather than reservoir rock) requires horizontal drilling and fracking or fracturing of the rock to extract a small fraction of the gas contained in the otherwise impermeable shale. With most of the good sites already used for dams to produce hydroelectric power, less ideal locations, with greater associated upfront energy costs and lower returns, are being considered. In many places in the U.S., dams are actually being removed to restore ecosystem function. As concerns for air quality lead to restrictions on emission of pollutants, coal-burning power plants must put into place systems that remove sulfur, for example, from flue gas. While this was critical to reducing the severity of acid rain in the industrial U.S. northeast (and Europe and Asia), it further reduced EPR for these power plants. Now, with powerful political industrial agriculture lobbies, corn-derived ethanol is being added as a contaminant to gasoline, further reducing EPR. (If diluting gasoline with 10% ethanol reduces efficiency as measured by gas mileage by 10%, then even if the ethanol were completely free, it would still not be worth it energetically. The ethanol is far from free and thus can be considered a contaminant in gasoline.) As high-EPR oil traditionally used in the process of industrial agriculture transitions to lower EPR oil or other low-EPR sources, the EPR of corn-derived ethanol may decrease further to well below one.

Now, in the early twenty-first century, the suite of energy sources available for electricity production as well as vehicular transportation can be assessed in terms of the energy involved in the life cycle of the energy production system, relative to the energy produced by that system over its lifetime. While corn-derived ethanol may actually use more energy to produce than is derived from burning it, it is not the only potential energy source that uses more than it provides. Nuclear fusion has been investigated for decades by isolated research groups globally, yet little emphasis has been placed on developing it as a viable means for generating electricity. Consequently, it still requires more energy in magnetic containment of plasmas and other supporting infrastructure to create a fusion reaction than is produced in the reaction. (For this reason, there are no operating fusion reactors commercially producing electricity.) Some conventional energy sources, when adjusted to reduce environmental impacts, also lose EPR to the point of veritable worthlessness. Coal-fired power plants, for example, once flue gas scrubbers are applied, lose some EPR, but when carbon capture and sequestration measures are added, can drop EPR to near one. Other fossil fuels are in the same situation. At the same time as more energy is needed to extract dwindling reservoirs, environmental concerns are calling for carbon capture and sequestration, which, when taken to the logical limit of zero emissions, may bring EPR for fossil fuels to near (or even below) one.

At present, photovoltaic electricity production has a low EPR (under 10) due to mostly monochromatic solar radiation utilization, but current research is leading to more efficient photocells, and there is every expectation that EPR will be greatly improved in the coming years. Consider the burning of biomass waste that results from forestry or urban waste. Since the material is already being produced for other reasons (paper, consumer goods, etc.), the waste already exists and can be burned for the mere cost of collection and transportation to the power plant. This enhances EPR to the twenties. Wind also has high EPR, as the energy cost of producing a wind turbine, while significant up front, when amortized over its operating lifetime, is relatively low, and thus EPRs of 20 to over 30 result. The highest EPR appears to be from hydropower from dammed rivers. The energy cost of building a dam is large, but because the operational energy is so small for hydropower and the facilities last for so long (over 100 years in most cases), the EPR reaches into the hundreds. Keep in mind that many of the best sites are already taken, and that there are numerous environmental concerns regarding building new dams (large or small) that will need to be taken into account before siting of new major dams is considered (e.g., Three Gorges Dam in China).

As we consider sustainable energy portfolios for the future in various parts of the world, an initial assessment of Energy Payback Ratio will be essential in determining the optimal mix of sustainable energy sources. Some regions are full of rivers and topography ideal for dams. Others are very sunny. Some have strong and steady winds, while others have very large tides. In each region, the mix of sources will need to be designed specially to take advantage of the greatest return and maximization of EPR (Figure 11.21).

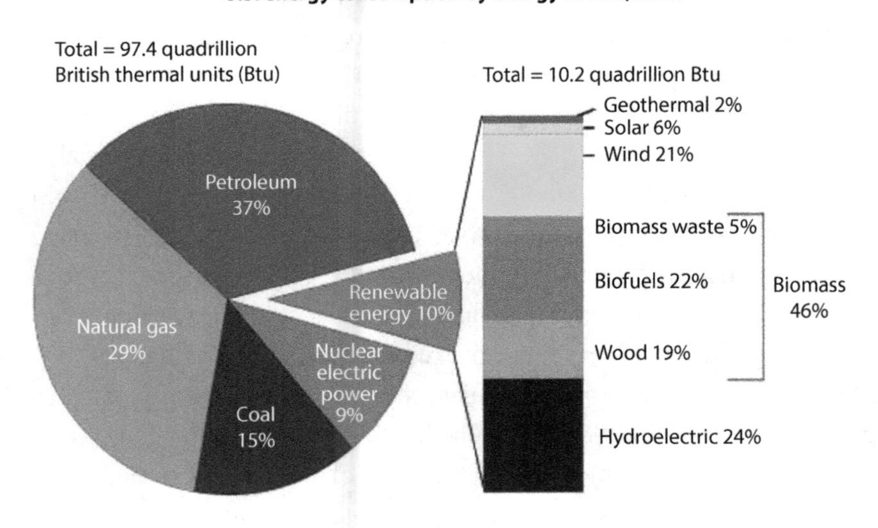

FIGURE 11.21 Energy Consumption 2016
In 2016, in the U.S., we still used more oil than anything else for our energy needs. Natural gas is gaining, however, as "fracking" delivers more cheap gas to the market, especially in replacing coal as a fuel source for electric power generation. Renewables (including "inexhaustibles") are also on the rise, but have a long way to go. (1 BTU is slightly more than 1 kilojoule - 1.055 kJ)
U.S. Energy Information Administration, *Monthly Energy Review*, Table1.3 and 10.1, April 2017, preliminary data

FOSSIL FUELS VS. SUSTAINABLE ENERGY SOURCES: A RATIONAL PERSPECTIVE

It appears to be relatively cheap to get energy from burning oil, coal, and natural gas, while it appears on the surface to be more expensive to get energy from sustainable sources such as wind, solar, or hydropower. The reason for this may be simple. Fossil fuels represent the concentration of many millions of years of solar energy (that made the plants or plankton, etc., that became hydrocarbons). This service of concentrating all this energy was done by geologic processes over the eons. The only thing that oil or mining companies have to do is take that already created material (that they did not create) and deliver it to the consumer. The consumer pays for only the cost of delivery, which is a small cost. The concern is that we can use up this concentrated energy in a relatively short time compared to what it took to create. (This is a separate concern from greenhouse gas emissions that result from burning fossil fuels.)

In the case of sustainable energy sources, energy must be harnessed from solar power or the things that the sun does (like make wind or flowing rivers) **directly in real time.** There is no mechanism to concentrate that solar energy over millions of years for the solar panel or windmill companies. So energy can be provided only at the rate that the sun brings it. Clearly, it is thus much more expensive per kilowatt, because all that concentration of solar energy found in fossil fuels cannot be simply taken for free by these sustainable energy producers.

A simple analogy to help understand how fossil fuels are different from sustainable energy sources is the case of buying a stereo, cell phone, or any other device, for example. When we go to the store to buy one, we deal with a company that collected the raw materials, refined and produced the components, assembled the parts in a factory (that they built for the purpose), packaged it, shipped it, and delivered it. This costs lots of labor, energy and materials, in addition to permits and taxes, and the item is priced accordingly, and the price you pay will be used, at least in part, to make another one.

Now consider buying a device from someone who stole it off a delivery truck. The (illegal and criminal) seller did not incur costs of production, assembly, packaging, shipping, taxes, etc., so can sell it much cheaper than an actual store could, and still make a profit. Since the seller did not create (or pay for) the product, the sale price is practically all profit. The thief has no intention of using the money to make another device—they will simply go steal another one as long as they can. This is just like fossil fuels, in which oil and coal companies simply take material out of the ground that they did not create, and sell it at a profit for the mere cost of delivery (and some processing). Further, like the thief, they have no intention of using the price you pay to create more oil, coal, or natural gas! They did not concentrate the energy—they externalized that cost to the geologic processes over prior eons. Sustainable energy producers do not have that advantage and must harness solar energy limited to the rate at which it arrives in real time. They cannot benefit from any externalized costs of energy production.

So it should be no wonder fossil-fuel-derived energy is cheaper than sustainable energy! It is like stolen goods—always cheap. The problem is that (like electronic goods) if no one produces energy in real time, when the stolen concentrated stuff runs out, there will be no more energy for anyone. As a society, how can we get sustainable energy to be produced when it can't compete for price with the "stolen goods" of fossil fuels? Can we jump-start an entire industry when it is undercut from the start? This is one of the most pressing issues in America, and indeed the world, and requires carefully designed energy policy over the coming years. Internalization of externalized costs involved in fossil fuels are unpopular with those who just want cheap energy now without regard to the future, and political structures are not designed to respond to long-term needs. The current generation of students will confront this issue head-on during their lifetimes.

References

Banarejee, D. (2014). Acid drainage potential from coal mine wastes: environmental assessment through static and kinetic tests. *International Journal of Environmental Science and Technology*, 11(5), 1365–1378.

Chu, S., & Majumdar, A. (2012). Opportunities and challenges for a sustainable energy future. *Nature*, 488, 294–303.

Davis, K. M., Nguyen, M. N., McClung, M. R., & Moran, M. D. (2018). A comparison of the impacts of wind energy and unconventional gas development on land-use and ecosystem services: An example from the Anadarko Basin of Oklahoma, USA. *Environmental Management*, 61(5), 796–804.

Eisenberg, A. M. (2015). Beyond science and hysteria: reality and perceptions of environmental justice concerns surrounding Marcellus and Utica shale gas development. *University of Pittsburgh Law Review*, 77(2), 183–234.

Gagnon, L. (2008). Civilisation and energy payback. *Energy Policy*, 36, 3317–3322.

Gong, J., Darling, S. B., & Fengqi, Y. (2015). Perovskite photovoltaics: Life-cycle assessment of energy and environmental impacts. *Energy & Environmental Science*, 8(7), 1953–1968

Huge, J., Waas, T., Eggermont, G., & Verbruggen, A. (2013). Impact assessment for a sustainable energy future—Reflections and practical experiences. *Energy Policy*, 39(10), 6243–6253.

Johnson, J. (2017). Climate change may impact future wind energy generation. *Chemical & Engineering News*, 95(49), 17.

Kammen, D. M. (2006). The rise of renewable energy. *Scientific American*, 295, 84–93.

Kaygusuz, K., & Kaygusuz, A. (2004). Energy and sustainable development. Part II: Environmental impacts of energy use. *Energy Sources*, 26(11), 1071–1082.

Lande-Sudall, D., Stallard, T., & Stansby, R. (2018). Co-located offshore wind and tidal stream turbines: Assessment of energy yield and loading. *Renewable Energy*, 118, 627–643.

Lee, J. C., Fthenakis, V. M., Morris, S. C., Goldstein, G. A., & Moscowitz, P. D. (1997). Projected photovoltaic energy impacts on US CO2 emissions: An integrated energy environment economic analysis. *Progress in Photovoltaics*, 5(4), 277–285.

Mock, J. E., Tester, J. W., & Wright, P. M. (1997). Geothermal energy from the earth: Its potential impact as an environmentally sustainable resource. *Annual Review of Energy and the Enviroment*, 22, 305–356.

Pimental, D. (2003). Ethanol fuels: Energy balance, economics, and environmental impacts are negative. *Natural Resources Research*, 12(2).

Requejo, A. G. (1994). Maturation of petroleum source rocks .2. Quantitative changes in extractable hydrocarbon content and composition associated with hydrocarbon generation. *Organic Geochemistry*, 21(1), 91–105.

Requejo, A. G., Gray, N. R., Freund, H., Thomann, H., Mechior, M. T., Gebhard, L. A., . . . & Hsu, C. S. (1992). Maturation of petroleum source rocks .1. Changes in kerogen structure and composition associated with hydrocarbon generation. *Energy & Fuels*, 6(2), 203–214.

Ritzen. M. J., Vroon, Z. A. E. P., Rovers, R. F. M., Lupisek, A., & Guerts, C. P. W. (2017). Environmental impact comparison of a ventilated and a non-ventilated building-integrated photovoltaic rooftop design in the Netherlands: Electricity output, energy payback time, and land claim. *Solar Energy*, 155, 304–313.

Robinson, J. (2012). Reducing environmental risk associated with Marcellus shale gas fracturing. *Oil & Gas Journal*, 110(4), 88–91.

Rozell, D. (2012). Water pollution risk associated with natural gas extraction from the Marcellus shale. *Risk Analysis*, 32, 1382–1393.

Rozell, D. J., Reaven, S. J. (2012). Water pollution risk associated with natural gas extraction from the Marcellus shale. *Risk Analysis*, 32(8), 1382–1393.

Tissot, B., Durand, B., Espitali, J., & Combaz, A. (1974). Influence of nature and diagenesis of organic-matter in formation of petroleum. *AAPG Bulletin*, 58(3), 499–506.

Figure Credits

Fig. 11.1: Copyright © 2015 Depositphotos/mrhighsky.

Fig. 11.2: Copyright © Hautala (CC BY-SA 3.0) at https://commons.wikimedia.org/wiki/File:Household_radiator.jpg.

Fig. 11.4: Copyright © 2009 Depositphotos/Iryna Volina

Fig. 11.5: Copyright © Mikenorton (CC BY-SA 3.0) at https://commons.wikimedia.org/wiki/File:HydroFrac2.svg.

Fig. 11.7: Copyright © 2009 Depositphotos/Iryna Volina

Fig. 11.8: https://commons.wikimedia.org/wiki/File:US_coal_production_by_coal-producing_region,_2010.png

Fig. 11.9: Copyright © Black Tusk (CC BY 3.0) at: https://commons.wikimedia.org/wiki/File:Northland_waste_water.jpg.

Fig. 11.10: Copyright © David J. Strozzi (CC BY-SA 2.0) at: https://commons.wikimedia.org/wiki/File:Binden_wiki.png.

Fig. 11.11: Copyright © Lothar Neumann, Gernsbach (CC BY-SA 2.5) at https://commons.wikimedia.org/wiki/File:Philippsburg2.jpg.

Fig. 11.12: https://commons.wikimedia.org/wiki/Nuclear_energy#/media/File:Fission_chain_reaction.svg

Fig. 11.13: Gallup, http://news.gallup.com/poll/190064/first-time-majority-oppose-nuclear-energy.aspx. Copyright © 2016 by Gallup, Inc.

Fig. 11.14: Copyright © OI-B-i.fernandez02 (CC BY-SA 3.0) at https://commons.wikimedia.org/wiki/File:Energiaberriztagarriak.jpg.

Fig. 11.15a: Copyright © eggi (CC BY-SA 3.0) at https://commons.wikimedia.org/wiki/File:Akosombo_hydropower.png.

Fig. 11.15b: Copyright © 2013 Depositphotos/sergeydolya.

Fig. 11.16: https://commons.wikimedia.org/wiki/File:Operation_of_a_basic_photovoltaic_cell.gif

Fig. 11.17: Copyright © 2014 Depositphotos/welcomia.

Fig. 11.20: Luc Gagnon, http://tc4.iec.ch/FactSheetPayback.pdf, pp. 2. Copyright © 2005 by Hydro-Québec.

Fig. 11.21: https://www.eia.gov/energyexplained/?page=us_energy_home

CHAPTER 12

The Atmosphere

The air was as pure as the white driven snow.
Then came our emissions and next thing you know,
Photochemical smog,
Cities left in a fog,
Destruction of ozone let UV below.

PV = nRT: Ideal gas law, where P is pressure, V is volume, n is the number of moles of gas molecules, R is the ideal gas constant

$$8.314 \frac{J}{(K \bullet mol)}$$

and T is temperature (always in kelvins!).

Troposphere: The bottom layer of the atmosphere (up to about 12 kilometers [km]), where we live, that contains most of the atmospheric air and also moisture. It is also where weather occurs.

Stratosphere: The next layer above the troposphere, from about 12 to 50 km, that contains an important ozone layer that absorbs UVB.

WE LIVE AT the bottom of a very thin layer of air, composed mostly of nitrogen (78%), oxygen (21%), argon (0.9%), and carbon dioxide (0.04% and rising rapidly). This accounts for about 99.94% or so of the atmosphere. The rest is composed of numerous trace gases such as methane (also rising rapidly), many of which are contaminants such as chlorofluorocarbons (CFCs), oxides of sulfur and nitrogen (NO_x and SO_x), and sulfur hexafluoride (SF_6). In addition (and not counted in the above percentages), there is a variable amount of water vapor (1%–4% by volume) depending on time and place. Most (90%) of the mass of the atmosphere is in the bottom twelve km. This is because it is a gas, and the weight of the gas above exerts pressure on the gas below, causing it to compress and become dense (and warm up): Recall from high school that **PV = nRT**. The atmosphere is not inert. It moves and it undergoes many chemical reactions driven by sunlight and temperature changes, interactions of different air masses, emissions of human-made chemicals, etc.

The atmosphere is divided into several layers (Figure 12.1). The bottom-most is the **troposphere**, where we live, and where most of the mass is concentrated, most water vapor is found, and most weather occurs. The bottom of the troposphere (the earth's solid and liquid surface) is the warmest part, averaging about 15°C or so (globally over all seasons), and cools with altitude to about -60°C at the top of the troposphere, an interface called the tropopause.

The next layer above is the **stratosphere**, extending from 12 km to about 50 km above the ground. An important layer within the stratosphere is the ozone layer, where the concentration of ozone is elevated due to photochemical reactions with sunlight that create ozone by dissociation of oxygen molecules by ultraviolet light (UVC)—(more on that later). All life on Earth, after the onset of an oxygen-rich atmosphere, evolved in the absence of UVB and UVC, so has no defense against cellular damage caused by these. This is why we hear so much about protecting the ozone layer, and also about wearing sunscreen when going outdoors.

Pressure controls the movement of the atmosphere, and temperature (in part) controls pressure at the ground. Because the atmosphere is a gas, and the overlying weight of air exerts pressure on the underlying air, as you rise in elevation from the ground, the pressure decreases from about 1 bar (1013 millibars) at sea level (also expressed in MKS units as 105 N/m^2 and in English units as 14.7 lbs/in^2 and by some as 760 mm of mercury), to about one-fifth to one-tenth of that at the tropopause (less pressure in the tropics, and more at the poles).

How much cooler would it be at the top of a 5500 m-tall mountain, where the air pressure is only half of what it is at sea level?

Answer: While one would be tempted to simply use the ideal gas law to calculate this, and say that the temperature would be half of what it is at sea level (in Kelvins, if it was 300 K at sea level this would be 150 K on the mountain, or -123°C, or -190°F, an absurdly low temperature). However, as P decreases, V expands, so it is not so simple. For dry air, we use an adiabatic lapse rate of 10°C/km of elevation, so the temperature would actually be about 55°C cooler. The real atmosphere is not dry, however, so a moist adiabatic lapse rate of about 6°C/km should be used, because condensation of water vapor as air rises (to make clouds, and even rain) releases heat to the air.

Atmospheric Circulation

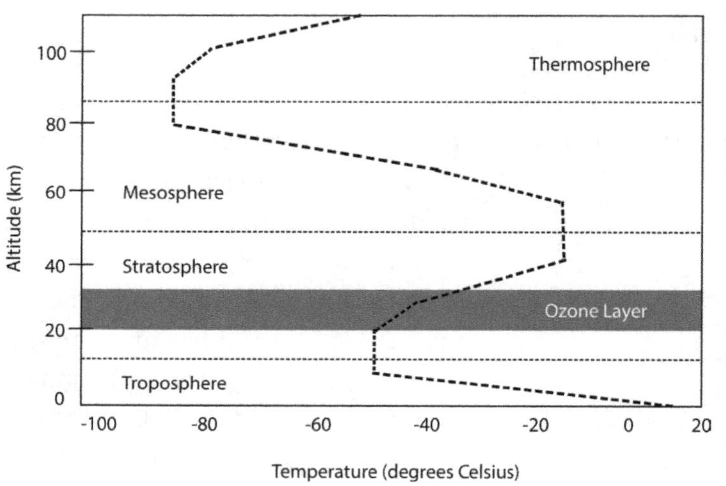

FIGURE 12.1 Structure of the atmosphere
Most of the mass of the atmosphere is in the troposphere, the lowest layer, due to the pressure exerted from the weight of the atmosphere overlying it. Most weather patterns are limited to the troposphere, 99% of the mass of the atmosphere is below 30 km, and 90 % of it is below 12 km. Bold dashed line indicates temperature. Note the position of the critical ozone layer in the stratosphere.

The atmosphere is far from stationary. It moves from place to place (what we feel on the ground as wind), driven by differential pressures caused by uneven net heating of the land surface, ocean surface, and atmosphere itself by sunlight by day and net cooling by infrared radiation out to space by night. In general, the sun shines most strongly in the tropics, as this is where it is most directly overhead, with the most energy impacting the surface per area. At higher latitudes, the angle of "impact" is tilted, so less energy is available per area. As tropical air is heated, it expands (PV = nRT) and rises buoyantly. Meanwhile, at the poles, extremely cold, dense air seeks to sink. This leads to a dipole forcing to make a loop with tropical air rising and moving out toward the poles and polar air sinking and moving along the ground back to the tropics (Figure 12.2). If the earth were not turning (and we had sunlight shining down all around the equator all the time), this is just what would happen. However, the earth IS turning (one rotation per day), so as tropical air tries to move poleward, because it is roughly a sphere, it finds that the earth is moving more slowly under it (imagine moving toward the center of a spinning disk). As polar air moves equatorward, it finds the earth moving faster and faster relative to the air. This is best imagined from a polar perspective.

Coriolis effect: An apparent deflection of the path of an object moving straight, when observed relative to a rotating frame of reference. It causes atmospheric circulation cells to deflect into a specific pattern.

Imagine a spinning disk (like an old-fashioned record player). Hovering above the center, you begin to move outward. You will notice that the surface of the disk moves more and more rapidly under you as you get to greater radii from the center. (If you were to draw a line with a marker to show your path, it would describe a spiral.) This is a simple way to imagine the **Coriolis effect**. Because the earth is a sphere, the same thing happens on the earth as air moves north or south. As such, the air gets deflected—toward the west if it is moving equatorward and toward the east if it is moving poleward (because the earth rotates west to east with the tropics moving much faster than the high latitudes). So air cannot simply rise at the equator and sink at the poles in a single loop that has equatorward-blowing wind at the ground and poleward-blowing wind aloft. It starts out that way, but gets deflected by the rotation of the earth.

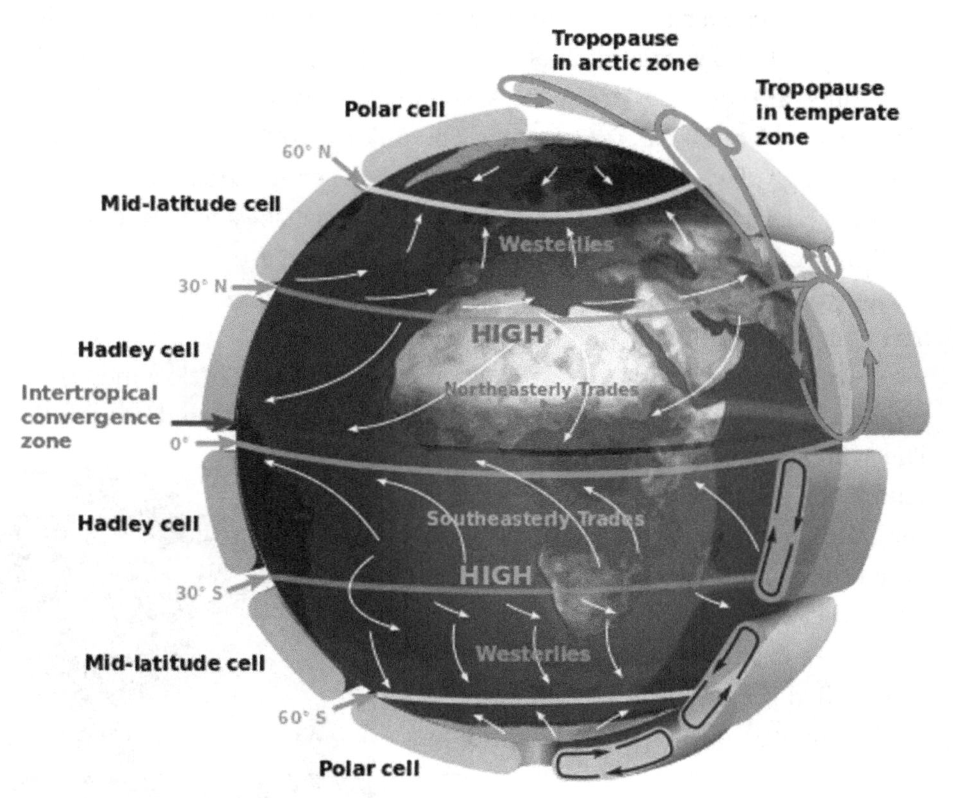

FIGURE 12.2 Global atmospheric circulation
Air is heated most in the tropics, so rises there and moves toward the poles, where it sinks again. However, because the earth is spinning on its axis, the Coriolis effect causes the air to deflect and sink again at ±30° where most deserts are found. Rising again at ±60°, it finally sinks again at the poles. In three dimensions, the air spirals around the planet in the various cells of the global atmospheric circulation system.

By the time tropical air reaches about 30° north and south in its poleward motion aloft, it has deflected sufficiently to the west, and cooled sufficiently at high altitude, that it sinks back down to the earth's surface. By the time polar air reaches about 60° north and south in its equatorward motion along the ground, it has deflected to the east and warmed sufficiently to rise again. This makes a set of smaller loops in vertical cross-section, or atmospheric circulation cells, with air rising at the equator, sinking at 30°, rising at 60°, and sinking at the poles (Figure 12.2). These loops have names—the one between the equator and 30° is the **Hadley Cell,** and the one between 30° and 60° is the mid-latitude, or **Ferrel Cell.** Between 60° and the pole is the **Polar Cell.**

This circulation system creates some general climate belts around the earth. At the equator, warm, moist air rises and cools. Because cool air cannot hold as much water vapor as warm air can, the vapor precipitates out as rain. This leads to the tropical rain forests. At 30° north and south, cool, dry air from high altitude sinks down to the surface and warms up. Because warm air can hold (and evaporate) more water vapor than cool air can, the already dry air becomes even farther from saturation, and maintains the low relative humidity of the deserts found at these latitudes. At 60° north and south, rising air again cools and precipitates, and we find the temperate rain forests of the Pacific Northwest, for example. The sinking, cold, dry air at the poles is exceedingly dry, so precipitation rates there (as snow) are extremely low. The only reason we find glaciers in Antarctica is that the snow never melts, and hundreds of thousands of years of accumulation are found in a single glacier (to the delight of paleoclimatologists who drill ice cores to determine past climate variability).

As a result of the rising air at the equator, there is very little wind there, and sailors call this region the doldrums, where their sails would hang slack. The same is true at 60°, but somewhat less sailing occurs there. At 30°, descending air also does not cause wind, and these are called the horse latitudes. However, in the regions between 30° and 60°, air moves along the surface in a remarkably steady (reliable) wind called the westerlies, while from 30° to the equator blow the trade winds, so useful for merchant sailing ships of previous centuries.

Because the sun warms the tropics most, the air there is warmest and thus least dense. This makes a low pressure area at the ground surface, into which the wind flows (trade winds). The cool dense air that sinks at 30° causes atmospheric pressure to be high at these latitudes. Low pressure is found again at 60° and high pressure at the poles.

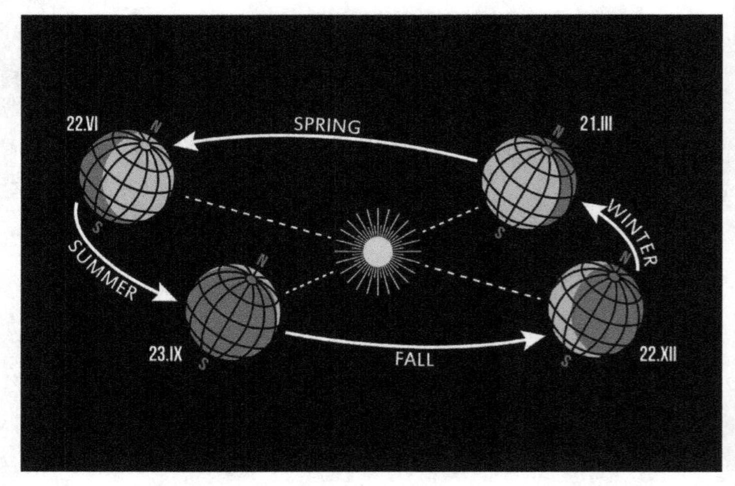

FIGURE 12.3 Tilt of Earth's axis
The tilt of the earth's axis as it revolves around the sun makes the seasons. In northern hemisphere summer, the North Pole points toward the sun (and has 24-hour daylight), while in winter, it points away (24-hour dark). At mid-latitudes, the days are longer in summer, and thus warmer with more time for solar heating, and less time for radiative cooling at night.

A slight complication on top of all this is that the earth's rotation axis is tilted about 23.5° relative to its orbit around the sun, so the sun shines directly down on the equator twice a year during the equinoxes, while the point of direct overhead sunlight moves north and south between the tropics of Cancer and Capricorn (at 23.5° north and south, respectively). This causes all the climate zones driven by the atmospheric circulation system to move north and south and provides us with the seasons (Figure 12.3).

Earth's Radiation Balance

The earth maintains a constant balance between radiant energy coming in from the sun (mostly as visible light) and going out from the earth (as infrared). To understand this, consider the electromagnetic spectrum (Figure 12.4).

At the lowest frequencies (longest wavelengths) are radio waves (including AM, FM, TV, etc.). These carry the least energy. The next higher range on the spectrum is microwaves, like the ones in your kitchen (microwave ovens are tuned to the frequency that most effectively excites harmonic frequencies of

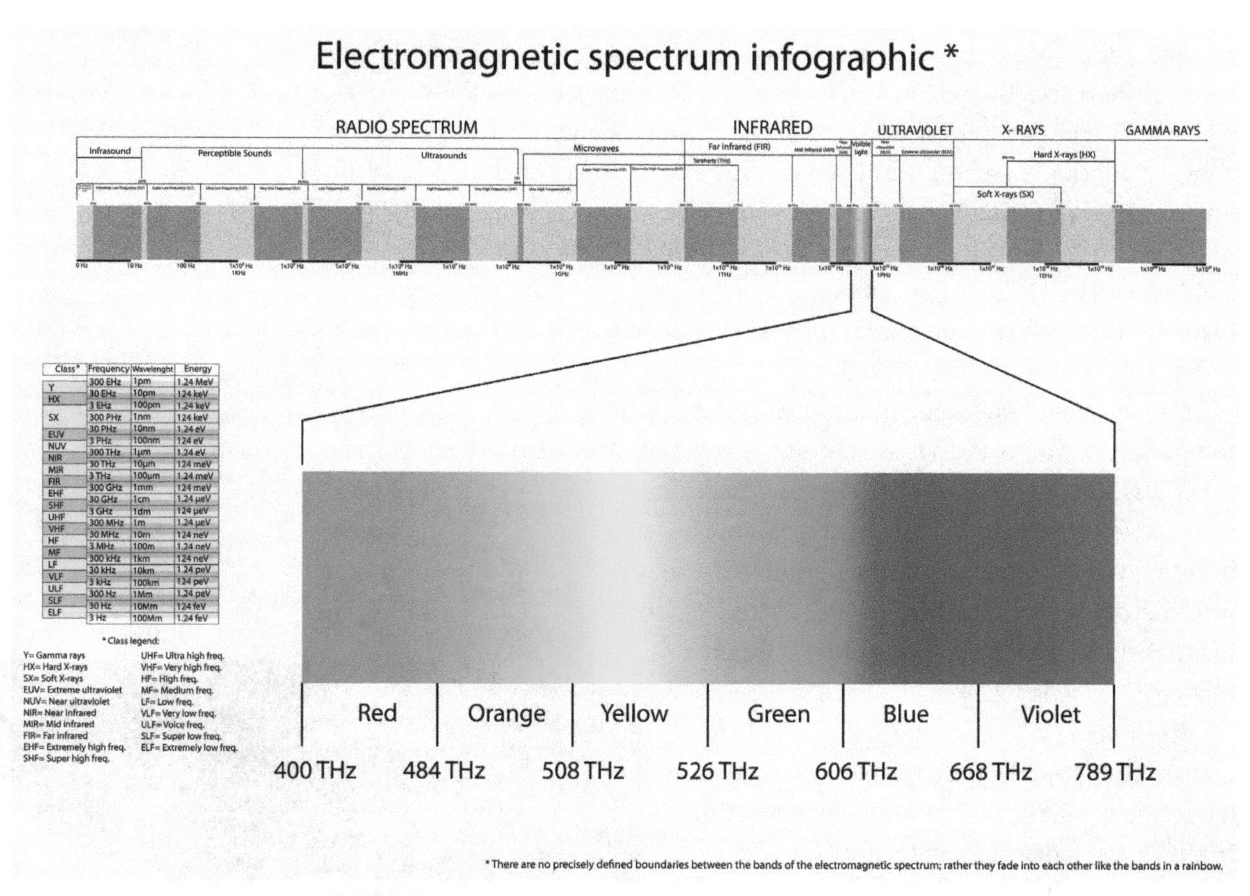

FIGURE 12.4 Electromagnetic spectrum

These waves travel at the speed of light (through a vacuum) and carry energy according to their frequency. The higher the frequency (shorter the wavelength), the more energy. Visible light is merely a very narrow range of frequencies, and is at the peak of the solar spectrum, meaning that the sun radiates most intensely at these frequencies.

the liquid water molecule). Next comes infrared, with wavelengths of about 1 to 100 microns. This is what the earth radiates to space. Beyond that is a very narrow range of visible light, with wavelengths of 0.4 to 0.7 microns. These wavelengths are at the peak of the solar spectrum. At yet-higher frequency is ultraviolet, generally divided up into UVA, UVB, and UVC. Then come x-rays, with wavelengths of 0.01 to 0.0001 microns. Finally, the shortest wavelength, and highest frequency and energy is gamma radiation, commonly given off in nuclear reactions and radioactive decay. All matter emits electromagnetic radiation of some sort. The hotter the body of matter, the higher the frequency (and energy) of the emitted radiation.

The most important negative feedback on the planet is black body radiative energy exchange, as defined by the Stefan-Boltzmann law:

$$E = \sigma T^4$$

where σ is a constant with a value of 5.67×10^{-8} Js^{-1} M^{-2} K^{-4} and T is temperature in Kelvin. E is the energy emitted per $meter^2$ per second, or the rate of energy emission per area of surface. This means that as the temperature of an object increases, the energy released (in units of joules per $meter^2$ per second) increases by the fourth power of the temperature. If a body is not a perfect "black body" that absorbs all incident radiation with an albedo of 0 at all wavelengths, the energy emitted is somewhat less, but still related to the fourth power of the temperature. This establishes a critical negative feedback in the Earth System, and is responsible for keeping Earth's climate relatively stable over the eons. If the temperature increases for whatever reason, the planet radiates away more energy, and this tends to cool it so that a relatively stable temperature is maintained.

As such, greenhouse gas–induced warming of the earth can increase the temperature, but as it does so, the earth becomes MUCH more efficient at shedding heat by infrared radiation to space. This most basic law of physics has enabled the earth to remain at a reasonably steady temperature, largely between the freezing and boiling points of water, for most of its 4.6-billion year history. However, the various positive feedbacks operating within the Earth System are pitted against this fundamental negative feedback, so temperature variations (within limits) do occur, and even small global temperature (and related climate) variations have led to major disruption of existing ecosystems. In the modern day, human technological society may be also vulnerable to climate changes, even within the moderate ranges constrained by the Stefan-Boltzmann law.

Considering the earth as a black body, how much would it warm up if it absorbed 10% more solar energy? Answer: Using the Stefan-Boltzmann law, 1.1 times the energy is balanced by T^4. So the fractional increase in T is the fourth root of 1.1, or 1.02 times the previous temperature. In Kelvins, if it started out at 288 K (15°C), it would increase to 295 K (22°C). While warming by 7°C seems little, it would cause a huge alteration of the climate system and life on Earth.

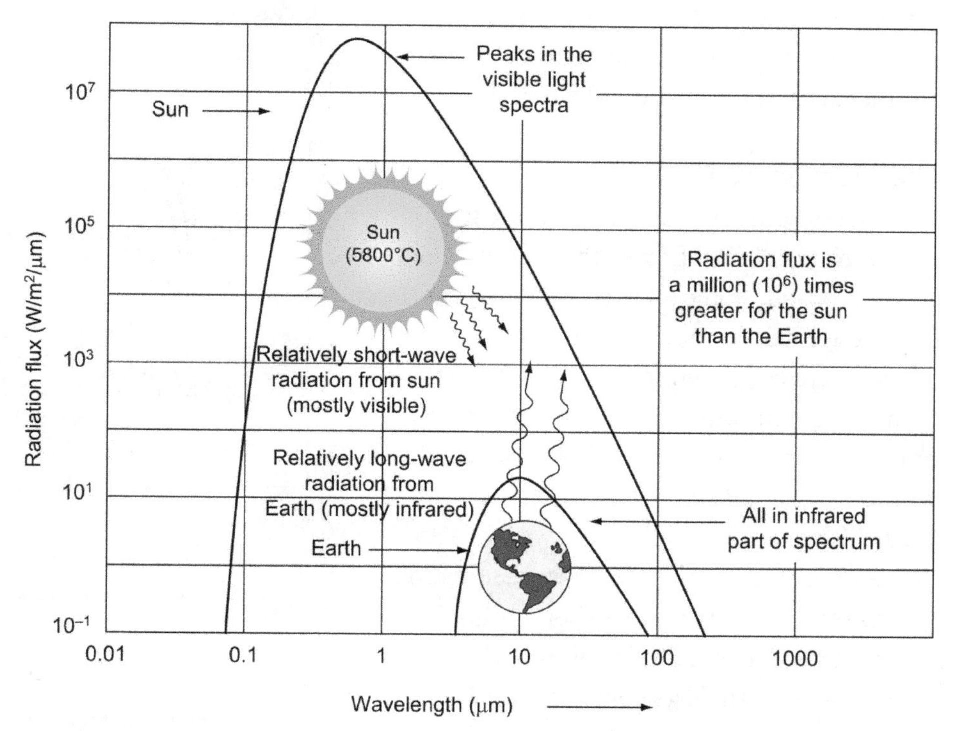

FIGURE 12.5 Radiation from the sun and the earth
The surface of the sun is almost 6000°C, so it radiates visible light (as well as infrared and ultra-violet). The earth is much cooler (only 15°C) so radiates in the infrared. The earth's atmosphere is fairly transparent to visible (which is why we can see a long way through it), but it is not so transparent to infrared. So visible light reaches and heats the earth's surface, but the reradiated infrared gets trapped in the atmosphere. This is the greenhouse effect.

The peak of the solar spectrum is in the visible, being produced by a radiating temperature of about 6000˚C (Figure 12.5). This should come as no surprise, as our eyes (and those of all animals), evolved to be sensitive to the range of frequencies that were most available. The sun also emits infrared, and its spectrum is much broader and deeper than the earth's. The earth emits only infrared because it is much cooler than the sun. With a radiating temperature of about 15˚C, it keeps a balance between the energy entering the atmosphere from the sun (mostly visible), and leaving the earth (as infrared).

The earth is not a simple black body, however. It is surrounded by an atmosphere that complicates the radiative balance of the planet. Visible light from the sun can penetrate the atmosphere partially, but some of the radiation is reflected back out to space by clouds (about 30%). Another 25% is absorbed by the atmosphere, warming it throughout. Finally, about 45% reaches the ground and warms the surface. In response to the surface warming, the earth radiates infrared back out into the atmosphere. However, while visible light from the sun can readily penetrate the relatively transparent atmosphere (It is transparent, or we couldn't see each other through it!), infrared does not as easily penetrate, being absorbed by certain gases in the atmosphere—the greenhouse gases (Figure 12.6). These are primarily water and carbon dioxide (CO_2), but include a whole host of other trace gases.

The Greenhouse Effect

The atmosphere contains gases that are transparent to visible light from the sun, yet are partially opaque to the infrared radiation emitted by the earth (Jerez et al., 2018).

These are the greenhouse gases. The two primary greenhouse gases are **water and carbon dioxide.** If not for these, the earth would be about 33°C colder on average—a very chilly place indeed! While water vapor in the atmosphere is an important greenhouse gas, its residence time is very short, as it rains out frequently, depending on weather conditions locally. As such, its concentration in the atmosphere is determined by atmospheric temperature distribution, evaporative moisture sources, and condensation nuclei for the development of precipitation. Carbon dioxide, on the other hand, has a very long residence time in the atmosphere, as there are few means for removal. Key sinks for modern anthropogenic CO_2 emissions are uptake by the surface ocean (solubility pump) and uptake by northern hemisphere forest regrowth after deforestation of previous centuries.

Water and CO_2 absorb outgoing infrared radiation by increasing their molecular vibrational modes, thus heating up as a result of absorbing the infrared photons. Because these molecules have a specific mass, structure, and bond strength, they can absorb only a specific set of frequencies of infrared. In the earth's infrared radiation spectrum there is thus a gap, or atmospheric window, in absorption by water and CO_2, between eight and twelve micron wavelength, through which radiation escapes, and lessens the strength of the overall greenhouse effect as a result. However,

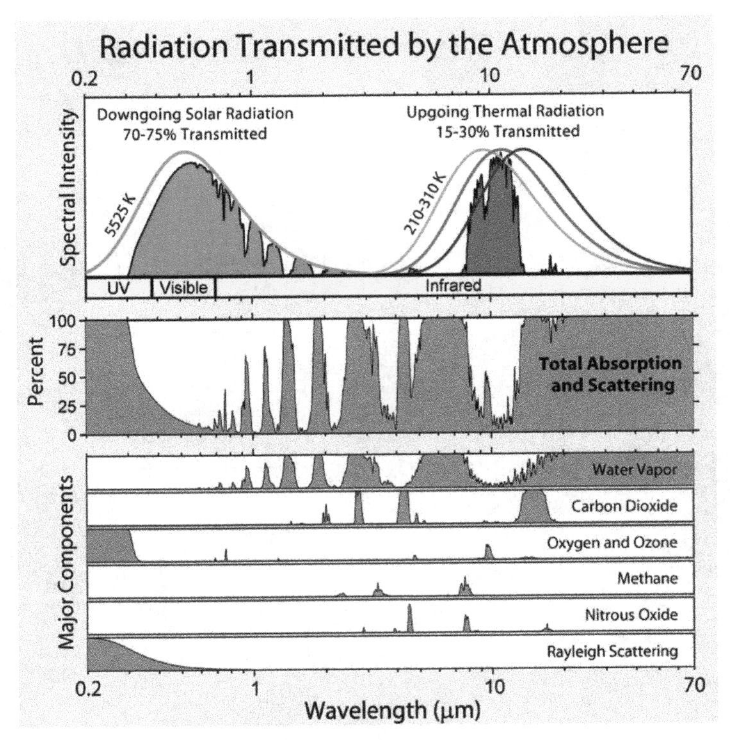

FIGURE 12.6 Outgoing atmospheric transmission
Outgoing infrared radiation from the earth, and absorption of various wavelengths by greenhouse gases according to their absorption spectra. Note that water absorbs the widest range of wavelengths, making it the most important greenhouse gas. CO_2, however, plays a critical role in partially closing a gap near the peak of the infrared radiation spectrum, around 11 microns (μm). As CO_2 concentration increases, the breadth of absorbed radiation wavelengths increases, and more is absorbed.

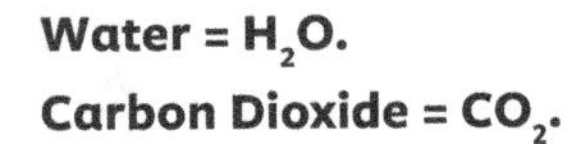

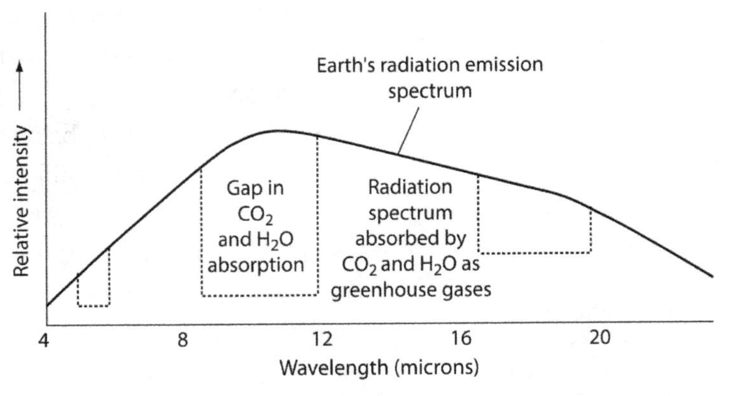

FIGURE 12.7 Outgiong atmospheric transmission in critical 5-20 micron range.
The earth radiates in the infrared, and CO_2 and H_2O, the main greenhouse gases, absorb radiation in that range, but leave a gap where some infrared can escape. Other gases such as CFC, ozone, and others absorb in the gap, but there is a relatively low concentration of these gases in the atmosphere at present.

other gases, such as anthropogenic CFCs, DO absorb in that frequency range, and thus serve to close the gap and increase the greenhouse effect. CO_2 already absorbs most of the outgoing infrared radiation that falls within its absorption spectrum. As more CO_2 is added to the atmosphere, additional absorption is mostly done around the "edges," or wings of the absorption curve, as wavelengths in the core of the CO_2 absorption spectrum are already absorbing most of the outgoing infrared (Figure 12.7). Still this additional absorption is sufficient to cause additional warming that triggers a positive feedback with water vapor, the primary greenhouse gas. This will be explored further in Chapter 13.

It is important to note that the greenhouse effect has been and always will be a critical function within the Earth System, without which the earth would be essentially in a deep freeze, 33°C colder than it has been on average. Greenhouse gases not only keep the earth a comfortable temperature, but they also reduce the diurnal temperature variability. By re-radiating down from the atmosphere back to the earth's surface, nighttime temperatures are kept higher than they would be if radiational cooling were allowed to proceed fully (Figure 12.8). You have perhaps noticed that clear nights are generally colder than cloudy nights. This is because clouds accentuate this effect and re-radiate infrared back to Earth. Also notice that because of this, dew rarely forms on cloudy nights, while it is ubiquitous on clear nights when the ground surface radiates away heat so that its temperature falls to below the dew point. The dew point is the temperature below which the air can no longer hold the water vapor it contains, and it precipitates. When this

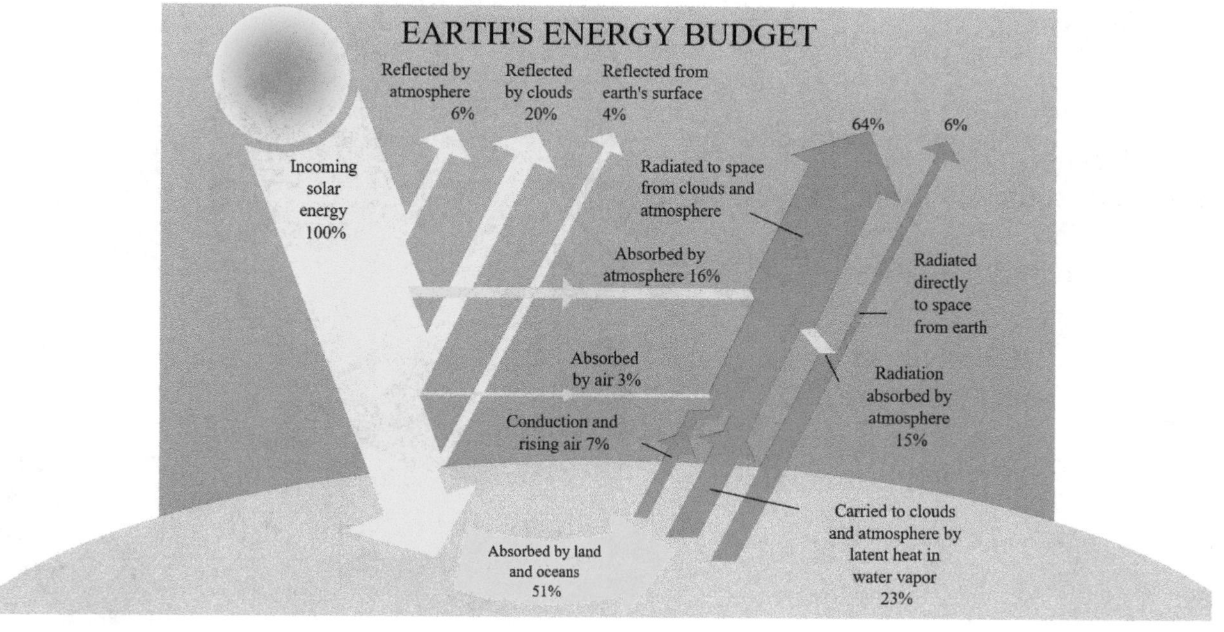

FIGURE 12.8 Earth's Energy Budget
The amount of incoming energy is always balanced by the amount of outgoing energy. If more outgoing energy is trapped by greenhouse gasses, atmospheric temperature must rise in order to emit enough energy to balance incoming radiation.

happens in the air, it makes rain. When it happens on the ground surface, it makes dew. While the greenhouse effect performs these critical functions, changes in greenhouse gases are of some concern, and are discussed in the context of climate change in Chapter 13.

Air Pollution

The atmosphere is not always as pristine as we might like it to be. Many solid, liquid, and gaseous pollutants are emitted into the atmosphere that have potentially deleterious effects on human (and other species) health. Pollutants can come from both anthropogenic and natural sources (Figure 12.9). Natural sources, however, are usually dispersed, like pollen or sea salt, or short-duration events like forest fires or dust storms, and seldom pose a serious health risk to humans. Anthropogenic sources, however, are often concentrated in urban areas and even indoors, where people spend a great deal of their time, and have proven to be serious health threats to many people. In the U.S., the EPA has established a number of classes of air pollutants, including nitrogen

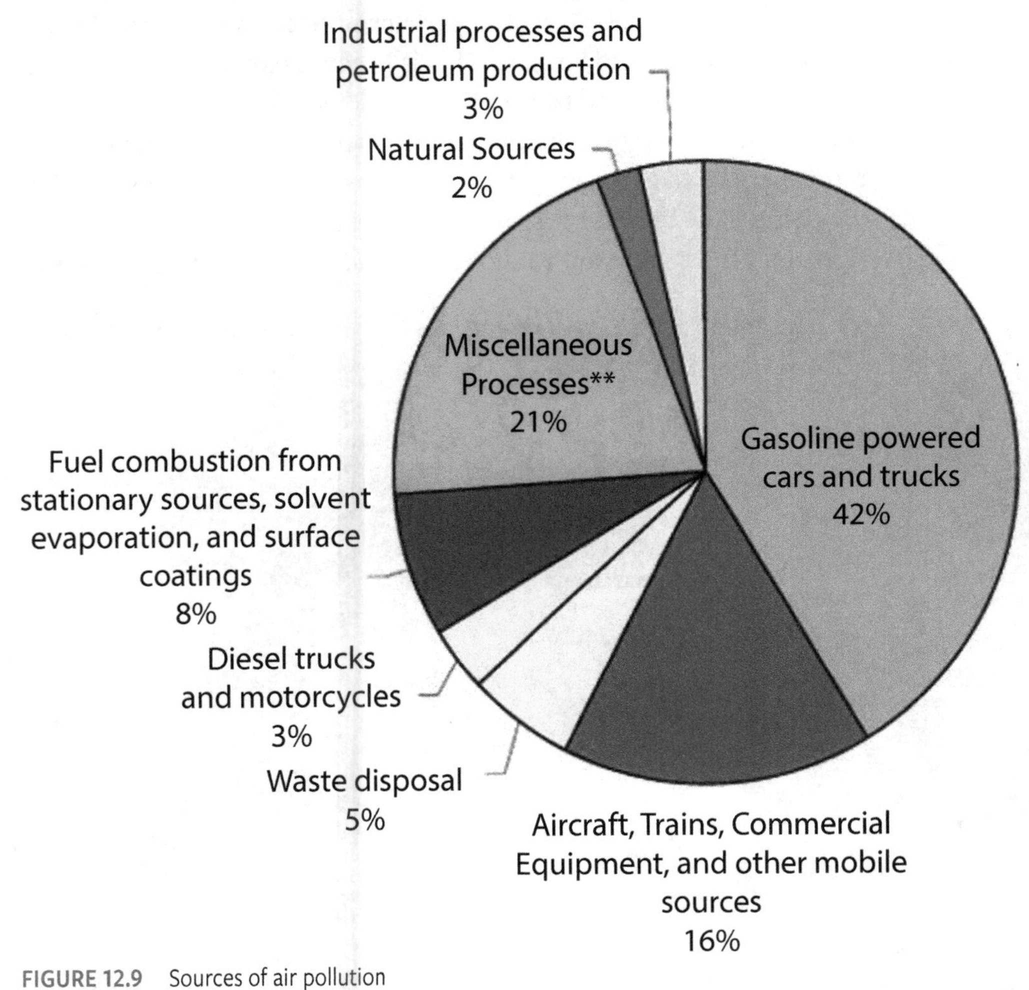

FIGURE 12.9 Sources of air pollution

oxides, sulfur oxides, ozone, hydrocarbons, carbon monoxide, and particulates. These stem mostly from motor vehicles, power plants, and factories, primarily from burning of fossil fuels. Nitrogen and sulfur oxides are emitted from fossil fuel burning, mostly in urban areas. Carbon monoxide stems from incomplete fossil fuel burning. Ozone (at ground level in the troposphere) is created by reactions between nitrogen oxides and oxygen (described below in "Photochemical Smog"). Finally, particulates are tiny solids emitted from a variety of processes, including fossil fuel burning, agriculture, mining, and dust raised by vehicles and other disturbance of the land surface. Solid particulates tend to partially block incoming sunlight, and thus tend to cool the lower atmosphere and Earth's surface from what it would otherwise be. In part, the acceleration of global warming is due to reducing emissions of particulates (for human and ecosystem health reasons) so that the actual rate of warming can be observed.

Try this at home! On a clear night when you expect dew to form on the ground or other surface, such as a picnic table, try "shading" part of the surface with a large umbrella or small tarpaulin to prevent IR radiation from escaping to space. Observe how the "shaded" part does not form dew even when the open part does.

In an extreme case of air pollution, an atmospheric inversion in 1948 in Donora, PA, prevented smokestack emissions from a zinc smelter from rising up away from ground level, and pollution accumulated along the ground, leading to the immediate deaths of 20 people. This was the first documented case of people actually dying from air pollution. More recently, in 1984, a Union Carbide chemical plant in Bhopal, India, released a cloud of methyl isocyanate, a deadly poison gas. With no escape to be had, 20,000 people died as a result of exposure. While these are extreme cases of pollution, most pollution acts over long periods, leading eventually to lung damage, respiratory diseases, and cancer.

Photochemical Smog

The brown haze sometimes seen, especially on summer afternoons, in some cities is smog produced by a number of chemical reactions between the products of fossil fuel burning and sunlight. All burning involves combination with oxygen (oxidation), and there is plenty of oxygen in air. However, 78% of the atmosphere is made of nitrogen. High-temperature combustion of fuels also causes nitrogen to combine with oxygen, leading to the formation of NO_2. If left unperturbed, the NO_2 in the atmosphere will dissolve with water and eventually make acid rain. However, sunlight (say, on a hot and sunny summer afternoon) can break down NO_2 and release individual oxygen atoms, which can then combine with oxygen O_2 molecules to form **ozone**. Ozone is a highly reactive oxidant, and irritates eyes, causes lung damage, and damages a broad spectrum of plants in croplands as well as terrestrial and aquatic ecosystems.

Ozone: A molecule of oxygen (O_3). The first two atoms are strongly bound, while the third is more weakly bound to the pair.

Morning traffic in urban areas burns hydrocarbons at high temperatures, and thus releases NO_2. By noon, the intense sunlight breaks down the NO_2 and leads to a build-up of ozone. By early evening, the reactivity of ozone leads to a reduction in ozone concentration, and without bright sunlight to create more, ozone concentration declines into the night, to start the cycle over again the next day.

Acid Rain

An acid is an aqueous solution that contains free hydrogen ions (protons). A pH of 7 is neutral on a log scale of 1 to 14 (Figure 12.10). The atmosphere has a great many chemical constituents along with water and water vapor, but one of the main ones is carbon dioxide. CO_2 dissolves in water (think of a carbonated beverage), and in so doing, forms a weak acid, carbonic acid. Therefore, all natural rain is slightly acidic, with a pH of about 5.7 or so. While increasing CO_2 concentrations in the atmosphere may make rain more acidic, there are more severe causes of acid rain. Fossil fuel burning emits large quantities of SO_2 and NO_2. Each of these dissociates in water to produce sulfuric and nitric acid, respectively. As a result, in highly industrialized parts of the world, such as the northeast U.S. and northern Europe, the rain had become so acidic (sometimes as low as 4 over large areas, and locally as low as 2), that otherwise pristine mountain lakes became acidic enough to kill all wildlife, and in fact all life to the microscopic level in the lake's ecosystem. Lakes in the Adirondack Mountains of

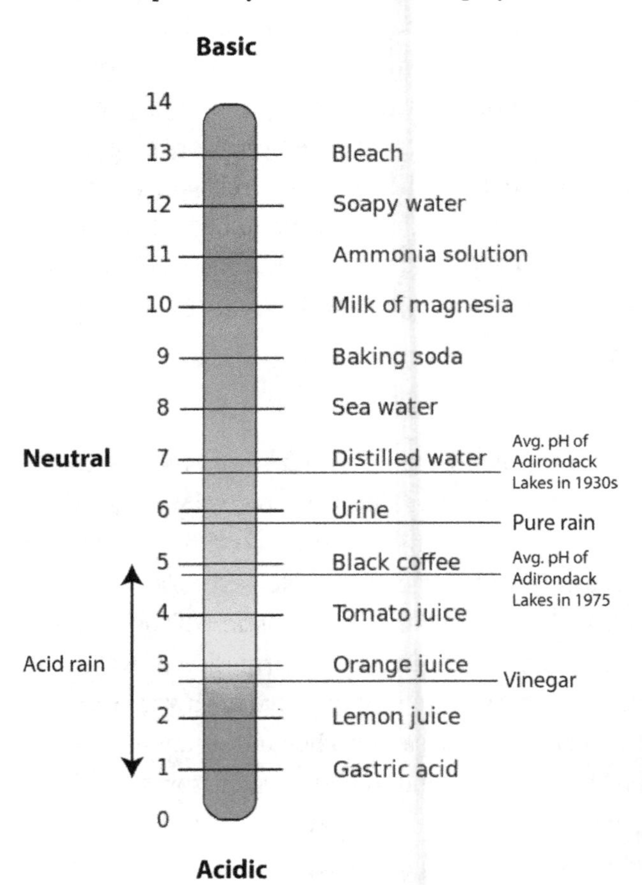

FIGURE 12.10 The pH scale of potential hydrogen in aqueous solutions. pH of 7 is neutral. Higher numbers are basic, while lower numbers are acidic.

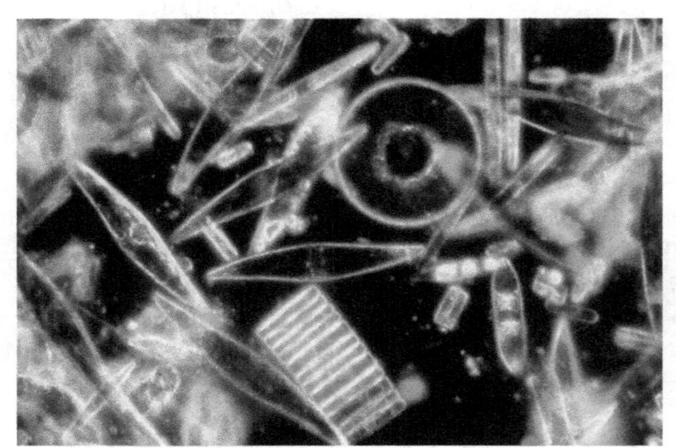

FIGURE 12.11 Phytoplankton

New York that were once teeming with fish and had water that was a murky green from phytoplankton and other aquatic microorganisms (Figure 12.11), became crystal clear (Lawrence et al, 2008; Pelly, 2003). There was no life remaining of any kind to cloud the water. This same problem occurred in Scandinavia and Canada.

The susceptibility of a lake to acidification depends on the acidity of the rain that runs into it and the buffering capacity of the water as determined by the surrounding soils and rock formations. In some localities, the geology is dominated by carbonate rocks such as limestone, which, when dissolved into water, serves to buffer acidification. These localities are resilient to acid rain and are not as negatively impacted as some other areas, and maintain pH above 6. Some areas have soils formed from weathering of granitic rocks, and these have very little buffering capacity. These areas, like the Adirondacks, are highly susceptible, and are places where the most acidic lakes are found, with pH sometimes below 5, a threshold for aquatic life.

Acid rain affects ecosystems globally, affecting all forms of life, from microorganisms to trees, birds, amphibians, and fish. Yet, we humans drink lemon juice (pH 2) and put vinegar (pH 3) on our salads. Swimming in an acid lake

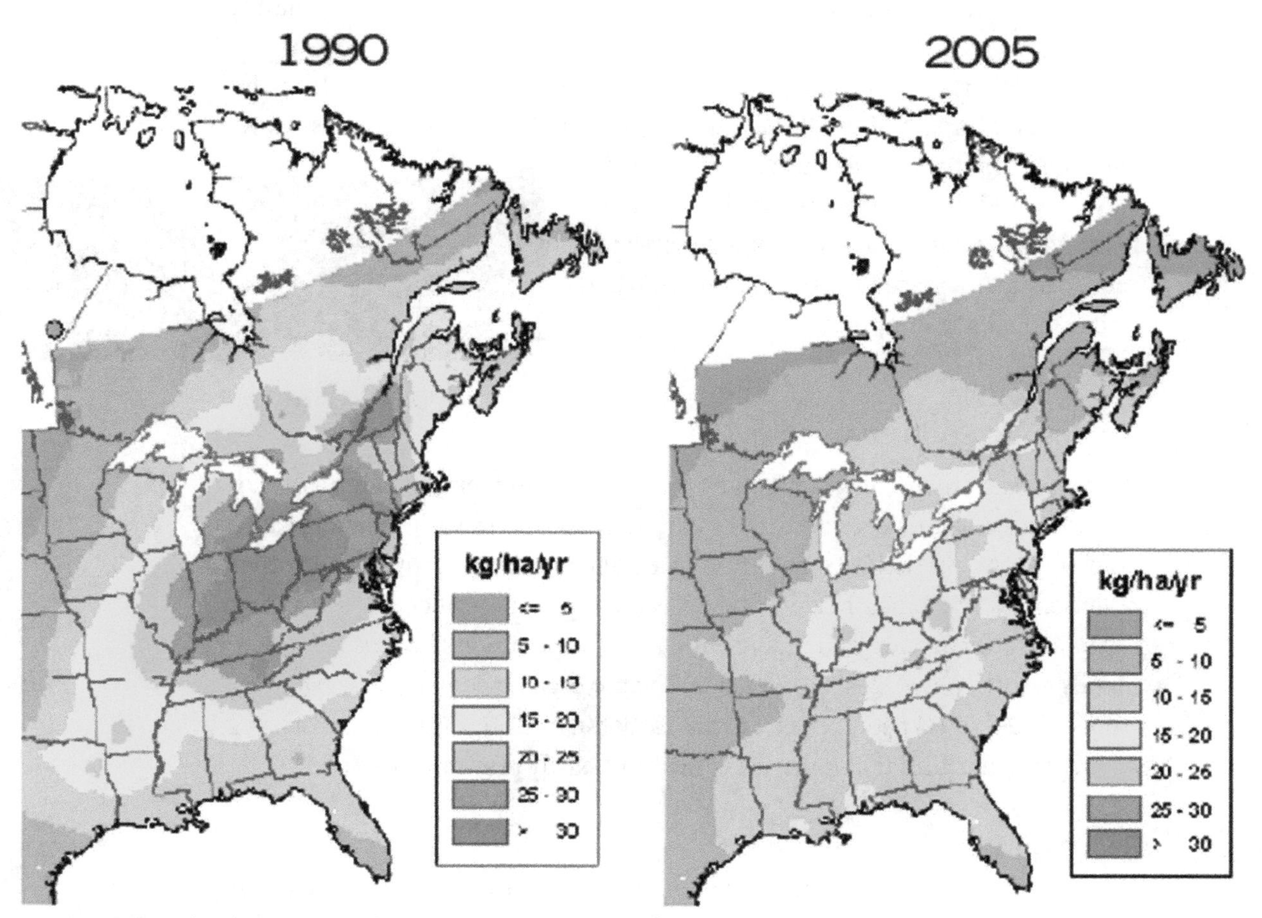

FIGURE 12.12 The Clean Air Act and consequent reductions in sulfur and nitrogen oxides (from coal-fired power plants and other fossil fuel sources) led to a dramatic reduction in the severity of acid rain in the eastern U.S. by the time most current college students were born. Scrubbers on smokestacks now remove sulfur from flue gases and combine with carbonate to make gypsum (used in construction wallboard), but CO_2 is still released to the atmosphere unabated.

would not affect us badly, but aquatic ecosystems are profoundly affected. As such, acid rain is a classic environmental problem caused by human activities, but that affects humans only indirectly (through loss of agriculture, parklands, and various ecosystem services).

In efforts to reduce acid rain caused by sulfur emissions from coal-fired power plants, scrubbers were introduced that serve to remove the sulfur from the exhaust in smokestacks (Figure 12.12). The scrubbers combine the sulfur with calcium carbonate from limestone, to make flue-gas desulfurization (FGD) gypsum. In detail, when the exhaust in the smokestack is put through calcium carbonate ($CaCO_2$), the sulfur gases (SO_2 that would produce acid rain in the atmosphere) react to make calcium sulfate ($CaSO_3$). This then gets combined with water (H_2O) to make gypsum ($CaSO_4 \bullet 2H_2O$), which is chemically identical to the gypsum mined from the ground. This gypsum is then used by wallboard manufacturers and saves both in mining costs and associated environmental impacts, and reduces the need for disposal of the FGD gypsum in landfills. While this is a win-win situation, it only slightly reduces the environmental impact of burning coal, and someday, when coal is no longer burned, if gypsum is still needed for building, it will need to be mined, or wallboard will need to be recycled far more effectively than it is today.

In general, there are many measures that could be taken to reduce the atmospheric impact of fossil fuel burning and other human activities. However, each of these measures can be readily counteracted by either increases in the rate of fossil fuel burning or the extent of other activities caused by an increase in human population and/or individual consumption.

The pH scale is logarithmic, so how much more acidic is water with a pH of 4.2 than water of 4.5?
Answer: $10^{4.2}$ = 158,489. $10^{4.5}$ = 316,228. Dividing one by the other yields about 2.5, so it is more than twice as acidic.

Indoor Air Pollution

Because most Americans spend most of their time indoors, the air quality of interior spaces has emerged in recent years as a major concern for human health. Some indoor pollution is from natural substances and organisms, while other pollution is from human-made products. Natural pollutants include pollen, dust mites, various bacteria, and radon. Human-made pollutants include **asbestos**, metals, ozone, pesticides, particulates, volatile organic compounds degassing from painted or carpeted surfaces, and carbon monoxide. Tobacco smoke is a particularly harmful pollutant, yet is the most easily preventable of all.

Radon

Of particular interest in recent years are the hazards posed by radon, a colorless, odorless radioactive gas that forms naturally in the uranium-238 decay series (Figure 12.13). Uranium occurs naturally within the earth, and fits

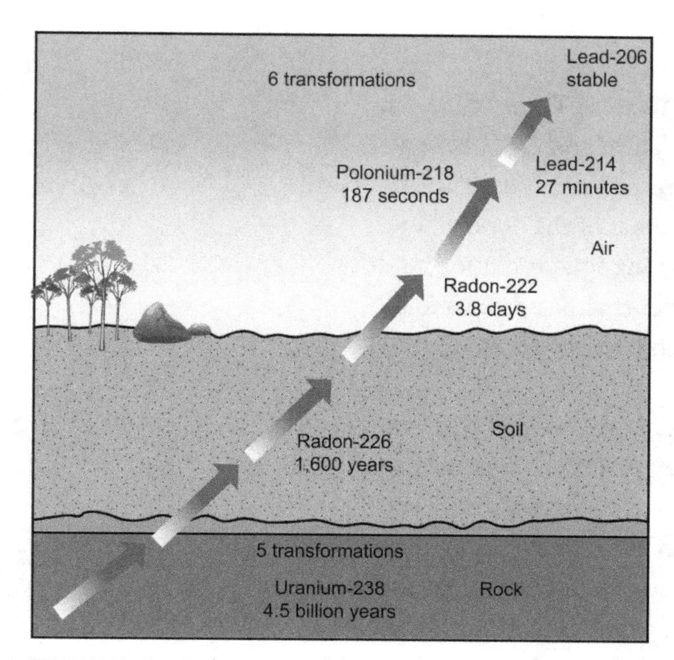

FIGURE 12.13 Radon is part of the U-238 radioactive decay series. When radon enters basements, it can accumulate and lead to a radiation hazard. Half-life for each isotope is indicated.

nicely into the lattice structure of minerals such as potassium feldspar in place of potassium. As it decays (half-life 4.5 billion years, or about the age of the earth) it goes through a series of transformations through alpha and beta decays, eventually making radon-222, a gas that commonly enters basements through the soil. In previous decades, this was not a problem because basements and houses were very leaky, causing drafts that blew away the radon so that it never accumulated into high concentrations. With the growing concerns regarding energy efficiency, buildings (especially houses) were better insulated, and virtually hermetically sealed to the point that if all windows were closed, slamming an exterior door would cause ears to pop in the house due to the pressure wave. This sealing of airflow enables radon seeping into a basement through foundation cracks, sumps, or even well water, to build up to potentially high concentrations. How high is high? The EPA threshold for concern is four picocuries per liter.

The health risk posed by radon is that with a half-life of a few days, it decays into polonium-218, which in just a few minutes decays, emitting an alpha particle (a helium nucleus). It is this alpha particle that can, in close quarters (such as the interior of lungs), cause cell damage by crashing through and breaking up strands of DNA. Statistically, the risk of getting lung cancer from a four picocurie concentration of radon, if exposed for an entire lifetime (70 years), essentially never emerging from a basement, is 0.2%. (In other words, of a thousand people in such an environment, two people would get cancer.) Cigarette smoking, however, greatly increases the cancer risk from radon, as the synergistic effect increases the risk to many times that of either alone.

If the half-life of the precious yet undiscovered element, Absurdium, is 1 year, and you have 1 kg of it initially, how much will be left after 5 years?
Answer: After what amounts to 5 half-lives, you would essentially have to divide by 2 five times, meaning $1/2^5 = 1/32$, so if you started with I kg, you would have only 0.031 kg, or 31 grams left after 5 years.

While the newly emergent risk posed by radon accumulations is an unwanted health risk, people are not ready to return to drafty, energy-wasting homes to remediate the problem. Rather, a more effective means is to install a blower fan that draws air from under the basement slab so that it does not enter the basement, and vent it to the atmosphere above the roof of the house, where it dissipates naturally into the atmosphere, completing its radioactive decays series that finally ends in lead-206 (Steck, 2012).

Stratospheric Ozone Depletion

Ozone is a form of oxygen that contains three oxygen atoms. Think of it as an O_2 molecule with an extra "third wheel" oxygen weakly attached. O_2 is strongly bound together, but the third oxygen is much more weakly bound. In the stratosphere (25 km altitude), far above most of the mass of the atmosphere, there is just enough air to begin to interact with incoming solar radiation, and this is where ozone is formed in a very important layer that has protected all life on Earth for billions of years from harmful ultraviolet radiation emanating from the sun.

In the electromagnetic spectrum, there are three energy ranges of **ultraviolet (UV)**. UVA is the weakest, UVB intermediate, and UVC the strongest, involving the highest-energy photons in the UV part of the spectrum. UVC has such high energy that it can break up the strong bond holding the O_2 molecule together. This means that as oxygen in the atmosphere (there is plenty at 21% of the total) and UVC interact, the breaking of the molecular bonds absorbs (destroys) UVC, such that no UVC at all reaches the planet surface. Instead, it is all absorbed in the stratosphere, where free individual oxygen atoms are left floating around. They can most readily find the ubiquitous O_2 molecules and attach weakly, making ozone. This is how ozone forms—basically as a result of UVC absorption. (Why don't the individual oxygen atoms recombine with each other? With 21% of the atmosphere a molecular oxygen, it would be exceedingly unlikely to actually find the lost partner atom rather than finding a nearby O_2.)

Meanwhile, UVB is streaming down from the sun as well. UVB has a much lower energy than UVC—not nearly enough to break up an O_2 bond, but just right to break off the weak bond that holds the third oxygen in ozone. Thus the interaction of ozone and UVB annihilates both, with the breaking of the weak bond absorbing the UVB photon. Thus, in sum, UVC makes ozone in the stratosphere, and UVB destroys it, so there is an equilibrium in which as long as they work in tandem, a constant concentration of ozone is maintained in a critical layer of the stratosphere, and little UVB reaches the ground surface, where people (and other organisms) live. This is the natural situation in which all life evolved and lived on Earth since free oxygen was added to the atmosphere about 2.5 billion years ago.

In the twentieth century, the industrialized world began building refrigerant systems, propellants, and industrial blowing agents made from various **chlorofluorocarbons (CFCs)**. These are ultimately released to the atmosphere and some made their way to the stratosphere and into the ozone layer, where the chlorine is broken from the rest of the CFC molecule by UVC and UVB. Now free for other chemical reactions, the Cl atom strips off the third O from ozone to make ClO, as it is energetically favorable, Cl-O being a stronger bond than O-O_2. So Cl has destroyed a single ozone molecule. This is not the end of the story, however, as there are more reactions.

Chlorofluorocarbons: A class of non-flammable, non-toxic chemicals used as refrigerants and propellants in the twentieth century until it was discovered that they led to destruction of the stratospheric ozone layer. Examples are Freon and CFC-11 (CCl3F).

When a ClO encounters a single O atom (created by UVC dissociating O_2), the O attached to Cl finds it even more energetically favorable to split off and reattach to the single O to make a new O_2. This frees the Cl atom to find another ozone to strip off the O and continue to destroy ozone (Figure 12.14), behaving as a catalyst to deplete the ozone layer.

$$Cl + O_3 \rightarrow ClO + O_2$$
$$ClO + O \rightarrow Cl + O_2$$

As such, Cl in the stratosphere is a catalyst to destroy ozone, and each Cl atom can continue to destroy ozone indefinitely. This is the essence of the ozone-depletion problem (Larin et al., 2016; Nair et al., 2013). There is a frightening aspect to this. Had the industry chosen to use bromofluorocarbons (bromine instead of chlorine), the ozone depletion problem would have been ten to 100 times worse, and may have become so severe before it was noticed, that extreme ecosystem damage and human health problems may have ensued. In a sense we dodged a bullet without knowing it. Still, the problem (even with Cl) is sufficiently severe that all are advised to use sunscreen, especially for lighter-skinned individuals.

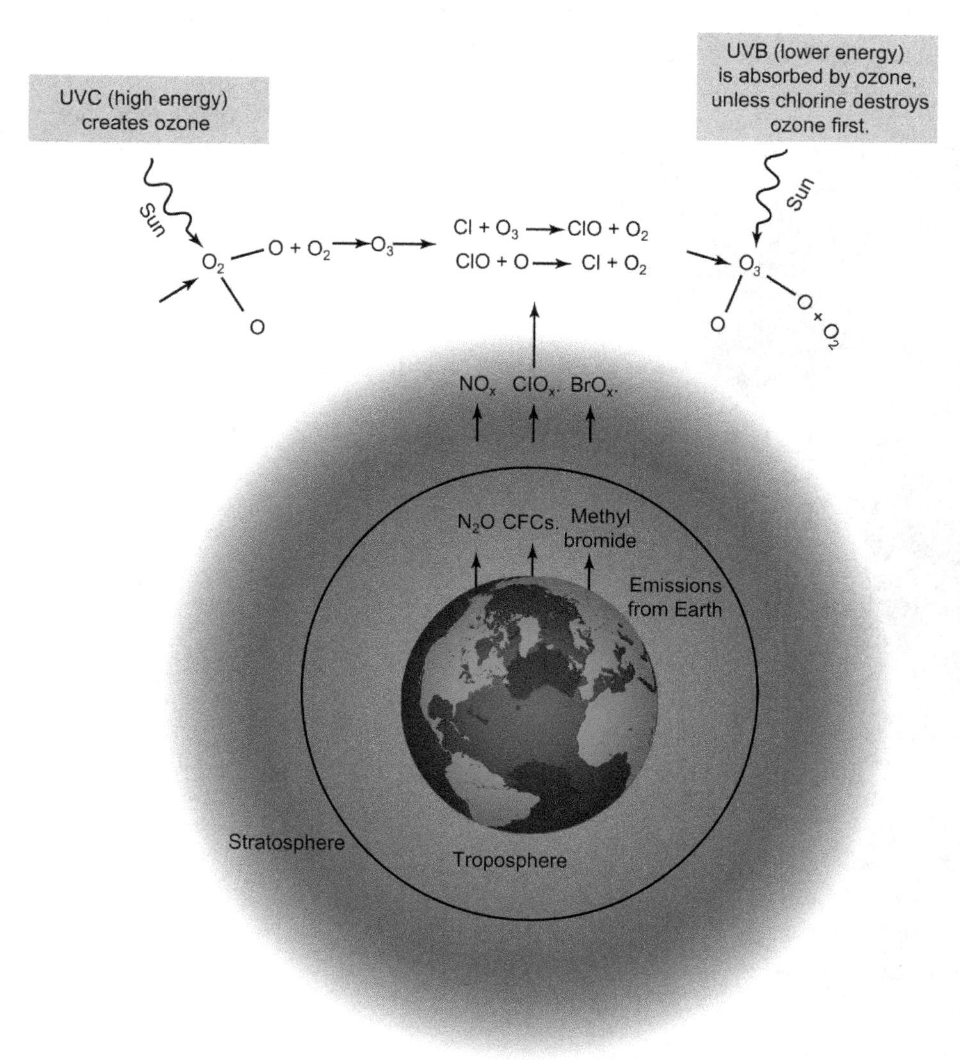

FIGURE 12.14 UVC from the sun dissociates O_2, freeing O to combine with another O_2 molecule to make ozone (O_3). The third O is weakly bound to the O_2, so lower energy UVB can break the bond, freeing the third O to combine with another single O to make O_2. This is the natural ozone cycle. Essentially UVC makes ozone and UVB destroys it in a dynamic equilibrium. When CFCs are emitted by industrial processes, UVB AND UVC dissociates the CFCs in the stratosphere, liberating Cl to react with ozone, stripping the third O to leave O_2. This makes ClO. When encountering another single O atom, the Cl loses its O, making O_2 again, and the Cl is free to destroy another ozone molecule.

Antarctic Ozone Hole

A region with a particularly extreme depletion of ozone

is over the Antarctic (Figure 12.15). While the Antarctic has no real source of CFCs, it nevertheless is the region with the most extreme ozone depletion and here is why:

In the severe cold and dark of the Antarctic winter, with temperatures plunging to the -80s° Celsius, a strong system of winds called the **polar vortex** develops that cuts off the Antarctic from air masses to the north. The stratosphere becomes so cold that sulfuric acid freezes out, nitric acid condenses, and a special set of chemical reactions takes place in polar stratospheric clouds. Chlorine remains in the stratosphere, tied up in reservoir species such as HCl (hydrochloric acid) and $ClONO_2$ (chlorine nitrate), which, in the presence of ice particles in the **polar stratospheric clouds**, slowly convert to chlorine gas (Cl_2), nitric acid (HNO_3), and water. However, when the sun returns in

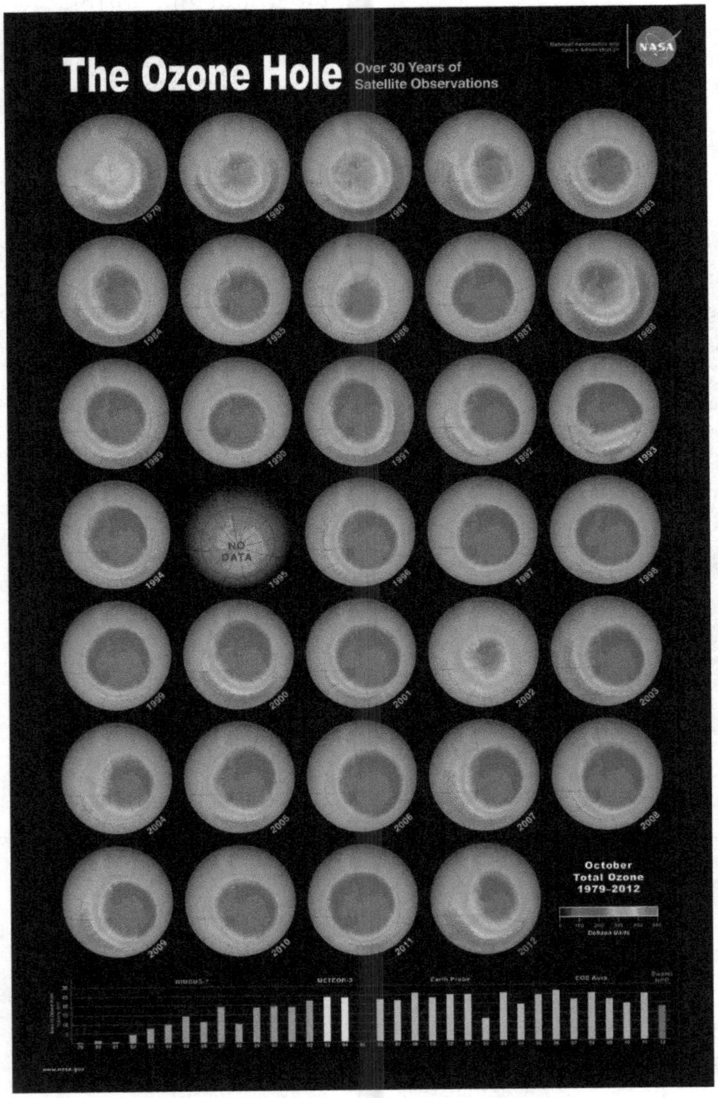

FIGURE 12.15 The ozone hole
The development and evolution of the ozone hole over Antarctica. It was discovered through observations made by the British Antarctic Survey, and the chemical explanation for it was subsequently developed, leading to a Nobel Prize for Molina, Sherwood, and Crutzen. There is some indication that the hole is now "healing" as a result of the industrialized world ceasing production of chlorofluorocarbons (CFCs).

Polar vortex: A system of strong, cold stratospheric winds that circle the poles (west to east), bounded by a jet stream that separates the polar vortex from lower latitude air masses.

Polar stratospheric clouds: Clouds made up of water, nitric acid, and sulfuric acid that occur in very cold polar winters in the lower stratosphere. The nitric acid precipitates out and is no longer available to combine with chlorine, so when the sun comes back out in the spring, chlorine bonds with oxygen, destroying ozone.

the spring, the sunlight dissociates Cl_2, and because there is little NO_x left after the cold winter to make chlorine nitrate and thus act as a sink for chlorine, Cl is freely available to destroy ozone. This rapidly brings the ozone level in the Antarctic stratosphere to extremely low concentrations, and essentially creates a seasonal hole in the protective ozone layer. Scientists from around the world have been monitoring the variations of the ozone hole from year to year, and while it has been both growing in area and deepening in severity, the expectation is that with cessation of CFC use (and release), the hole will begin to heal in the coming decades. This healing could have some effects on Antarctic climate, which has been held artificially cold by the ozone hole and its effect on enhancing the circum-Antarctic winds, which provide more sea spray and associated aerosols into the atmosphere, thus increasing summer cloudiness and helping keep Antarctica colder. Without this effect of the ozone hole, the actual warming of Antarctica in keeping with overall global warming is expected to become evident in the coming decades. This is an area of current research and new developments are expected in our understanding of the relation between ozone, regional climate forcing, and Antarctic ice budgets in the coming years.

References

Jerez, S. Lopez-Romero, J. M., Turco, M., Jimenez-Guerrero, R., Vautard, R., & Montavez, J. P. (2018). Impact of evolving greenhouse gas forcing on the warming signal in regional climate model experiments. *Nature Communications*, 9(1304).

Larin, I. K., Aloyan, A. E., & Ermakov, A. N. (2016). Chlorine activation of the lower stratosphere at mid-latitudes: Impact on the ozone layer. *Russian Journal of Physical Chemistry*, 10(5), 860–864.

Lawrence, G. B., Roy, K. M., Baldigo, B. P., Simonin, H. A., Capone, S. B., Sutherland, J. W., . . . & Boylen, C. W. (2008). Chronic and episodic acidification of Adirondack streams from acid rain in 2003–2005. Journal of Environmental Quality, 37(6), 2264–2274.

Nair, P. J., Godin-Beekmann, S., Kuttippurath, J., Ancellet, G., Florence, G., Pazmino, A., . . . & Pastel, M. (2013). Ozone trends derived from the total column and vertical profiles at a northern mid-latitude station. *Atmospheric Chemistry and Physics*, 13(20), 10373–10384.

Pelley, J. (2003). Adirondack lakes recovering from acid rain. *Environmental Science & Technology*, 37(11), 202A–203A.

Steck, D. J. (2012). The effectiveness of mitigation for reducing radon risk in single-family Minnesota homes. *Health Physics*, 103(3), 241–248.

Figure Credits

Fig. 12.1: Source: https://commons.wikimedia.org/wiki/File:Profil_temperature_atmosphere.png

Fig. 12.2: Copyright © Kaidor (CC BY-SA 3.0) at: https://commons.wikimedia.org/wiki/File:Earth_Global_Circulation_-_en.svg.

Fig. 12.3: Copyright © 2015 Depositphotos/eurovector.

Fig. 12.4: Copyright © 2012 Depositphotos/AlexCiopata.

Fig. 12.6: Copyright © Dominic (CC BY-SA 3.0) at: http://en.wikipedia.org/wiki/File:Atmospheric_Transmission.png.

Fig. 12.8: Source: https://commons.wikimedia.org/wiki/File:Earth_energy_budget.svg

Fig. 12.9: Source: http://www.bar.ca.gov/80_bar-resources/02_smogcheck/air_pollution_sources.html

Fig. 12.10: Copyright © Edward Stevens (CC BY 3.0) at: https://commons.wikimedia.org/wiki/File:PH_Scale.svg.

Fig. 12.11: Source: https://en.wikipedia.org/wiki/Phytoplankton#/media/File:Diatoms_through_the_microscope.jpg

Fig. 12.12: Source: http://www.epa.gov/usca/acid_2008.html

Fig. 12.15: Source: https://aura.gsfc.nasa.gov/ozoneholeposter/

CHAPTER 13

Climate and Climate Change

We didn't perceive that there was any danger
In being a one-species Earth re-arranger.
Clearing land for our food,
Fossil fuels we thought good,
Turned out as an overall world climate changer.

Weather vs. Climate

WEATHER IS THE day-to-day atmospheric conditions, including hourly temperature, precipitation, cloudiness, relative humidity, fog, and the various other atmospheric phenomena we encounter in any given day at any particular point at the earth's surface. **Climate** is the long-term averaged conditions for a given locality or region. Over periods of decades to centuries, climate controls the nature of vegetation, soils, morphology of the land surface, and the interaction between the land surface, groundwater, surface water, and the atmosphere (Figure 13.1). Climate in a given locality is controlled most generally by latitude (hot, humid, and rain at the equator; dry with large diurnal te perature variations at 30° north and so latitudes; cool, cloudy, and rainy at 60 itude; and very cold and dry at the p

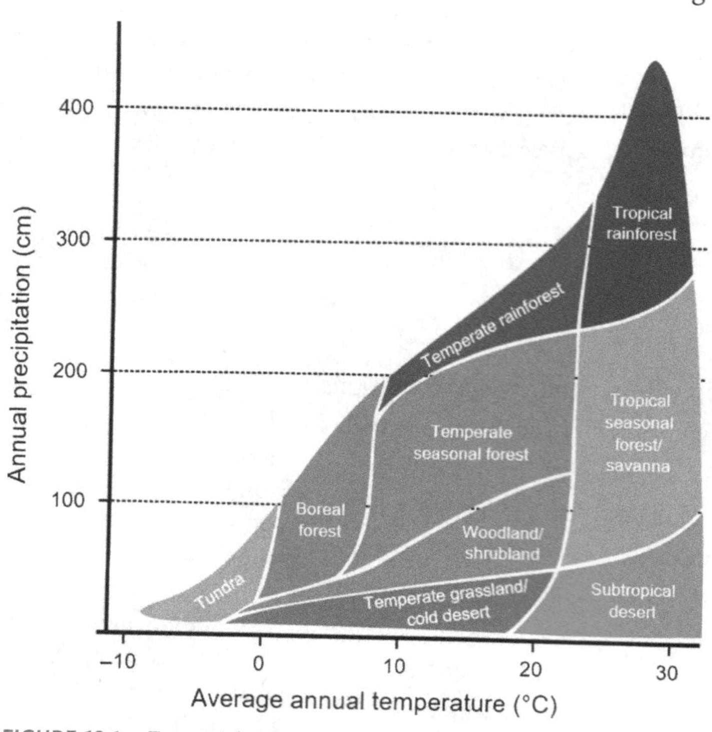

FIGURE 13.1 Terrestrial climate and associated biomes

It is also controlled by elevation, mountains are much cooler than tudes. Prevailing winds also can d a region's climate, as hillslopes wind are generally rainier facing away from the preva (This is because warm, mo

it rises up a slope, thus losing capacity to hold moisture while heading down the other side, cool dry air gets warmer and thus drier, creating the **orographic effect**). Hawaii is an excellent example of very moist tropical climate on one side of the islands, and very dry desert-like conditions on the other. Yet another factor controlling climate is proximity to an ocean or a large lake. Water, due to its great heat capacity, takes much longer to warm up (in summer) or cool off (in winter) than the land surface, so if the ocean is upwind, seasonal temperature variations are greatly moderated. The British Isles are a prime example of the moderating effect of the ocean. In the other extreme, central Siberia and the northern plains of the U.S. experience very large seasonal variations in temperature, with exceedingly hot summers, and very cold winters.

In order to understand climate and climate change, it is necessary to explore the earth's radiation balance, carbon and other biogeochemical cycles, the hydrologic cycle, all the feedbacks and interactions between Earth's subsystems, and the human impact the climate system. In this chapter, for both the ogic past (pre-human) and post-industrial, we vestigate climate change on the basis of obser- mechanisms that control climate change, ns of future climate, and expected impacts change on natural and social systems.

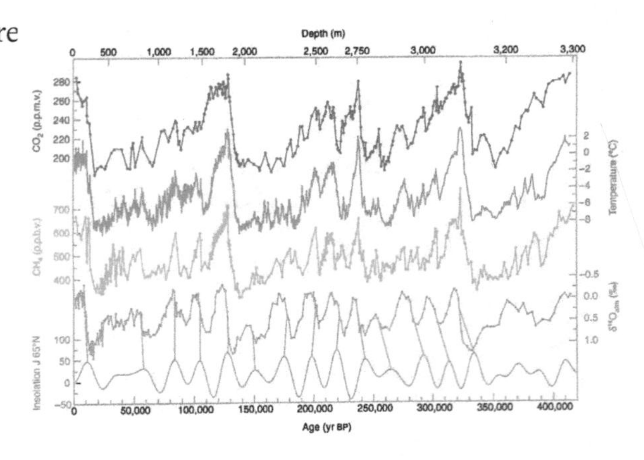

FIGURE 13.2 Ice Age Temperature
The Vostok ice core drilled from Antarctica preserves a record of climate changes for the latter part of the Pleistocene (Petit et al., 1997; Bonacci and Lacaze, 2018; Pol et al., 2014). Air preserved in bubbles in the ice can be recovered and analyzed for CO_2 concentration, isotopic composition, and other characteristics to reconstruct the changing climate over the last 400 thousand years. (This has now been extended back to almost a million years with the same pattern.) Note that the period of oscillation is about 100,000 years. This is understood as driven by Earth's orbital variations. Atmospheric CO_2 concentration varied along with temperature, oscillating between about 180 ppm and 280 ppm, never rising higher or dropping lower. Apparently, some feedbacks within the Earth System served to emit CO_2 into the atmosphere once it fell to 180 ppm, and draw it back out once it rose to 280 ppm. The details of this are not yet fully understood, and it is an area of active research. However, it is important to note that due to twentieth century and current fossil-fuel burning, we have raised CO_2 concentration to over 400 ppm, far outside the envelope of stability of the last million years or so.

...nate

...e Change

be characterized as the long-term atmospheric conditions ..., climate constantly changes at a range of timescales ...of millions of years. This has been well documented ...s proxies, including glacial ice extent, tree rings, ...n caves), ocean and lake sediment cores, isotopes ...es, and many other means that geologists use to ...2).

...rs, global climate (averaged over the entire ... from about 12°C to 22-°C and back again ...idered an **icehouse world**, while the warm

Icehouse world: Globally cold conditions, during which extensive continental glaciers (ice sheets) waxed and waned (glacial-interglacial cycles within icehouse times). We are in an icehouse world now, in an interglacial period, characterized by a cold deep ocean.

Greenhouse world: Warm conditions globally, during which there was little polar ice, resulting in a warm deep ocean.

times are a **greenhouse world**. Over the last 10,000 years or so, the global average temperature has been about 15°C, and we have been in an interglacial interval of an icehouse world. Throughout Phanerozoic time (last 550 million years), the earth has oscillated between icehouse and greenhouse conditions about five times, with the most recent greenhouse time being the Cretaceous period, until about 65 million years ago. Since that time, the earth gradually cooled, culminating in the major glaciations of the Pleistocene Epoch in the last couple of million years. Glaciation waxed and waned throughout the Pleistocene (Figure 13.2), and the most recent major ice sheets that covered northern North America and Eurasia 18,000 years ago essentially melted away by 10,000 years ago. Remnants remain today only in Greenland and Antarctica.

The Holocene Epoch of the last 12,000 years or so has been a time of remarkable climate stability. (There is a great deal of discussion among scientists about how this time of stable climate may have facilitated agriculture and modern society.) However, even during this stable time, minor climate variations have been recorded. In the last thousand years (Figure 13.3), a Medieval Warm Period occurred in the twelfth and thirteenth centuries AD, during which the average global temperature was almost a degree warmer than the centuries previous or since, until the twentieth century. In the fifteenth through eighteenth centuries, the Little Ice Age brought temperatures down to about half a degree below the long-term Holocene average, and Londoners were able to skate on the Thames River, an activity that is quite impossible these days (it doesn't freeze anymore). In the middle of the nineteenth century, temperatures began to rise and continue to rise at present. The rate of climate warming in the present day is greater than that in the paleo record, and this may be because the temporal resolution of the proxies used to reconstruct past temperatures are insufficient to record such rapid changes, and may be because global warming in the post-industrial era, during which great quantities of anthropogenic greenhouse gases have been released into the atmosphere, is actually more rapid than past natural changes.

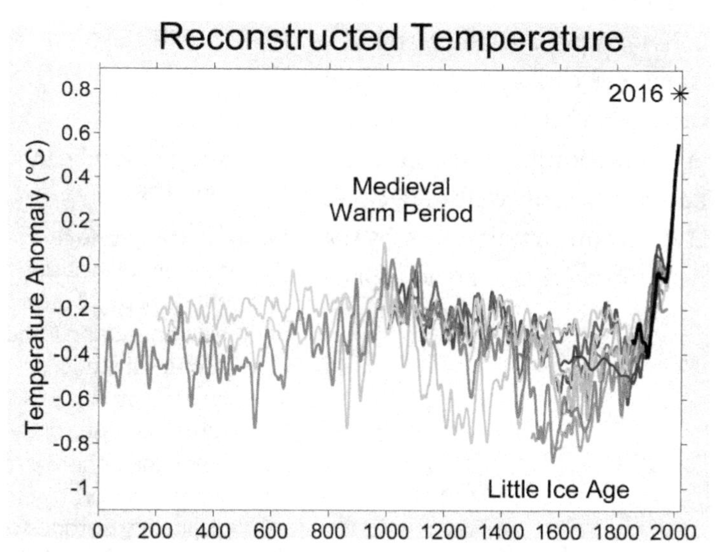

FIGURE 13.3 Earth's temperature record resembling a hockey stick, with various global and northern hemisphere reconstructions since the year 0. The Medieval Warm Period and Little Ice Age were minor compared to modern warming. The rate and magnitude of 20th and 21st century warming are unprecedented in the last 2,000 years, and temperatures are projected to continue to rise throughout the 21st century and beyond, depending on the rate of fossil fuel burning and associated greenhouse gas emissions.

Observing Evidence of Past Climate Change

We were not around to measure climate changes in the ancient past, so we must rely on **proxies** that, by their response to climate change, record these changes and preserve them in the geologic record and other records that scientists can observe and measure (Fig 13.4) (Scheff et al., 2017; Saraswat et al., 2017). A key observational record comes from ice cores, most notably in Greenland and Antarctica. When snow compacts and recrystallizes to ice, the air trapped between the snowflakes is enclosed in the ice and cannot escape. This air has the composition of the atmosphere at the time the ice formed. By drilling out long ice cores that represent hundreds of thousands of years of ice accumulation, scientists can carefully analyze the air in those bubbles and determine past atmospheric composition. This is how we know how CO_2 and methane changed over time, for example.

Proxies: Records of past environmental conditions preserved in sediments, rocks, trees, ice, and any other remnants from the time of interest. The records may be chemical, biological, isotopic, or physical, and could be preserved in any number of ways. Scientists use these records to infer past conditions, because they understand that the processes that create the records depend on the past environmental conditions that created them.

Another means for observing evidence of past climates is the analysis of isotopic composition of various organisms, such as marine microorganisms (e.g., foraminifera—**forams**). The carbon and oxygen isotopes in their shells reflect the isotopic composition of the ocean surface water from which they made their shells. The oxygen isotopic composition of the water is determined by the amount of glacial ice stored, because high-latitude snow and ice has a much lighter isotopic composition than low-latitude rain has. This is because as water precipitates from vapor to liquid, the heavy isotope (O-18) changes phase into the liquid earlier than the lighter phase (O-16). (The more sluggish heavy molecules want to move more slowly, as in the liquid, rather than energetically bouncing around in the gaseous vapor phase.) This is called **isotopic fractionation.** Because heavy isotopes rain out first, and most evaporation occurs in the tropics, by the time a moist airmass reaches high latitude (or high altitude), it has already rained

Forams: (short for foraminifera). A single-class phylum of the kingdom Protista that primarily lives at the ocean bottom or surface, and makes its shells from calcium carbonate, whose isotopic composition reflects the temperature as well as isotopic composition of the ocean at the time that organism lived.

Isotopic fractionation: The preferential inclusion of a light or heavy isotope of an element during phase change of a substance. The heavier isotope "prefers" liquid over gas, and solid over liquid. Therefore, when water evaporates, for example, heavy isotopes of O and H are preferentially left behind in the ocean, and when the water rains back out of the atmosphere, any heavy isotopes that had evaporated are the first to precipitate, leaving rain and snow at high latitudes and altitudes isotopically very light.

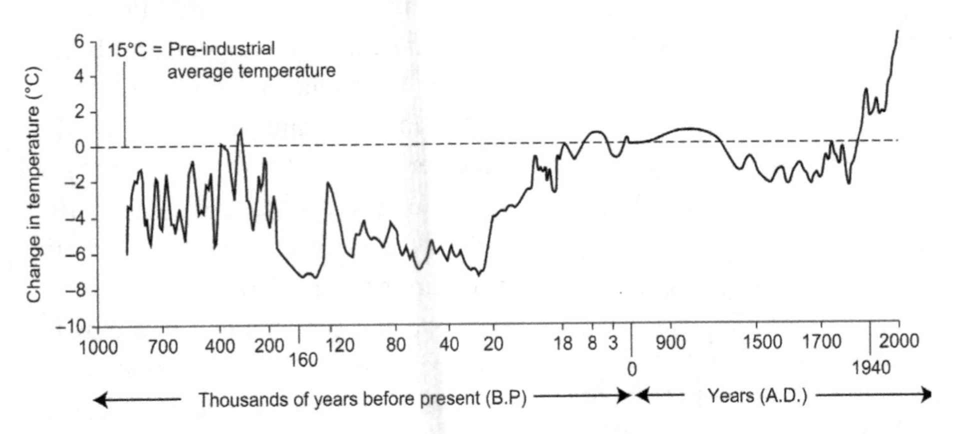

FIGURE 13.4 History of climate change for last 800,000 years. Note glacial-interglacial cycles, and modern climate warming. Note also that the scale is non-linear. The ancient past is compressed and the recent is expanded. This has the effect of reducing the apparent rate of recent warming, but it stands out nevertheless.

out the heavy isotopes, and the resulting snow and ice is light, leaving the

Biological fractionation: Isotopic fractionation by organisms during growth. Photosynthetic pathways (both C3 and C4) strongly prefer the light isotope of carbon (both in land plants and marine phytoplankton). Photosynthetic phytoplankton thus deplete the surface ocean in C-12 (relative to C-13) "enriching" the water in C-13, so that when planktonic heterotrophic organisms make their shells from the remaining inorganic C, they are enriched (isotopically heavy) in C-13 relative to the deeper ocean.

Tree rings: Annual growth rings in trees. They are normally wider (faster growth) during warm and especially wet times.

remaining ocean heavy, so marine organisms during glacial times have isotopically heavier water to work with. Temperature also affects how organisms incorporate the different isotopes into their bodies, depending on how they do it. Phytoplankton, the base of the marine food chain, use photosynthesis, which strongly fractionates carbon in favor of the light isotope in a way that is sensitive to temperature. In cold water, when there is plenty of dissolved CO_2, phytoplankton can be pickier about preferring the light C isotope, and there is greater fractionation, leaving more heavy C-13 in the water. When zooplankton and larger organisms make their shells from dissolved carbonate, they do not fractionate much, so reflect the composition of the seawater. As such, the degree of **biological fractionation** depends on the temperature of the water and this is reflected in the composition of marine shells. By measuring the isotopic composition of the shells preserved in deep sea cores and rocks, scientists can thus calculate the temperature of the water at the time that the organisms were living.

Yet another proxy for past climate is the nature of **tree rings**. During warm and wet periods, trees grow faster and make thicker rings. Some trees grow very old, and some living Bristlecone pines are up to 5,000 years old, so there is a substantial climate record that can be recovered. By correlating ring patterns from living trees with older, dead trees, the record can be extended even further.

Similar to tree rings but in the marine realm are the growth patterns of **corals**. Annual growth rings can be observed and interpreted regarding the factors that control growth rate in corals (marine chemistry, nutrients, temperature, etc.). In addition, as done for forams and other microorganisms, corals preserve isotopic ratios that reflect the temperature of the water at the time of growth-ring formation.

Water dripping into caves precipitates calcium carbonate and forms stalagmites, or **speleothems** (Wong and Breeker, 2015; Frappier et al., 2007). Because the water comes from rain and flows through the land surface and rocks above the cave, the isotopic microstructure of the speleothems reflects the oxygen composition of the rain water as well as the carbon fractionation processes of the terrestrial ecosystem overlying the cave, with temporal resolution that can be as fine as weeks. It has even been discovered that some speleothems can record the passing of individual hurricanes in the tropics, and used as a measure of hurricane frequency in the past.

Natural Drivers of Climate Change

At the longest timescales of tens of millions of years, climate can be affected by the motions of the earth's tectonic plates, moving land areas to more polar or more equatorial positions. In more polar positions, glaciation and the associated ice-albedo feedback can lead to global cooling, while equatorial land and polar oceans lead to more moderate general climates. On shorter timescales, however (less than a million years), too short for tectonic motions to matter, there are other mechanisms for climate change. The most notable of these is the variations in the earth's orbit around sun. These were discovered in the 1920s by Milutin **Milankovitch**, who was able to relate orbital variations to geologic evidence of the timing of Pleistocene glaciations. There are three timescales of orbital variability (Figure 13.5). The first and longest, with a periodicity of roughly 100,000 years, is the shape of the orbit, which is always elliptical, but with varying **eccentricity**—sometimes more circular, sometimes less. The second is the tilt of the axis relative to the plane of the orbit (known as **obliquity**) with a period of about 41,000 years, varying between 21.5 and 24.5 degrees (current obliquity is 23.5 degrees, the latitude of the tropics of Cancer and Capricorn). The third, **precession**, relates to the interaction between the former two, because the rotational axis wobbles so that sometimes northern summer occurs when the earth is closest to the sun (in its eccentric orbit), and sometimes, as at present, northern summer occurs when it is farthest from the sun, thus reducing seasonality (recall that most of the earth's land surface is in the northern hemisphere). The period for precession is about 23,000 years. All three Milankovitch cycles affect global climate, but sometimes one frequency more strongly controls glacial–interglacial cycles, and sometimes another frequency dominates. It is not entirely clear why this should be so.

On shorter timescales, **solar cycles** can influence climate, as they control solar output, reflected in the number of sunspots. There have been times, such as the **Medieval Warm Period**, when there were an abundance of **sunspots** (solar storms) that emitted more solar energy, and times such as the Maunder Minimum that corresponds with the **Little Ice Age**, when there were few sunspots, but there were likely other contributors to these climate variations. There is a short-term cyclicity that results in variations in sunspot numbers, which oscillate between 200 and just a few in a cycle that has been about 11 years in period ever since it has been recorded, but it appears to be shortening to about ten years recently.

Other natural variations occur within the earth's climate system, and **El Niño Southern Oscillation** (ENSO) is a prime example. Every three to seven years, the warm pool of equatorial Pacific surface water moves east and west, altering the global atmospheric circulation patterns, and while not significantly altering global average temperatures, marked changes in regional climates are observed throughout the world. Other oscillations include the North Atlantic Oscillation, and the Pacific Decadal Oscillation, each of which affects regional and global climates in a cyclic fashion with a period of several years to decades.

Eccentricity: The non-roundness of an ellipse (or in this case, elliptical orbit), as measured in terms of the ratio of major and minor semiaxes as

$$\sqrt{1 - \frac{b^2}{a^2}}$$

—circle has eccentricity 0, and a very elongated ellipse has eccentricity approaching 1. Earth's orbital eccentricity waxes and wanes between about 0.004 (almost circular) and 0.06 with a period of about 100,000 years.

Obliquity: The tilt of the earth's rotational axis relative to the plane of its orbit around the sun. It increases and decreases with period of about 41,000 years.

Precession: The relation between eccentricity and obliquity, pertaining to whether the earth is nearer the sun during the northern hemisphere's summer or winter. Presently, it is closer to the sun in northern hemisphere winter, thus slightly reducing seasonality.

Solar cycles: Periodic variability in the number of sunspots on the sun, with a period of about 11 years. Sunspots produce bursts of energy that temporarily provide greater insolation to the earth (warming).

Medieval Warm Period (eleventh to fourteenth centuries, AD): A time of many sunspots that led to a warmer than usual climate on Earth.

Little Ice Age (seventeenth to nineteenth centuries, AD): A time of few sunspots, leading to a cooling climate.

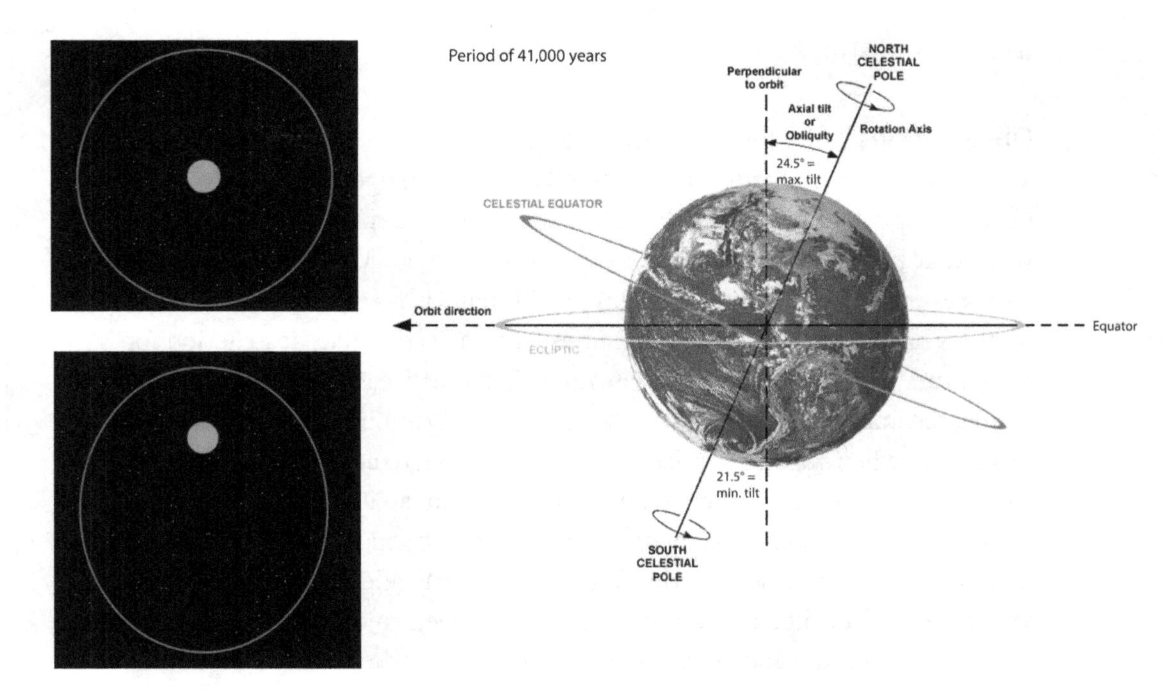

El Niño Southern Oscillation (ENSO): Oceanographic and atmospheric conditions caused by a shift of a warm pool of equatorial Pacific surface water from the western side of the Pacific to the eastern side, closer to South America. This occurs every 3–7 years and suppresses upwelling of nutrient-rich deep waters off the coast of South America. Without nutrients, phytoplankton do not bloom as profusely, thus affecting the higher trophic levels, reducing the utility of the fishery off South America.

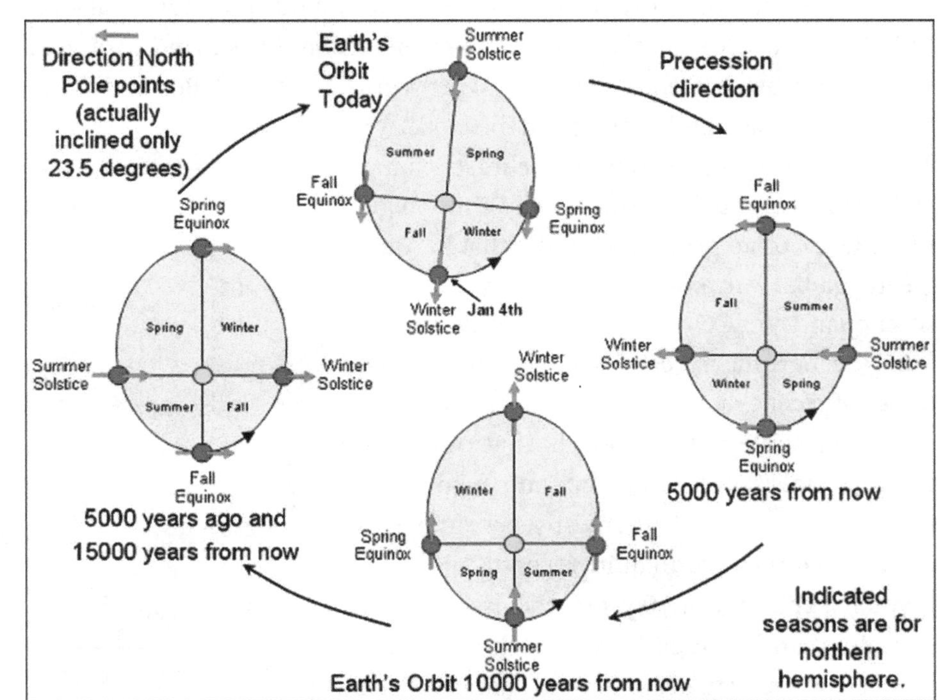

FIGURE 13.5 Milankovitch cycles. Long-term climate is affected by Earth's periodically varying orbital parameters. Eccentricity is the "roundness" of the orbit around the sun (the sun remains at one focus of the elliptical orbit). At present, the orbit is fairly round. The tilt of the Earth's rotational axis also changes, defining the tropics of Cancer and Capricorn as well as the Arctic and Antarctic circles. At present, the tilt is a moderate 23.5 degrees. Finally, the interaction between the eccentricity and tilt changes (precesses), so that the earth is closer or farther from the sun during northern hemisphere summer. At present, we are farther from the sun in northern hemisphere summer, thus moderating climate slightly. The last several glacial-interglacial cycles were driven at the eccentricity period of about 100,000 years.

Modern Climate

Observations of Recent Climate Change

The issue of rapid recent and projected climate change has risen from an obscure topic for specialized academic scientists to explore, to a household word that has come to the fore of the political arena. Many consider it the most serious environmental problem that humanity has ever faced. Although humans have been conducting activities, such as ecosystem destruction for agriculture, hunting, and biomass burning, that alter the Earth System for millennia, the impacts on global systems were relatively minor until the industrial revolution, when huge quantities of fossil fuels began to be mined and burned. As a result, various gases were emitted into the atmosphere, including oxides of sulfur and nitrogen, and most notably, carbon dioxide. While the sulfur and nitrogen led to acid rain that was subsequently alleviated by scrubbers and other end-of-pipe technologies, there has been to date no satisfactory means for capturing and permanently storing carbon dioxide. Consequently, the atmospheric concentration of CO_2 has risen from the pre-industrial level of 280 ppm, to a 2018 level of over 410 ppm, and it continues to increase at an accelerating rate, currently about 4 ppm per year (Figure 13.6). Because CO_2 is a greenhouse gas, this has triggered the modern era of global warming and associated climate changes and impacts throughout the world. While there is a long list of other greenhouse gases that have even greater global warming potential per molecule, the sheer quantity of CO_2 emitted to the atmosphere by fossil fuel burning has caused it to be the dominant driver of recent rapid climate change.

A remarkable observation is that of average global temperatures since the industrial revolution. If you sort the 20 warmest years since 1890, you see that they are all quite recent (Table 13.1). If you list the coolest 20 years, they are all in the late nineteenth and early twentieth century. In fact, 17 of the warmest 18 years (since typical college freshmen were born) ever recorded were in the 21st century. All 20 of the warmest years have been since 1997. In 2016, for the first time in recorded history, the all-time warmest year record was broken in three years in a row. The post-industrial history of global temperature shows dramatic warming (Figure 13.7). This observation settles any question about whether or not there is any global warming.

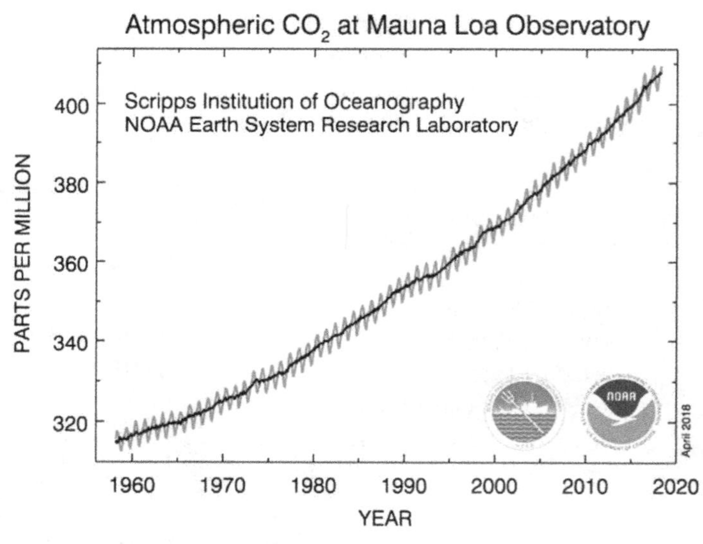

FIGURE 13.6 Atmospheric CO_2 at Mauna Loa
The record of atmospheric CO_2 concentration on top of Mauna Loa, on Hawaii in the middle of the biggest ocean on Earth, far from most industrial or other sources of anthropogenic CO_2. Note that the rate of increase has been increasing since 1958, when recording began. The thick line is the annually averaged CO_2 concentration. The thin line varies annually as the northern hemisphere plants bloom in the spring and decay in the fall and winter. This can be likened to the "breathing" of the northern hemisphere, where most deciduous land plants live. In 2017 a threshold was passed—we will not see CO_2 concentration below 400 ppm in our lifetimes again.

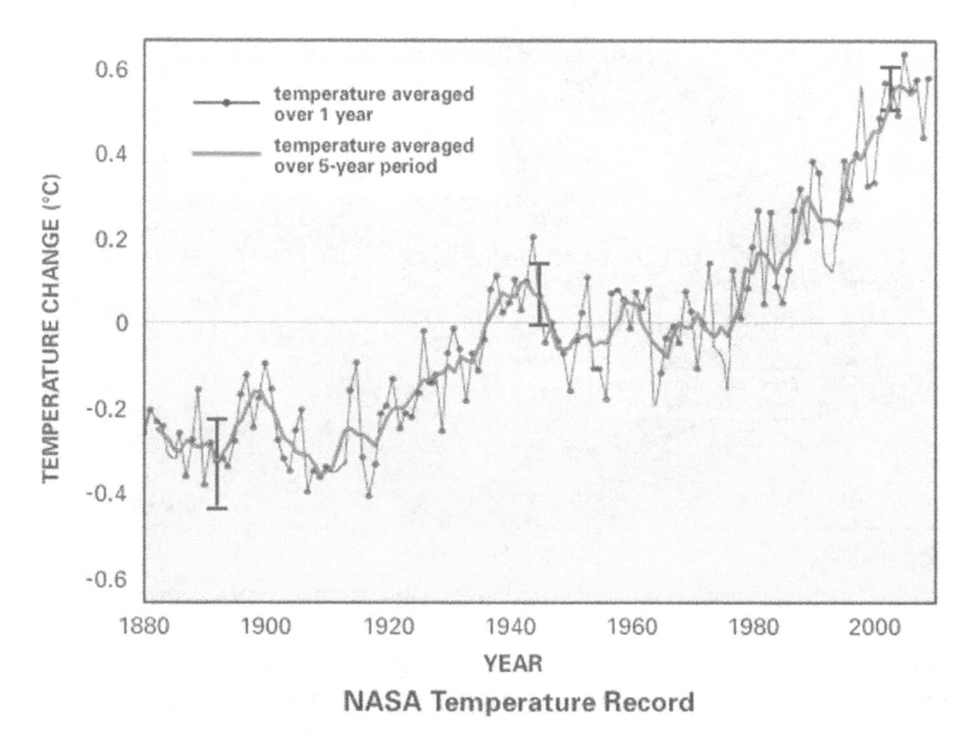

FIGURE 13.7
Observed global averaged temperature as a departure from the 1951–1980 average.

In response to observed climate changes, there has been an outcry from a number of international organizations as well as the scientific community to reduce and shortly thereafter curtail fossil fuel burning. This may, at least at first, involve necessary reductions in energy consumption, until sufficient non-carbon-based energy sources are developed.

There has been a great deal of confusion in the public realm, particularly in the U.S., about the causes of observed twentieth-century warming (and twenty-first), and the role of human activities in the climate system. This confusion has arisen from the fact that there are a great many factors that control observable climate, including, but not limited to, sunspots, ocean currents, cycles such as ENSO, other regional effects, volcanic activity, land use changes, marine biological activity, and many others. These complexities are the realm of Earth System Science, a very complicated business, indeed. It has taken generations of scientific investigation just to begin to understand what is involved, but some headway is being made.

Arctic Ice

An alarming observation in recent years is the deterioration of sea ice in the Arctic Ocean. While it has been long understood that global warming would affect the Polar Regions more profoundly than the tropics, no prediction even came close to predicting the rate of Arctic ice disintegration that has been

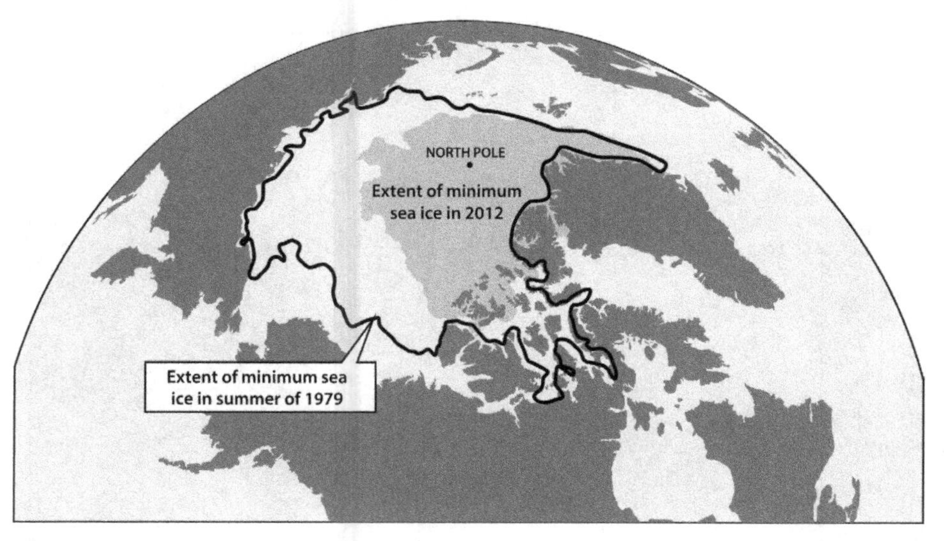

FIGURE 13.8
Arctic sea ice has dramatically declined in the last several years. This triggers a strong positive feedback from ice albedo (near 1) and exposed ocean water (albedo near 0). The more that ice melts to expose water, the more solar energy is absorbed, melting more ice, exposing and warming more water, etc. The rate of Arctic sea ice loss was severely underestimated by the scientific community as reflected in the Fourth Report of the Intergovernmental Panel on Climate Change (IPCC).

observed in the last several years. The famous ice minimum of late September, 2007, was observed just a month before the monumental volume on climate change was published by the Intergovernmental Panel on Climate Change (IPCC), such that the volume was out of date before it was even released. It did not predict the huge decrease in Arctic ice that was observed a month before it was released, after years of research and production. The minimum ice in 2011 came close, but 2012 surpassed all (Figure 13.8), setting a new record for minimum ice (Figure 13.9). In 2012, there was very little multi-year ice remaining. In March, 2017, the winter (maximum) extent of Arctic ice fell to its all-time minimum breaking records for the third year in a row, and at the same time, Antarctic sea ice fell to its all-time minimum, making 2017 the year of the least (by far) sea ice globally ever observed. Almost the entire Arctic will now be covered by first-year ice, meaning new, thin, recently formed ice, rather than thick leftover ice that has persisted for many years. This has never been observed in recorded human history.

The global impact of declining Arctic ice stems from the strong **ice-albedo effect**, causing incident sunlight to be absorbed by an ocean with albedo near zero rather than reflected by snow and ice with albedo near one. This positive feedback has already set in, and we may see further declines in summer ice and an extended ice-free season in the Arctic, with thinner and thinner winter ice in future years.

Ice-albedo effect: The strong feedback between snow/ice cover and solar heating. The sun reflects away from high-albedo snow and ice, thus not heating it, so snow makes it colder, keeping things cold. If there is a reduction in snow/ice cover, this heats the low-albedo land surface or ocean water much more, thus heating more, reducing the snow/ice cover, and heating more. This positive feedback creates an instability that can lead to rapid loss or gain of ice. Currently we are losing Arctic ice very quickly.

TWENTY WARMEST AND COOLEST YEARS ON RECORD (°C ANOMALY FROM 1901–2000 MEAN)

The tables below are sorted by temperature anomaly, or the amount by which the average temperature of each year exceeds the average for the entire century. Note that the warmest years (left) are recent, while the coolest years (right) were a century ago. 2015 shattered the record for the warmest year on record by the greatest margin in instrumental history. This was in turn exceeded by a similar amount in 2016. This table is updated every year, and every year, the most recent year must be added to the list of warmest years, deleting an older twentieth-century cooler year, but the list of coolest years has never needed to be updated. Data from NOAA.

TABLE 13.1 **A) 20 HOTTEST YEARS. B) 20 COLDEST YEARS.**

a		b	
2016	0.99	1909	-0.4179
2020	0.98	1908	-0.4144
2019	0.95	1911	-0.4131
2015	0.85	1904	-0.3999
2017	0.84	1910	-0.3999
2018	0.83	1907	-0.3702
2014	0.69	1912	-0.3544
2005	0.6183	1893	-0.3459
2010	0.6171	1903	-0.3453
1998	0.5984	1913	-0.3340
2003	0.5832	1917	-0.3333
2002	0.5762	1894	-0.3095
2006	0.5623	1892	-0.3087
2009	0.5591	1890	-0.3082
2007	0.5509	1916	-0.2949
2004	0.5441	1905	-0.2759
2001	0.5188	1891	-0.2654
2011	0.5124	1898	-0.2585
2008	0.4842	1887	-0.2516
1997	0.4799	1895	-0.2503

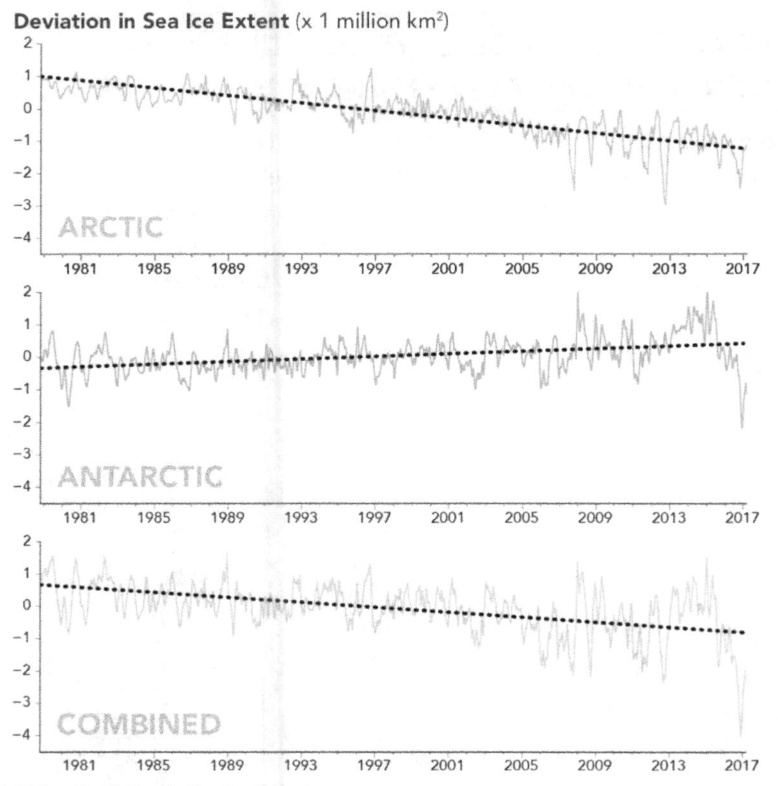

FIGURE 13.9 Deviation in Sea Ice Extent
Historical variation of sea ice in the Arctic Ocean and surrounding Antarctica. The minimum of Arctic sea ice cover each year is observed in September, and a dramatic minimum occurred in 2007. In 2008, some thought that a partial recovery may have occurred, but in 2010, there was another dramatic decline, and in 2011, the areal extent was almost as small as in 2007. The all-time minimum so far was in September, 2012, with a slight increase in 2013. It is noteworthy that record lows have never been set two years in a row (after each record low, there is a slight rebound). This may be caused by the albedo of open water vs. ice. Water's very low albedo absorbs almost all energy (like a "**black body**"), but it also emits energy much more readily than high-albedo ice. So in the dark winter of a minimum ice season, all that open water radiating to space loses more energy than would be lost by water insulated conductively, and especially by radiation, by ice. Note that the minimum in September is followed by an uptick in March ice cover the following calendar year. This would lead to greater ice cover the next September as well, so you don't see two record-low Septembers in a row. However, the long-term trend continues to decline. This has led to projections of an ice-free Arctic during the summers in as little as 10–15 years. Due to the severity of the ice-albedo positive feedback, such projections are difficult to quantify, and have been found to underestimate rates and magnitudes when applied to other Earth System parameters. This feedback may also be causing the average age of Arctic ice to be declining, with more first-year ice, and less multi-year ice each year. On that basis, we would expect the more rapid decline in summer ice than in winter ice, as can be discerned from these figures from the National Snow and Ice Data Center. However, on February 25, 2018, in the deepest and darkest dead of polar winter, there was a thaw at the North Pole, during which it reached 35°F after not seeing the sun since last September. The all time lowest maxium sea ice extent (so far) was in March of 2018.. Note that until 2007, the time of maximum ice extent was in February, but since 2007, the (declining) annual maximum has been occurring later, in March. The situation in Antarctica had been less severe until 2017, when the trend of gentle increase dramatically reversed, making 2017 the year with the least global sea ice ever observed since reliable satellite observations started in 1979. From NASA (https://www.nasa.gov/feature/goddard/2017/sea-ice-extent-sinks-to-record-lows-at-both-poles).

Black Body: An object that absorbs and emits every wavelength of electromagnetic radiation perfectly (0 albedo). Black bodies heat in the presence of incident radiation (and cool in its absence) more quickly than objects that do not absorb and emit radiation as efficiently. The ocean is close to a black body, while floating ice is close to its opposite, a white body.

Drivers of Recent (and Near Future) Climate Change

Greenhouse Gas Emissions

The primary driver of recent (20th–21st century) climate change is anthropogenic emissions of greenhouse gases. These emissions have been dominated by CO_2 resulting from fossil fuel burning, but also include methane, and numerous other trace gases that contribute to climate warming due to trapping of outgoing infrared radiation.

The greenhouse effect has already been described in Chapter 12, and for the purposes of this chapter, it will suffice to say that when greenhouse gases are added to the atmosphere, a temperature increase cannot be avoided by any means. The correlation between observed greenhouse gas emissions and observed climate change is no accident. Despite delays in the response of the climate system to greenhouse gas forcing, most of the observed twentieth-century warming trend has been found to be due to anthropogenic emissions (Figure 13.8). Some climate contrarians maintain that greenhouse gas emissions have not altered global climate. Clearly this is wrong, but more importantly, the delayed response of the climate system to the twentieth-century emissions is not so severe that we have yet to observe associated warming. If warming had not yet started due to delayed response, all impact of greenhouse gas emissions would occur in the future, above and beyond what we have observed so far, leading to far more severe climate impacts than actually expected.

In 1996, the Second IPCC Report stated that "The balance of evidence suggests that there is a discernible human influence on global climate." In 2001, the Third Report stated that "There is new and stronger evidence that most of the warming observed over the last 50 years is attributable to human activity." In 2007, the most definitive report to date, the Fourth Assessment Report stated that "Most of the observed increase in globally averaged temperatures since the mid-20th century is very likely due to the observed increase in anthropogenic greenhouse gas concentrations." Most recently, in 2013, the Fifth Report stated that, "It is *extremely likely* that human influence has been the dominant cause of the observed warming since the mid-20th century." While the scientific conclusions have remained essentially unchanged since the early 1990s, the mounting scientific evidence clearly demonstrates the magnitude of the human influence on climate changes through greenhouse gas emissions, as exacerbated by land use changes.

CO_2-Water Feedback

A major area of confusion among the American public is the role of water vapor as a greenhouse gas. Anthropogenic greenhouse gases are being emitted to the atmosphere as a result of human activity—mostly agricultural and industrial. The greenhouse warming caused by these gases serves to trigger an increase in the most important greenhouse gas—water vapor. Water vapor is responsible for most of the 33°C of the total greenhouse effect on the earth. Because warmer air can hold more water vapor (think of hot, humid weather relative to cold, dry days), there is a strong positive feedback that forms a vicious cycle of warming, once the atmosphere is warmed a little by other means. During the glacial–interglacial climate oscillations, this external driver was orbital variations that warmed the atmosphere slightly, allowing the water vapor effect to take over and cause much more warming. Since industrial times, the trigger has been the anthropogenic greenhouse gases described here (mostly CO_2). By causing a very slight amount of warming directly, these gases enable the atmosphere to hold more water vapor, which in turn causes a great deal more warming, which, in turn, allows the atmosphere to hold yet more water vapor, leading to additional warming, etc. Some "climate contrarians" misunderstand the role of water and CO_2 and contend that because CO_2 is a mere trace gas, anthropogenic emissions have no significant effect on climate. When climate contrarians claim that CO_2 has no effect on climate because during glacial–interglacial cycles warming preceded CO_2 increase, they are missing the point that the initial warming was caused by orbital variations and then greatly amplified by the water vapor greenhouse effect. The warming of the atmosphere then warmed the ocean, which, when warmed, could not hold as much CO_2 in solution, so CO_2 exsolved from the water and went into the atmosphere, and this is recorded in ice cores. The additional direct greenhouse effect from the CO_2 itself would have been small relative to water vapor. However, the CO_2 NOW being added to the atmosphere (from fossil fuel sources) is sufficient to trigger the water vapor effect and lead to marked global warming and related climatic changes. Water has a very short residence time in the atmosphere, as it literally rains out frequently. CO_2, on the other hand, has a very long residence time, on the order of millennia, so its effects cannot be removed quickly, and we have only begun to feel the warming triggered by CO_2.

Some who have not studied the details of planetary energy balance may also fall victim to the **CO_2 saturation fallacy.** Some climate contrarians claim that CO_2 already absorbs all of the outgoing energy within its absorption spectrum (see Chapter 12), and thus promote the fallacy. In fact, absorption of infrared by CO_2 can only happen for discrete wavelengths, according to the energy levels of the vibrational modes of the molecule. Thus there is not enough CO_2 in the atmosphere to absorb all the outgoing infrared in the broad range of absorption. Near the ground the high temperature and pressure of the lower troposphere causes intermolecular collisions that serve to broaden absorption bands, but only slightly, still leaving large gaps for plenty of infrared wavelengths to slip through. Further, not only is CO_2 not saturated in the Earth's atmosphere (as it actually is on Venus), but even if it were saturated for infrared absorption as a whole, it is only the top portion of the atmosphere (that is even further from saturated) that emits radiation to space. Lower in

CO_2 saturation fallacy: The impression (by some who know just a little about science) that there is enough CO_2 in the atmosphere to absorb ALL outgoing infrared radiation in the entire range of the CO_2 absorption spectrum, so thinking that adding more CO_2 wouldn't make a difference. However, a molecule absorbs radiation of a specific set of wavelengths, and even in the high pressure of the lower troposphere, where the specific absorption bands are spread out due to collisions and interactions between molecules, there are wide gaps for radiation of not-quite-the-right-wavelengths to escape. So even the lower atmosphere is not saturated in the sense of absorbing all the infrared absorbed be that can by CO_2. Furthermore, the atmosphere is not like a sheet of greenhouse glass—rather, it absorbs and re-radiates infrared in all directions at every level of the atmosphere. So it is the concentration of CO_2 at the top of the atmosphere (where there is no water, which is limited to troposphere) that matters, and if you add CO_2, it will increase, thickening the "insulating" blanket of greenhouse gases surrounding the planet, necessarily causing warming at all levels of the atmosphere.

the atmosphere, CO_2 (and water) molecules absorb and then radiate infrared radiation in all directions (up AND down). So when you add CO_2 to the atmosphere, you get a thicker "blanket" for absorbing and radiating, with warming at all levels, while the level from which radiation escapes to space rises higher. Moreover, although water vapor is the most important greenhouse gas near the ground, the upper atmosphere is dry so water does not play a role there. However, CO_2 is well mixed throughout the entire atmosphere, and thus dominates the radiative balance at the top of the atmosphere where it controls planetary radiation out to space.

The problem is that the atmosphere should not be imagined as a single piece of matter, like the glass of a greenhouse for which the effect is named. As such, the "greenhouse effect" is not really an accurate view of what happens. When the ground (and water) surface of the earth emits infrared radiation, much of it is absorbed by the greenhouse gases in the air near the ground. These gases then re-radiate infrared in all directions, including up to overlying air and back down to the ground. This sets up a chain reaction of absorption and re-radiation that eventually reaches the top of the atmosphere. By adding CO_2 to the atmosphere, a deepening amount of it engages in absorption and re-radiation, thus effectively thickening the "blanket" of greenhouse gases on the planet and moves the location that the atmosphere radiates out to space to a higher altitude, leading to greater warming at lower levels in order to maintain radiative equilibrium.

Twentieth-Century Aerosols

In the mid-twentieth century, the industrialized nations of North America and Europe burned a great deal of high-sulfur coal, thus emitting large amounts of sulfur dioxide into the atmosphere. This formed tiny droplets, or aerosols, of SO_2 that served to reflect incoming solar energy back into space before reaching the ground. Ultimately, this cooled the lower troposphere (where thermometers reside on the ground) in these large regions, in addition to causing severe acid rain and surface waters in particularly susceptible areas. In response to the acid rain problem, technologies were developed to scrub out the sulfur from coal-fired power plant smokestacks, greatly reducing sulfur emissions. This removed the cooling effect of the aerosols as well. Many climate contrarians who wish to obfuscate the scientific discussion of climate change have cited the mid-twentieth century cooling during a time of rapid fossil fuel burning as "proof" that climate change is not caused by CO_2 emissions from fossil fuel burning. However, they do not understand (or admit) the local cooling effect of aerosols, which was temporary in any case. Now that we do not have that cooling effect, climate is warming everywhere, including on the ground in industrialized nations.

An additional source of aerosols is specific volcanic eruptions. In a notable 1991 eruption of Pinatubo, tons of sulfur (as SO_2) erupted into the atmosphere, and this had a temporary cooling effect on global climate. For a couple of years, temperatures were measurably cooler than in previous years. However, like industrially produced aerosols, they quickly rained out, and climate was restored to the normal of the time. Although sulfur aerosols have a temporary cooling effect on climate, this should not be misconstrued to suggest that globally, we can effectively add aerosols to the atmosphere in order to cool surface temperatures, even locally. Nevertheless, there is some controversial discussion regarding "geoengineering", to be explored in Chapter 14. The acid rain produced is a deadly impact of sulfate aerosols in the troposphere, as experienced by the many dead lakes in the Adirondack Mountains of New York and many other places. Aerosols in the dry stratosphere do not rain out as quickly, so alter climate a bit longer and globally, but still temporarily. Further, the cooling effect of aerosols functions only during the daylight hours, while the global warming effect of CO_2 operates 24/7, as the earth emits infrared all the time. Consequently, aerosols cannot be used to effectively mitigate the climate changes triggered by anthropogenic CO_2 emissions.

Carbon and Climate

Because anthropogenic CO_2 plays such an important role in the climate system, the carbon cycle has become a critical area of research for climate scientists. As discussed in Chapter 6, the natural carbon cycle includes short- and long-term sources and sinks, such as annual northern hemisphere spring bloom, weathering of silicate rocks, the solubility and biological marine pumps, subduction of carbonate sediments, and CO_2 emission from volcanic eruptions. These processes, while cycling vast quantities of carbon, maintain a steady state, such that the CO_2 that resides in the atmosphere is kept at reasonably constant concentrations, varying between glacial and interglacial times between 180 and 280 ppm. With the advent of fossil fuel burning, we have removed enough geologically stored carbon from deep reservoirs and thus released enough CO_2 to raise atmospheric concentration to 410 ppm, and while it had been rising at a rate of about 2 ppm per year, it has been accelerating and is currently closer to 3 ppm per year. This has suddenly brought atmospheric chemistry out of equilibrium with global climate, ocean temperature and circulation, biome distribution (although this is also perturbed by direct land use), and ice cover. The earth may not have ever seen such a severe level of disequilibrium—certainly not as long as humans have existed as a species, and the process of re-equilibration will involve some fundamental shifts in the operation of the Earth System, including (in addition to general warming), storm intensification, glacial melting, precipitation changes (with wet places getting wetter and dry places drier in general), and perhaps most significantly, **ocean acidification**, and **sea level rise** as discussed in Chapter 9. Such rapid CO_2 increase with such cold ocean water makes it possible for the ocean to dissolve greater concentrations of CO_2 than in previous times of high

Ocean acidification: The reduction of pH of ocean water by solution of excess atmospheric CO_2 through the solubility pump. Historical pH was 8.2, but it is declining. If it falls below 8.0, marine organisms will have difficulty making calcium carbonate shells, thus threatening the biological pump, as well as the marine ecosystem. With a cold ocean, as present, and with sudden high atmospheric CO_2 concentrations (as we are tending toward due to fossil fuel burning), more CO_2 can be dissolved in the ocean than would naturally be possible, leading to concerns regarding the marine ecosystem.

atmospheric CO_2 because the ocean has not had time to warm up yet in response to global warming, and thus can hold more carbon, and thus become more acidic than perhaps ever before. Coastal communities have recently noticed the problem of sea level rise, and adaptation measures are already being taken, yet mitigation measures are lagging behind. This will be explored in Chapter 14.

The burning of fossil fuels removes ancient carbon from geological storage and transfers it to the atmosphere. Carbon incorporated into living tissue includes all isotopes of carbon, including C-14. While C-12 and C-13 are isotopically stable, C-14 is created in the atmosphere by absorption of a neutron by nitrogen N-14. These neutrons are created by cosmic rays (mostly protons and some alpha particles) interacting with the atmosphere (mostly near the tropopause). The neutrons then strike and are absorbed by a N-14 nucleus, in which a proton essentially turns into a neutron, and the nucleus becomes C-14. C-14 is not a stable nucleus, and decays back into N-14 with a half-life of about 6,000 years (actually 5,730 years). Consequently, after tens of millions of years underground, there is virtually no C-14 left in oil, coal, or natural gas, which remains composed only of C-12 (99%) and C-13 (1%). Enough fossil fuels have already been burned and C-14-free carbon released to the atmosphere to dilute the atmospheric concentration of C-14 which is become further reduced as more C-12 and C-13 are added to the atmosphere. This is called the "Seuss Effect."

Interestingly, another source of C-14 is nuclear bombs, and a major pulse of C-14 doubled the concentration of C-14 in the atmosphere in the 1950s and early 1960s during bomb testing, after which the nuclear test ban treaty went into effect. Enough C-14 was added at that time to serve as a marker in living tissue that incorporated it. The global ecosystem is incorporating that C-14 at a rate that halves the excess C-14 every 11 years or so. In several decades, atmospheric C-14 concentration will be back to the level that is in equilibrium with its formation by cosmic ray-generated neutrons.

A key issue regarding anthropogenic perturbation of the carbon cycle is the expected emissions in the coming decades. Regarding this, the International Panel on Climate Change (IPCC) developed a number of emissions scenarios and the climate impacts of each of these. They include low (B1), medium (B2), and high (A2) emissions. A discussion of what the world would look like under the various IPCC scenarios may become moot, however. The scenarios were defined almost 20 years ago, and sufficient time has passed already to observe that we are actually going along the path of the extreme emissions case of the business-as-usual A2 case. In 2019, about 37 billion metric tons of carbon dioxide were emitted by human activities (equivalent to about ten gigatons of just carbon—do the stoichiometry), near the old IPCC A2 scenario (or the more recent IPCC RCP8.5 scenario—see IPCC section below). If we continue along this path, only the higher temperature, sea level, and other projections will be relevant in the coming decades.

Impacts of Climate Change

Temperature

Temperature is the primary parameter one thinks of in discussions of climate change (Figure 13.10), but it is by no means the only change, and, in fact is of less real concern than various other parameters that are driven by changing thermal structure and energy balance of the planet. Indeed, temperature variations between glacial and interglacial cycles were only a few degrees. It changes more than that every day and night, and much, much more than that seasonally. People would not notice an average temperature change of the expected four or five degrees (F) over the course of the twenty-first century, when it changes much more than that every day and night. It is the impacts of other parameters that are very sensitive to temperature that are of greater direct concern.

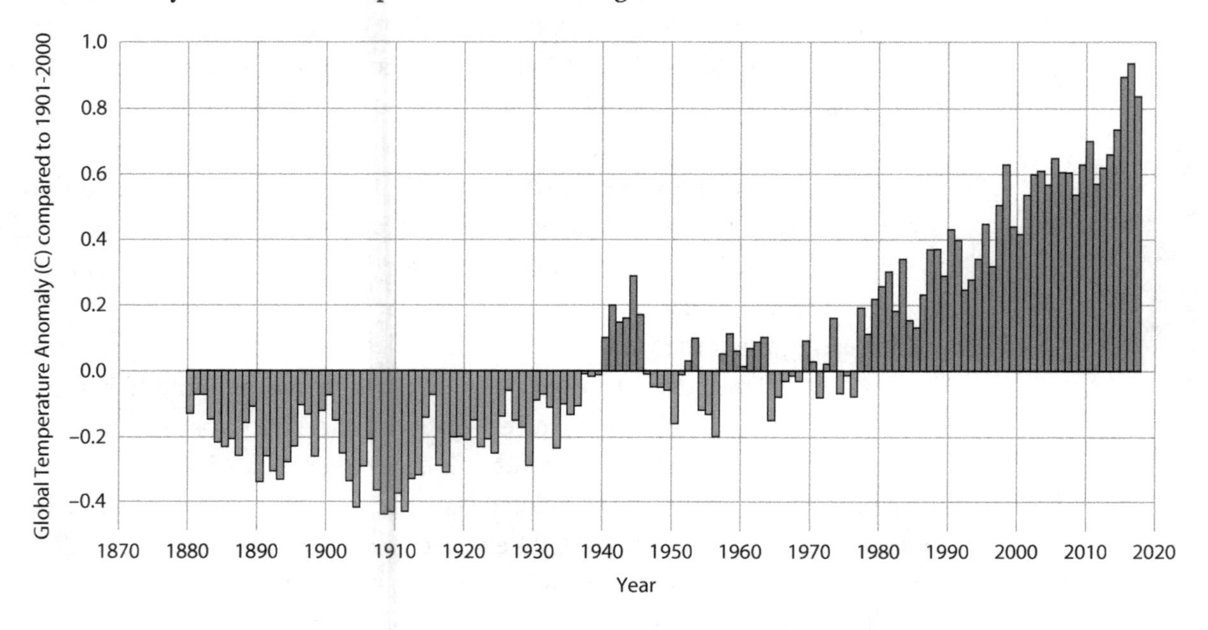

FIGURE 13.10 Average global termperature anomalies
Post-industrial temperature record, relative to the twentieth-century average. Note rapid warming in the late twentieth century. The warming trend that started in the early 1900s was interrupted by the emission of sulfur aerosols in mid-century that led to acid rain and were subsequently halted by using scrubbers in coal-burning power plants, after which warming quickly caught up to its twentieth-century trend. Data from NOAA.

Precipitation Patterns

Climate models indicate that the changes in precipitation patterns will depend on the extent of warming caused by different potential scenarios of anthropogenic greenhouse gas emissions. In a business-as-usual scenario (A2 or RCP 8.5 of IPCC), severe changes should be expected, with relatively dry places generally getting drier and relatively wet places getting wetter (Jiang et al., 2016; felzer and Sahagian, 2014). The implications of this are more droughts where agriculture may already be water-limited, and more floods where additional water is not needed. In a scenario in which emissions are rapidly curtailed (B1 or RCP 2.6 of IPCC), this effect would be greatly reduced (but still present). As precipitation patterns change in the twenty-first century, adaptation measures will need to be taken to ensure that food production can be maintained, and that areas subject to additional flooding and droughts bolster infrastructure and community organization to reduce vulnerability.

Storms

Hurricanes: Cyclonic (anticyclonic, spinning counterclockwise in the northern hemisphere) storms triggered by tropical atmospheric instabilities in the absence of wind shear, and fueled by the latent heat of evaporating seawater.

Large cyclonic storms such as **hurricanes** are driven by energy entrained by the evaporation of warm sea surface water (Figure 13.11). (This is why they happen most frequently in late summer and fall, when ocean temperatures are highest.) The warmer the water, the more energy there is to fuel the storm. Global warming involves heating of the ocean as well as the atmosphere (in fact the ocean water is far more important due to its heat capacity), so we should expect more serious storms as a result. Global warming does not make hurricanes. They occurred in pre-industrial times.

FIGURE 13.11 Storm

However, ocean temperature plays a strong role in determining the nature of hurricanes. There has been considerable discussion in the scientific community regarding the impact of climate change on large storm systems (e.g. Romero and Emanuel, 2017). It has now been reasonably well established that warming will not spawn more storms, but that the once the cyclonic flow is generated in the tropical North Atlantic, for instance, with increasingly warm water to fuel them, the storms will grow more powerful and maintain their strength longer as they travel farther north. Coastal cities will be braced for more severe impacts, as seen in New Orleans in 2005 (Katrina; Rita), New York in 2012 (Sandy), and Houston and Florida in 2017 (Harvey and Irma).

Sea Level Rise

One of the most troubling impacts associated with climate change is its effect on sea level (Jacobs and Sahagian, 1993). Sea level has varied by hundreds of meters over geologic time, and is rising rapidly today (Figure 13.12). Observations of twentieth-century sea level rise have been made from tide gauges in the world's harbors. Although originally installed to measure the diurnal and semi-diurnal times for shipping purposes in the ports, over the long term, it was noticed that the average level of the tides was rising—sea level rise. After correcting some locations for tectonic motions, including delayed vertical rebound of the earth's crust due to deglaciation, a twentieth-century rate of sea level rise was observed to average about 1.5–1.75 mm/yr. In the second decade of the twenty-first century, we now have a long enough satellite altimetry record to observe that the rate of sea level rise has increased to over 3 mm/yr (Nerem et al., 2018; Legeais et al., 2018). This is discussed in detail in Chapter 9, and is one of the most dramatic impacts of global climate change (Sahagian, 2000). With the rate of sea level rise increasing from between 1.5 and 1.75 mm/yr in the twentieth century, to more than 3 mm/yr and accelerating in the twenty-first century, there is concern for the security of many of the world's coastal cities. Problems will be felt first during storms, such as Hurricane Sandy that struck New York and New Jersey. There is a double whammy involved here. As storms become more intense (but not more frequent), the increased storm surge and wave heights will act on top of an already higher base level due to sea level rise. When the storm surge is over a couple of meters, every additional centimeter counts in terms of flooding large areas, inundating and filling tunnels, subways, and other infrastructure, and the rate and extent of transport of coastal sediments such as the sands of beaches.

FIGURE 13.12 Sea level rise

Ocean Acidification

Anthropogenic CO_2 emissions do more than contribute to the greenhouse effect and cause global warming. As explained in Chapter 9, in order to maintain chemical equilibrium between the ocean and the atmosphere, some of the excess atmospheric CO_2 enters solution in the ocean. This leads to lower pH and makes it more difficult for marine organisms to make shells out of carbonate dissolved in the water (Matear et al., 018; Carpenter et al., 2018). If this ocean sink (the biological pump) weakens, more emitted CO_2 will remain in the atmosphere, contributing further to greenhouse warming. As ocean water warms in response, CO_2 becomes less soluble, slowing (and potentially reversing) the solubility pump as well. There is evidence that the marine carbon sink is already weakening. There has not been a great deal of concern to date regarding ocean acidification, but will become a topic of increasing interest in the coming years.

Ecosystem Disruption

Climate change has a variety of impacts on ecosystem function and stability. The main perturbations involve changing storm tracks and intensity, shifting patterns of pestilence and disease vectors, altered precipitation, changing phenology, and ocean acidification. These disruptions are directly caused by climate change, and act in addition to deforestation (land use), invasive species, and pollution. Storms have always occurred, but have been confined to specific regions and tracks. Altered atmospheric circulation patterns shift the locations and paths of storms, and warming sea surface temperatures enable tropical storms to become more intense and travel to higher latitudes than previously in human history. The ecosystems that have not evolved in the presence of such storms are not resilient and suffer major damage from which it is difficult to recover.

As climate warms insects and other pests that have been confined to lower latitudes can migrate into regions whose plants have no defenses against them, leading to major biotic disruption such as in the case of the bark beetle in North America. This has already decimated millions of acres of norther pine forests in the U.S. and Canada.

Although climate change is often associated with global warming, the main concern for ecosystems is water availability through precipitation. As precipitation patterns change plants may no longer be suited to the alterations, thus reducing productivity, increasing vulnerability to invasive species, and reducing the ability to recover from other disruptions such as storms. Because plants are the base of the ecosystem (primary producers), all animals depend on their productivity. Climate change may be exacerbating the "sixth mass extinction" being caused now directly by human activities such as deforestation. While if you were a tree, you would probably be less afraid of climate change than you would be a chainsaw, the impact is more keenly felt in ecosystems when both climate change and saws are present.

Many plants in temperate climates lose their leaves in the fall, with new buds in the spring. The timing of these seasonal life cycles, or **phenology**, depends on a number of factors, including length of daylight (not changing on human timescales), and seasonal variations in precipitation and temperature (changing rapidly). Consequently, trees are holding onto their leaves longer in fall and budding sooner in spring. In the fall, this has already left trees full of leaves when heavy snow has fallen in late October, and caused extensive damage in the northeastern U.S. in 2011. (It also caused broken branches to break power lines with widespread power outages.) In spring, leaves bud out earlier as a result of warmer weather, and insects begin to graze on those buds and young leaves. However, the migratory birds which rely on those insects to eat and feed their young are not yet there, or eggs have not yet hatched by the time the insects have already moved on. This type of mismatch of timing disrupts ecosystem function.

Ocean acidification was discussed in Chapter 9, and is only mentioned here because it may become a fundamental cause for major marine ecosystem disruption as pH reduces further as a result of continued fossil-fuel-derived CO_2 emissions.

Societal Impacts

When you hear about climate change impacts in the news, they are typically talking about how it affects people through disruption of built infrastructure. (Preston, 2015; Crutzen, 2002). One place this is keenly felt is along coastlines as a result of sea level rise and exacerbated storm damage. As sea level rises, the damage caused by storms such as Hurricane Sandy that struck New York and New Jersey in 2012 is worsened as storm surge and wind-driven waves reach higher elevations, flooding homes, businesses, roads, and in particular, machinery and wiring in buildings, vehicles, subways, and underground that quickly degrades in the presence of salt water. This combines two impacts of climate change both related to warming of ocean temperatures. The first is the expansion of ocean water that contributes to sea level (along with glacial melting). The second is the provision of heat energy to hurricanes as they travel poleward from tropical oceans. Warmer temperatures enable tropical storms to become more intense and also travel farther toward high latitudes, impacting regions that have not seen such storms in human history.

A general trend in global climate change is that wet places are becoming wetter and dry places becoming dryer. In many parts of the world, agriculture depends strongly on precipitation, and control of water resources becomes a contentious issue. For example, the Colorado River was a point of contention between the western states and Mexico throughout the twentieth century, and the Colorado River Compact had to be negotiated to allocate the flow of the river between the states and still ensure that sufficient water reaches Mexico for their own needs. This semi-arid region is now experience reduction in both rainfall, and snowpack in the mountains, and the level in dammed reservoirs as well as the river is decreasing. This will have serious implications on the ability of the region to sustain agriculture that depends on the extraction of water from the river.

Meanwhile, in the Middle East, reduction in precipitation and streamflow is leading to mass migrations of people who can no longer grow sufficient food with the little remaining water available. This is leading to "climate refugees", most notably recently in Syria, but expected to intensify throughout Africa and the rest of the world. People typically first migrate from rural regions of subsistence agriculture to cities in hopes of finding work and food. Finding none there, they emigrate as refugees to other nations—mostly Europe in the case of Syrian refugees.

The impact of climate change on social systems is not limited to the direct effects of water, storms, and migrations. There are more broadly felt economic issues as well. As coastal cities suffer increasing damage, and are eventually inundated, extreme costs will be incurred throughout reginal and national economies. This brings into question the relative cost of reducing the rate of climate change (mitigation) and coping with its impacts (adaptation). This will be explored in detail in Chapter 14.

The Intergovernmental Panel on Climate Change (IPCC)

In response to the potential threats climate change poses not only to human systems, but also to the functioning of all aspects of the global ecosystem, the governments of the world formed the Intergovernmental Panel on Climate Change (IPCC). This body serves to bring together scientists from all over the world to address the difficult science of understanding the climate system and predicting climate change for the coming century and beyond. IPCC was established by the World Meteorological Organization (WMO) and the United Nations Environment Programme (UNEP) to assess scientific, technical, and socio-economic information relevant for the understanding of climate change, and its potential impacts and options for adaptation and mitigation. It is open to all members of the UN and of WMO. It bases its assessment on peer-reviewed and published scientific/technical literature. The goal of IPCC is to use scientific literature to evaluate the extent and understanding of climate changes, as well as the potential to adapt to or counteract climate changes.

The IPCC organizes and reviews the published findings of the world's leading scientists, convenes workshops, produces detailed reports on various topics, and has provided five major Assessment Reports starting in 1990. Although the scientific community's understanding of the climate system progressed a great deal in the intervening two decades, the projections and predictions have changed remarkably little since the early years, indicating that even then, enough was known about the climate system for governments to begin to take action to mitigate future climate change.

IPCC reports were published in 1990, 1995, 2001, 2007, and 2013. While earlier reports were less detailed and based on fewer years of scientific research, they all resulted in the same basic conclusions regarding observation, attribution, and projection of climate changes. However, most

BASIC QUESTIONS OF CLIMATE CHANGE

Despite the complexities of global climate change, there is one aspect of Earth's climate that is relatively simple to understand, and that is the balance of energy between the sun and the earth's surface. Sunlight comes through a fairly transparent atmosphere and heats the surface; that heat is radiated back to space as infrared radiation. However, the atmosphere is not quite as transparent to infrared, so absorbs this energy, warming the atmosphere. This process is the greenhouse effect. If we make the atmosphere more or less transparent to infrared radiation by adding or subtracting certain gases, we can cool or warm it directly. These effects are very simple physics, and are well understood, explained in introductory textbooks such as this one. Further, CO_2 is a trace gas that warms the atmosphere, thus enabling it to hold more water vapor, the dominant greenhouse gas. So CO_2 acts as a trigger to load more water vapor, which warms more, thus holding more water, warming more. On top of that, because warm water cannot hold in solution as much CO_2 as cold water can (try opening a warm seltzer!), warming atmosphere warms the ocean, which then could begin to exsolve some of its CO_2 into the atmosphere, warming further. This runaway feedback between CO_2, temperature, and water vapor is the primary concern regarding CO_2 and other greenhouse gas emissions. The details of these processes are explained in the comprehensive report of the Intergovernmental Panel on Climate Change (IPCC, 2007), which summarizes and double checks the scholarly publications of the international scientific community in area of climate change. While there may be a great deal of discussion within the scientific community about the details of the interactions between the various components of the Earth System, the basic physics of radiation and absorption has been settled for over a century. In sum, if CO_2 or other greenhouse gases are added to the atmosphere, it must get warmer.

Regarding anthropogenic climate change there are thus five basic questions:

1. **Is climate changing?** Yes. There are undeniable instrumental data that temperatures are rising, precipitation patterns are changing, and ocean and atmospheric circulation systems are changing throughout the twentieth and twenty-first centuries.
2. **Do people have anything to do with it?** Yes. Greenhouse gas emissions (primarily CO_2 from fossil fuel burning) have to warm the atmosphere—it is what they do. The consensus of model results shows that the global climate is sufficiently sensitive to historic anthropogenic CO_2 emissions to have already warmed by the amount measured over the last 150 years.
3. **Is climate change bad?** Yes. While this is a more normative question to be considered by philosophers and the general public rather than by scientists, history has shown that any change in the environment of stable civilizations is disruptive to those civilizations. Alterations in areas in which crops can be grown, changes in phenology (when plants bloom or flower and leaves fall, and when insects emerge, etc.), shifting storm tracks, and rising sea level may have devastating economic, social, and political consequences to modern societies.
4. **Can we do anything about it?** Yes. Because much of the warming caused by past emissions has already occurred, cessation of emissions can stabilize climate in the twenty-first century. Until they are overwhelmed, natural carbon sinks in the ocean and terrestrial ecosystems can continue to absorb previously emitted carbon and return global climate to the stable state in which civilization evolved over the last 10,000 years.
5. **Is it worth doing anything about?** Yes. Economic analyses indicate that the cost of adaptation to climate change in the form of agricultural disruptions, damage to coastal cities, and impacts of extreme events will be much greater than the cost of mitigation by transition to sustainable energy sources.

What we learn from the past is that nearly every major climate change in Earth's history has been accompanied by changes in greenhouse gases, with warming associated with more CO_2 and cooling associated with less. In the geologic past, before humans existed, climate and atmospheric CO_2 concentrations varied together, with CO_2 change not always predating climate change. This was due to the runaway feedbacks between temperature, CO_2 in the atmosphere and ocean, and water vapor in the atmosphere. However, now that we have devised a way to inject CO_2 directly into the atmosphere (fossil fuel burning), CO_2 is preceding climate warming, which is already responding to the additional greenhouse gases.

predictions of the previous reports have proven to be underestimates of both the rates and magnitudes of climate change and its impacts on ice cover, sea level, ecosystems and other aspects of the Earth System. The Fourth Assessment Report, published in 2007, included greater detail than any previous report, and gained the notice of more people than any report that preceded it. In addition, it served as the basis for the book and documentary film *An Inconvenient Truth*, which brought the science of climate change into more homes and discussions than had all the scientists in the world for the preceding 20 years. For this, and for the continuing exhaustive work of the IPCC and all the scientists involved, Al Gore and the entire IPCC were awarded the Nobel Peace Prize in 2007. (Al Gore also received an Oscar for the film in the same year.) In the past, some Nobel Prizes were awarded for work already completed and problems solved. In this case, however, the scientific community is still in the process of identifying the extent of the problems associated with climate change, and active research continues in the areas of understanding drivers, responses, and interconnections within the climate system, impacts of climate change on terrestrial and marine ecosystems, and implications for alterations in climate and ecosystems for human society.

The Fifth Assessment Report of the IPCC was released in fall, 2013. The basic conclusions are the same as previous IPCC reports, but uncertainties have been reduced, and some twenty-first-century predictions of rates and amounts of change for temperature, regional variability, and sea level rise have been increased as our models and observations have improved in recent years. In 2018 IPCC released an extra report indicating that the impacts of a global temperature increase of 2 °C would be significantly more severe than the impacts of a 1.5 °C increase, and that a revised goal for global climate change mitigation should be 1.5.°C. While the basic scientific conclusions may not differ greatly from the previous reports, the added evidence and reductions in uncertainty of predictions of changes and responses of the various parts of the Earth System may help steer the governments of the world who have organized the IPCC to take the most effective action possible to both mitigate climate change and adapt to its impacts. The general strategy is to mitigate what you can, and adapt to the rest. In the simple words of the UN Scientific Export Group Report on Climate Change and Sustainable Development, and later the Union of Concerned Scientists, "Avoid the unmanageable so that you can manage the unavoidable."

The following sections are provided as background for understanding the science of climate change. The most comprehensive and up-to-date treatment of climate change is the 2013 report of the IPCC. It was written in several parts, the most relevant of which is the report of Working Group I, "The Scientific Basis." IPCC also published a "Technical Summary," and a "Summary for Policymakers," both of which provide a more digestible synopsis of the findings of the international scientific community than the full report. The entire report is several thousand pages long; even the first part on "The Scientific Basis" weighs in at over 2000 pages, providing plenty of reading for those who wish to explore every detail of the science of climate change. Rather than reproducing this landmark work within this introductory textbook, because it is free for all to access online, the links are provided here for students to obtain the relevant parts of the reports directly from IPCC, starting with the "Summary for Policymakers" and the "Technical Summary" of Working Group I.

The first reading in this section is the "Summary for Policymakers" found at

http://www.ipcc.ch/report/ar5/

The summary provides a short introduction to the science of climate change as well as a few of the key results that are policy-relevant.

With the publication of the 2013 AR5, the 2007 report became out of date, although the basic science remains unchanged. For example, in the 2007 report, it was indicated that 11 of the 12 years between 1995 and 2006 ranked among the 12 warmest years since 1850, when instrumental records began to be reliably kept. In 2012, this was updated to "20 of the last 25 years (1987–2011) were the warmest years in the twentieth century." And now, in 2020, we see that of the warmest 18 years ever recorded, 17 of them were in the 21st century (and the 18th was in 1998!). The warmest year recorded so far was 2016, with 2019 a close second (Table 13.1). The warming trend continues unabated despite well-discussed but completely unimplemented measures to reduce greenhouse gas emissions and land use changes.

The next part of the IPCC to be read is the "Technical Summary," which provides the scientific basis for the conclusions summarized in the "Summary for Policymakers."

Those interested in a fully comprehensive treatment of the problem can obtain the full report of Working group 1 (and 2 and 3) at the IPCC website.

The National Climate Assessment

A more recent and equally authoritative (but not quite as comprehensive) report is the National Climate Assessment of the United States Global Change Research Program (USGCRP). This report covers similar topics as the IPCC reports and reaches similar conclusions, but is authorized by the U.S. government rather than an international collaboration. The report can be found, including a convenient Executive Summary, at

https://science2017.globalchange.gov/.

References

Bonacci, D., & Lacaze, B. (2018). New CO2 concentration predictions and spectral estimation applied to the Vostok ice core. *IEEE Transactions on Geoscience and Remote Sensing*, 56(1), 145–151.

Carpenter, R. C., Lantz, C. A., Shaw, E., & Edmunds, P. (2018). Responses of coral reef community metabolism in flumes to ocean acidification. *Marine Biology*, 165(4), 66.

Crutzen, P. J. (2002). The effects of industrial and agricultural practices on atmospheric chemistry and climate during the Anthropocene. *Journal of Environmental Science and Health Part A-Toxic/Hazardous Substances & Environmental Engineering*, 57(4), 423–424.

Felzber, B., & Sahagian, D. (2014). Climate impacts on regional ecosystem services in the United States from CMIP3-based Multimodel comparisons. *Climate Research*, 61(2), 133–155. doi:10.3354/cr01249

Frappier, A. B., Sahagian, D., Carpenter, S. J., Gonzalez, L. A., & Frappier, B. R. (2007). Stalagmite stable isotope record of recent tropical cyclone events. Geology. 35(2), 111–114.

Jacobs, D. K., & Sahagian, D. L. (1993). Climate-induced fluctuations in sea-level during non-glacial times. *Nature*, 361(6414), 710–712.

Jiang, M., Felzer, B., & Sahagian, D. (2016). Predictability of precipitation over the conterminous U.S. based on the CMIP5 Multi-Model Ensemble. *Nature, Science Reports*, 6(33618). doi:10.1038/srep33619

Legeais, J., Ablain, M., Zawadzki, L., Zuo, H., Johannessen, J. A., Scharffenberg, M. G., . . . & Benveniste, J. (2018). An improved and homogeneous altimeter sea level record from the ESA Climate Change Initiative. *Earth System Science Data*, 10(1), 281–301.

Matear, R. J., & Lenton, A. (2018). Carbon-climate feedbacks accelerate ocean acidification. *Biogeosciences*, 15(6), 1721–1732.

Nerem, R. S., Beckley, B. D., Fasullo, J. T., Hamlington, B. D., Masters, D., & Mitchum, G. T. (2018). Climate-change-driven accelerated sea-level rise detected in the altimeter era. *Proceedings of the National Academy of Sciences of the United States of America*, 115(9), 2022–2025.

Petit, J. R., Basile, I., Leruyuet, A., Raynaud, D., & Lorius, C. (1997). Four climate cycles in Vostok ice core. *Nature*, 387(6631), 359–360.

Pol, K., Masson-Delmotte, V., Cattani, O., Debret, M., Falourd, S., Jouzel, J., . . . & Stenni, B. (2014). Climate variability features of the last interglacial in the East Antarctic EPICA Dome C ice core. Geophysical Research Letters, 41(11), 4004–4012.

Preston, C. J. (2015). Framing an ethics of climate management for the Anthropocene. *Climatic Change*, 130(3), 359–369.

Romero, R., & Emanuel, K. (2017). Climate change and hurricane-like extratropical cyclones: Projections for North Atlantic polar lows and medicanes based on CMIP5 models. *Journal of Climate*, 30(1), 279–299.

Sahagian, D. (2000). Global physical effects of anthropogenic hydrological alterations: Sea level and water redistribution. *Global and Planetary Change*, 25(1–2), 39–48.

Saraswat, R., Nigam, R., Tiegang, L., & Griffith, E. M. (2017). Marine paleoclimatic proxies: A shift from qualitative to quantitative estimation of seawater parameters. Palaeogeography, Palaeoclimatology, *Palaeoecology*, 483, 1–5.

Scheff, J., Seager, R., & Liu, H. (2017). Are glacials dry? Consequences for paleoclimatology and for greenhouse warming. *Journal of Climate*, 30(17), 6593–6609.

Wong, C. I., & Breecker, D. O. (2015). Advancements in the use of speleothems as climate archives. *Quaternary Science Reviews*, 127, 1–18.

Figure Credits

Fig. 13.1: Source: https://commons.wikimedia.org/wiki/File:Climate_influence_on_terrestrial_biome.svg

Fig. 13.2: https://commons.wikimedia.org/wiki/File:Vostok_420ky_4curves_insolation.jpg.

Fig. 13.3: Copyright © Dragons flight (CC BY-SA 3.0) at https://commons.wikimedia.org/wiki/File:2000_Year_Temperature_Comparison.png.

Fig. 13.4: Source: https://climate.nasa.gov/vital-signs/global-temperature/

Fig. 13.5a: Source: https://commons.wikimedia.org/wiki/File:Eccentricity_zero.svg

Fig. 13.5b: Source: https://en.wikipedia.org/wiki/File:Eccentricity_half.svg

Fig. 13.5c: Copyright © Dna-Dennis (CC BY 3.0) at: https://commons.wikimedia.org/wiki/File:AxialTiltObliquity.png.

Fig. 13.5d: Copyright © Greg Benson (CC BY-SA 3.0) at: https://commons.wikimedia.org/wiki/File:Precession_and_seasons.jpg.

Fig. 13.6: Source: https://www.esrl.noaa.gov/gmd/ccgg/trends/full.html

Fig. 13.7: Source: http://climate.nasa.gov/interactives/warming_world/

Fig. 13.9: Source: https://www.nasa.gov/feature/goddard/2017/sea-ice-extent-sinks-to-record-lows-at-both-poles

Fig. 13.10: Source: https://www.climate.gov/news-features/understanding-climate/climate-change-global-temperature

Fig. 13.11: Source: https://commons.wikimedia.org/wiki/Storm#/media/File:Chaparral_Supercell_2.JPG

Fig. 13.12: Source: https://commons.wikimedia.org/wiki/Category:Sea_level_rise#/media/File:Sea_Level_Rise_(14227656790).jpg

CHAPTER 14

Global Environmental Change and the Earth System

On system components they pondered and stewed.
Their earth understanding was sketchy and skewed.
Separate parts, there were none.
They behaved all as one
And old Earth's emergent behavior ensued.

WE KNOW NOW that the natural environment behaves as a system, in which a fluctuation of one part affects many others. In this book so far, we have explored the various sub-systems of the earth that provide the environmental goods and services upon which society depends. What is the value of these goods and services? How would we fare in their absence? This depends on large part on what they are and how we use them. Some attempts have been made to quantify the value of the world's ecosystem goods and services. Bob Costanza, for example, calculated a total value of 33 trillion dollars per year for the total provided to humanity from the global ecosystem. This figure has raised some controversy, as has the concept of placing a dollar value on such goods and services in the first place. Those who object to the practice suggest that if you place a number on it, someone may feel entitled to "buy" it and thereby justify destroying that part of the ecosystem, along with any other parts that depend on it. It is unclear how to get the money used to "buy" (meaning destroy) ecosystem goods and services to the populations who do not even know they are "selling" their access to what the global ecosystem has always provided for their societies. As such, ecosystem valuation remains a contentious issue.

Anthropocene: A new and informal designation of a geological epoch to follow the Holocene, during which humans have had a major influence on Earth's environmental systems. It was suggested by Paul Crutzen at a conference, and then written up in an IGBP newsletter with Eugene Stoermer in 2000. It is suggested that the Anthropocene started at the time of the industrial revolution, but there are other options, and it is still an informal designation. (How do YOU think it should be defined, as a geological epoch?)

We also now realize that changes in the Earth System involve much more than merely climate change, and that global change encompasses alterations in climate, land cover, ocean chemistry, terrestrial and marine biome distribution and health, ice extent, soil composition, and perhaps now in a new **Anthropocene** era (Crutzen, 2002), a new understanding of humans and our role in the Earth System.

It is worth reviewing a few key observations of elements within the Earth System that have led to an appreciation of the importance of human activity in global change. One of the most telling is the CO_2 record from the Mauna Loa observatory, on the island of Hawaii in the middle of the earth's largest ocean, far from industrial sources.

Measurements since 1958 show an accelerating increase in atmospheric CO_2 concentration resulting from fossil fuel burning. However, due to the interactions within the Earth System, the full amount of emitted CO_2 has not remained in the atmosphere. Roughly a third has been dissolved into the ocean through the solubility and biological pumps, and a third has been absorbed by the regrowth of previously deforested northern hemisphere forests, leaving only about a third of the emitted CO_2 in the atmosphere to contribute to global warming and trigger the water-vapor positive feedback. As such, we have been enjoying fossil fuel burning at a discount in terms of greenhouse gases. The concern at a system level is how long this discount will remain. Observations of ocean uptake indicate that it is already beginning to weaken, as warming water cannot dissolve as much CO_2 as cold water. It is not clear that there is any reduction in the rate of terrestrial carbon uptake, but inevitably, as forests mature and regain carbon equilibrium, this sink will weaken as well, leaving a greater fraction of emitted CO_2 in the atmosphere to contribute further to accelerating greenhouse warming.

Another key observation is the climate record recovered from the Vostok ice core obtained from Antarctica. Because the bubbles in the ice lock in the atmospheric chemistry of the time at which the ice formed, a detailed climate history can be reconstructed for the last million years from Antarctic ice cores. It was found that during glacial–interglacial cycles that recur every 100 thousand years or so (driven by Milankovitch orbital variations), temperature, atmospheric CO_2, and atmospheric methane all varied together. CO_2 variations occurred in an envelope between 180 ppm (glacial) and 280 ppm (interglacial), and never strayed above or below. This stability was presumably maintained by interactions within the Earth System, perhaps in the marine carbon cycle, terrestrial ecosystems and soils, and perhaps other factors. However, in the last century, fossil fuel burning has brought the atmospheric concentration of CO_2 to more than 415 ppm, far outside the stability envelope. How the Earth System will respond to this disequilibrium is the subject of current research throughout the world. There has been no time in the past that the Earth System has been as far out of equilibrium as it is now, and its re-equilibration will surely bring on some "exciting" times.

Water is critical to all life on Earth, and the harnessing of water resources has been a key element in human civilization. Water is used mostly for irrigation for agriculture; in essence, we do not drink water resources—we eat them. In exploiting global water resources, society has appropriated more than half of all the available fresh water for the support and maintenance of the human population (Postel et al., 1996). This has been accomplished by digging or drilling wells, diverting surface water flows, and building dams. Because clean water is such a critical commodity, when we stress the availability of water resources, difficulties will ensue, not only for human civilization, but for global ecosystems as well.

FIGURE 14.1 Phytoplankton bloom

Another remarkable observation involves nitrogen fixation. This is the reduction of N_2 from the atmosphere by bacteria (often in root nodules of legumes and other plants, but also free-living in the soil), creating ammonium (NH_4), which is then oxidized by nitrifying bacteria, and through a series of biochemical reactions, converted back to N_2 and released to the atmosphere. The fixed nitrogen is very important for plant growth, and is a key component of agricultural fertilizers as a result. Through the **Haber-Bosch process**, atmospheric N_2 and hydrogen are combined at high temperature to make ammonia artificially. This is where industrially produced fertilizer comes from, and in the mid-1980s this process surpassed all of nature in all the world's soils and oceans in the amount of nitrogen fixed per year. As such, people now fix more nitrogen than all of nature. This excess fixed nitrogen is washed off of farms and into streams, leading to explosive algal and other growth. When such abundant life dies, decomposition uses dissolved oxygen and the resulting eutrophication kills fish and other animals in the waterways. The net effect is that current agricultural practice kills fish and otherwise disrupts not only terrestrial ecosystems, but aquatic and marine ecosystems as well (Figure 14.1).

In the development of human civilization during the early Holocene, an important food source transition was made from hunting-gathering to agriculture, enabling more sustainable and reliable support of an increasing human population. While agriculture has itself been developed to the point that it has profound environmental impacts at the Earth System level, any attempt to revert and feed the current world population by hunting and gathering would have immediately driven to extinction a large fraction of the world's animal and plant species with subsequent severe food shortages for the remainder of humanity. Yet, in the marine realm, we are still behaving

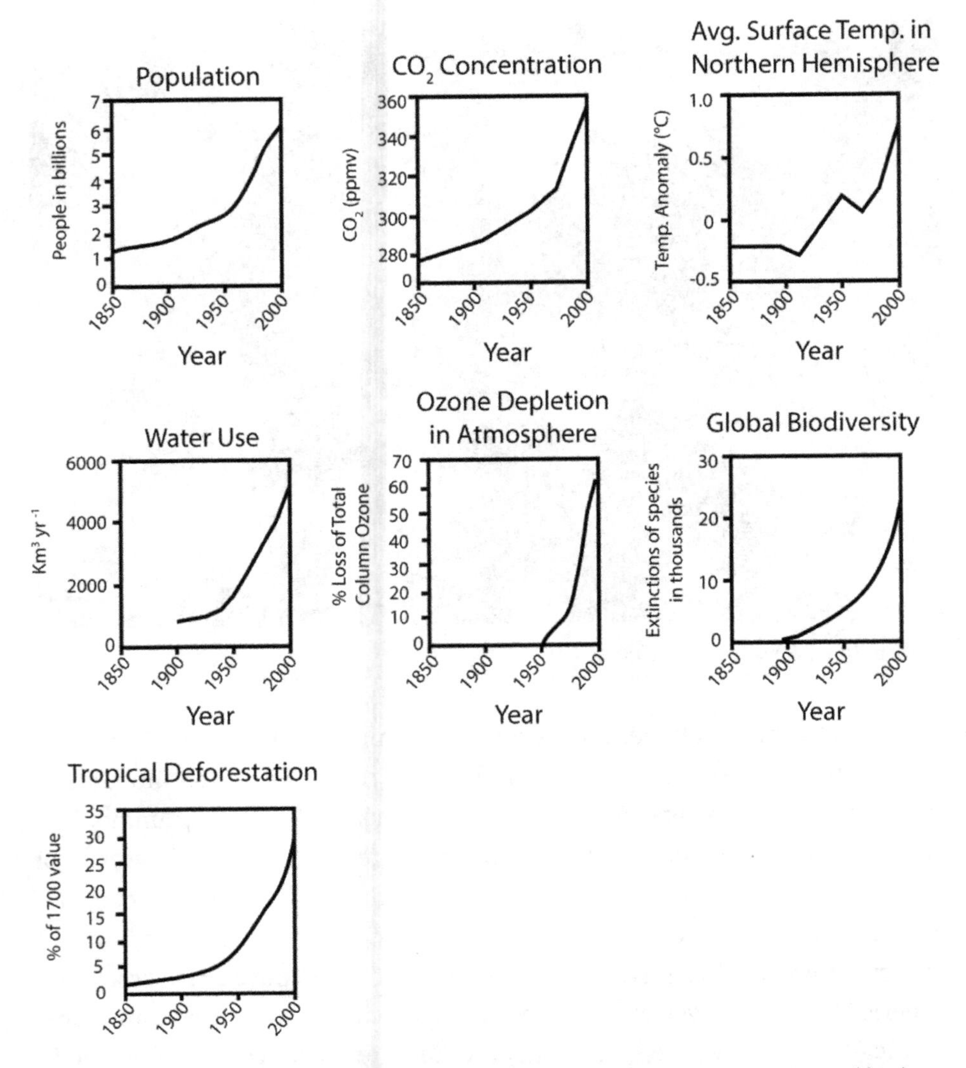

FIGURE 14.2 Increasing population and increasing per capita energy consumption and land use have driven dramatic changes in the Earth System.

as hunters and gatherers. We go to sea in boats and use increasingly sophisticated technology to find and capture fish to sell at market for human consumption. As pointed out by Daniel Pauly, former Canadian Minister of Fisheries and now professor at the University of British Columbia, "We are at war with fish ... And we are winning!" (Pauly, 2018a; b). We have sufficient technology to wipe out fish populations and have recently done so in many regions, destroying ancient fisheries and perturbing marine ecosystems globally. In stark contrast to terrestrial agriculture where crops are planted, animals are raised and fed, and there is a steady state between food production and consumption, in the ocean, we are mining food at a vastly greater rate than it is regenerated, and we make no significant attempt to replace what we remove. This is the classic tragedy of the commons. In partial response to the increasing unavailability of ocean fish, mariculture is being attempted in a number of places, but this is mostly for cash crops, and often involves hunting and gathering fish (that we could eat) to feed high trophic level fish (with high market value) with the consequent 90% loss between trophic levels. Thus, ironically, in many cases, mariculture is leading to faster depletion of ocean fisheries than simple hunting and gathering would do. How can such a result be avoided if we are to create a sustainable source of food from the ocean? This is discussed below.

Each observation of global change is based on changes caused by people, and it is axiomatic that the more people there are, the more impact there is (Figure 14.2). As such, the observation of human population explosion in the last century has great significance to the Earth System. If any other species of organisms underwent such a population explosion, we would call it an infection or a plague (depending on the size of the organism involved), and antibiotics or other population control measures would be applied. In any case, discussion of global change, both in attribution and in impacts, is impossible without accounting for the role and fate of humanity.

There is some question regarding the human invention that has most profoundly impacted the Earth System. Some would say the plow, as turning soil destroys its internal structure, exposes soil carbon to oxidation in the atmosphere, and makes it vulnerable to water and wind erosion. In the U.S., almost half of our topsoil is already gone as a result. Others might contend that the ax, and later, chainsaw, have been the most disruptive to the earth, as they are the means for massive deforestation, and thus ecosystem loss with the consequent species extinctions. Since we are now undergoing a mass extinction as a result, this could be considered as most serious. If one counts fire as a means of deforestation, then the discovery and control of fire could be considered man's most damaging invention. The harnessing of fire also led to the burning of fossil fuels and the resulting climate change we are now experiencing. Some might argue that CFCs, due to their profound impact on destroying stratospheric ozone, have been the most significant because in the last billion years, the earth has never seen an atmosphere so transparent to UVB, and all plant and animal life is now vulnerable to cell damage as a result. Others might suggest that harnessing of electricity was man's most perturbing invention, in that the generation of electricity has involved burning huge quantities of coal and emission of much of the anthropogenic greenhouse gases. Add the internal combustion engine to that, and, along with electricity, one has the primary cause of global warming, a major planetary disruption. One could generalize the entire conversation and claim that actually, antibiotics and other medicines have led to the greatest impact on the Earth System, as they have enabled the human population to explode, both reducing infant mortality and increasing life expectancy, thus increasing the impact of all human activities to unprecedented levels. Regardless of which particular invention one considers most significant, it is clear that humanity has become a critical component within the Earth System, and the first one (that we know of) that is in a position to control the Earth System *on purpose*. How effectively humanity manages the Earth System remains to be seen.

Emergent Behavior of the Earth System

When multiple internal interactions and feedbacks occur within a system and control the overall behavior of the system in ways that cannot be predicted by the behavior of any of the individual components, new modes of operation emerge, and this is called **emergent behavior** (Anderson et al., 2018). In recent decades, scientists have discovered that the Earth System is a prime example of such behavior, as the atmosphere, ocean, and land, through chemistry, physics, biology, and human activities, interact in ways that produce stability and instability in a highly non-linear fashion. This means that the system can behave in a way that is insensitive to perturbations within some of the sub-systems until a critical threshold is reached, after which the behavior of the system switches to something new and different, with a new quasi-stability, but a very different state.

Emergent Behavior: The operation of a system composed of many interacting parts, whose resulting behavior is driven by the rules of interaction between the parts, rather than the behavior of the individual parts themselves. The behavior of complex systems such as the earth thus cannot be predicted on the basis of individual component responses to drivers or perturbations, and thus must be analyzed on the basis of the full suite of couplings and feedbacks between components.

Some "classic" cases of emergent behavior may serve as simple examples. The first is a flock of birds. They fly together using simple rules: 1) Keep a certain distance from the birds around you (separation); 2) fly in the same general direction as the birds around you (alignment); 3) try to stay surrounded by birds around you and do not stray away (cohesion). The number of birds involved and the minor variations on these rules determine the details of the flocking behavior. For example, a dozen or so geese fly in the familiar "V" by following a leader and lining up, each next to, and behind the last. Starlings, on the other hand, fly in flocks of hundreds to thousands, and form a 3-dimensional body that moves and evolves in complex ways based on the simple flocking rules. The videos of this you can easily find on line are impressive. Schools of fish behave in a similar fashion to starlings, forming large 3-dimensional groups that minimize the external surface area available to predators. If a predator appears and perturbs the system, it responds in complex ways that cannot be predicted on the basis of the behavior of any single bird or fish. This is classic emergent behavior.

There are many examples of thresholds that have been surpassed in the earth's past, and which led to a new mode of system behavior on the basis of a much more complex set of rules than those that apply to bird and fish. Prior to 5,400 years ago, what is now the Sahara Desert was covered by a thriving savanna ecosystem (Hopcroft et al., 2017), with grasses, large animals, and a society of cave dwellers who drew images of their environment on cave walls (which is how we know what was there). During that time, the Milankovitch cycles of orbital variation were leading to decreasing northern hemisphere summer **insolation** (solar energy reaching the ground) (Figure 14.3). The hydrologic cycle of the Sahara included an internal cycling of water, maintained by the grasses and trees that covered the land, decreasing the albedo relative to bare ground. This kept it warm in summer and helped maintain the hydrology, even with declining summer insolation. However, a critical threshold was passed and the ecosystem could no longer sustain itself. As soon as the fraction of vegetation cover declined a bit, the albedo increased, leading to further summer cooling, and further decline, so that the ecosystem rapidly collapsed about 5,400 years ago and became the Sahara Desert that persists at present.

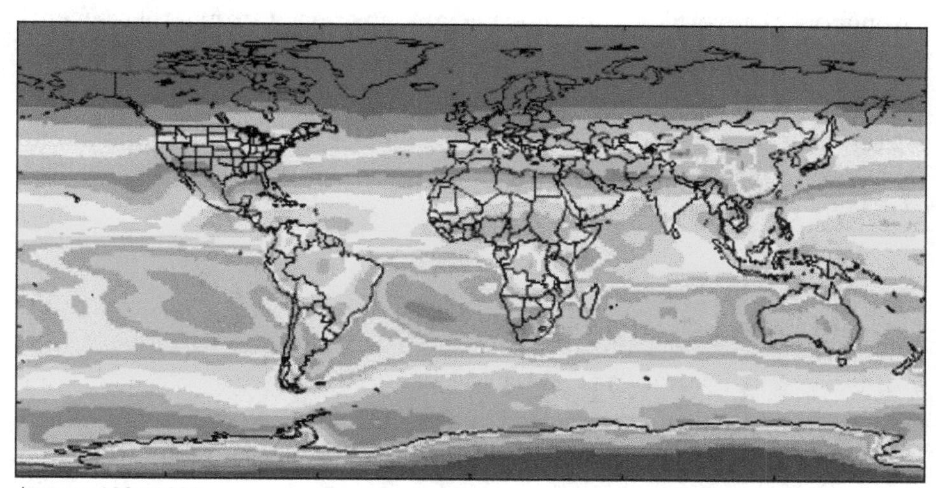

January 1984-1993

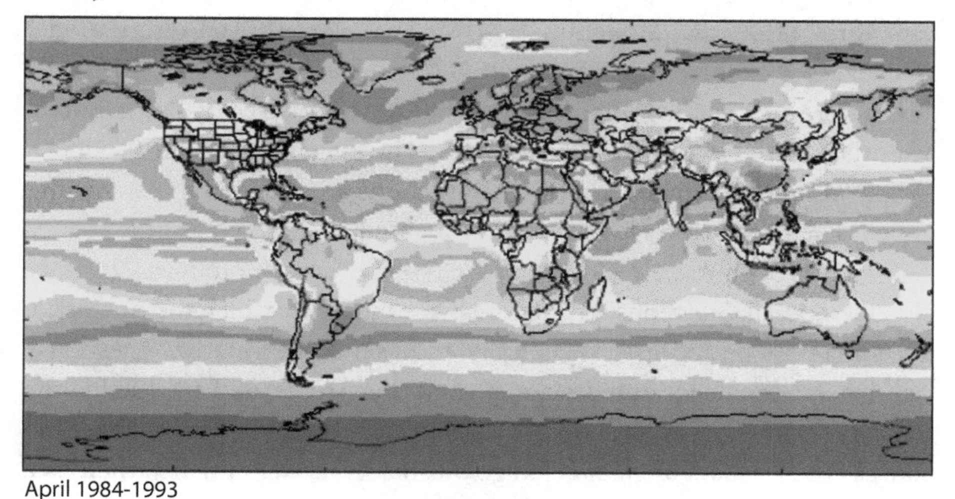

April 1984-1993

Solar Insolation (kWh/m²/day)

0 >8.5

FIGURE 14.3 Seasonal variation in solar insolation

Dansgaard-Oeschger events: 25 episodes of rapid warming during glacial times, followed by a gradual cooling back to "normal" glacial conditions.

Another example of a threshold in the Earth System that can lead to changes in behavior is the global thermohaline circulation driven by North Atlantic bottom water formation. This has been associated with abrupt climate changes during glacial times, known as **Dansgaard-Oeschger (D/O)** events and Heinrich events (Rabassaand Federico Ponce, 2013; Jennerjahn et al., 2004). During glacial periods (we are now in an interglacial), D/O events start out with rapid warming in Greenland, and then gradual cooling back to cold, glacial conditions. During these cold times, **Heinrich events** occur, during which the large North American ice sheet surges ahead, releasing armadas of icebergs into the North Atlantic, which melt and leave a layer of light, fresh water at the top of the ocean. This inhibits the formation of bottom water, and temporarily slows the global thermohaline circulation system. Because the global ocean circulation system controls climate throughout the world, and most especially northern Europe, these events alter climate in ways not predictable by considering the ice or ocean by itself. Only by accounting for the complex network of interactions within the Earth System can the overall behavior of the system be understood, and can responses of the system to perturbations, be they anthropogenic or natural, be predicted.

Heinrich events: Extreme cold events between some of the Dansgaard-Oeschger events, during which large armadas of icebergs melted in the North Atlantic, dropping ice-rafted sediment farther south than it normally extends and adding lots of fresh water to the surface ocean, thus reducing the strength of the ocean conveyor and making regional climate even colder.

The Hilbertian Questions

In 1900, a mathematician named David Hilbert posed a set of 23 fundamental mathematical problems to be solved by the scholars of the twentieth century, and most were indeed solved by the close of the century. In 2000, a set of questions pertaining to the Earth System was posed by the Global Analysis, Integration, and Modelling (GAIM) Task Force of the International Geosphere-Biosphere Programme (IGBP), modeled after those of Hilbert. These were dubbed the "Hilbertian Questions" to be solved by Earth System scientists of the twenty-first century (See http://gaim.unh.edu/Products/News/Summer2002/index.html). The questions are grouped into four categories. The "Analytical Questions" are the fundamental scientific problems to be solved with respect to the Earth System. These are followed by a set of "Operational Questions" that address the means by which the Analytical Questions can be addressed. Then there is a set of "Normative Questions" that involve the needs and desires of societal systems, and are outside the realm of science, yet control the anthropogenic drivers of global change, and address the impacts of global change on society as well. Finally, there is a group of "Strategic Questions" that address ways to achieve the needs and desires determined by the Normative Questions. In essence, we have 1) How does the Earth System work? 2) How do we figure it out? 3) What do we want? and 4) How do we get what we want?

Answering each of these questions may involve years or decades of research by numerous groups around the world, and scholars are already working on various aspects of each. Perhaps the most difficult to address are the normative questions, as the answers cannot be derived scientifically, but depend on human choices and preferences, many of which vary across societies and cultures, and even between individuals. A trivial, yet specific, example of the latter is Americans' view on windmills to generate electricity. Some consider them a blight on the landscape, and don't want to have to look at one anywhere near their homes or travels. Others see them as remarkable constructs that blend technology and art, majestic monuments to humanity's ability to solve problems at the planetary scale. Still others think they just look cool. A simple poll of your classmates may reveal some very contrasting opinions.

HILBERTIAN QUESTIONS REGARDING THE EARTH SYSTEM

Analytical Questions:

1. What are the vital organs of the ecosphere in view of operation and evolution?
2. What are the major dynamical patterns, teleconnections and feedback loops in the planetary machinery?
3. What are the critical elements (thresholds, bottlenecks, switches) in the Earth System?
4. What are the characteristic regimes and time-scales of natural planetary variability?
5. What are the anthropogenic disturbance regimes and teleperturbations that matter at the Earth-System level?
6. Which are the vital ecosphere organs and critical planetary elements that can actually be transformed by human action?
7. Which are the most vulnerable regions under global change?
8. How are abrupt and extreme events processed through nature-society interactions?

Operational Questions:

9. What are the principles for constructing "macroscopes," i.e., representations of the Earth System that aggregate away the details while retaining all systems-order items?
10. What levels of complexity and resolution have to be achieved in Earth System modelling?
11. Is it possible to describe the Earth System as a composition of weakly coupled organs and regions, and to reconstruct the planetary machinery from these parts?
12. What might be the most effective global strategy for generating, processing and integrating relevant Earth System data sets?
13. What are the best techniques for analyzing and possibly predicting irregular events?
14. What are the most appropriate methodologies for integrating natural-science and social-science knowledge?

Normative Questions:

15. What are the general criteria and principles for distinguishing non-sustainable and sustainable futures?
16. What is the carrying capacity of the earth?
17. What are the accessible but intolerable domains in the co-evolution space of nature and humanity?
18. What kind of nature do modern societies want?
19. What are the equity principles that should govern global environmental management?

Strategic Questions:

20. What is the optimal mix of adaptation and mitigation measures to respond to global change?
21. What is the optimal decomposition of the planetary surface into nature reserves and managed areas?
22. What are the options and caveats for technological fixes like geoengineering and genetic modification?
23. What is the structure of an effective and efficient system of global environment and development institutions?

Let's explore just one of the key Hilbertian Questions. For example, "2. What are the major dynamical patterns, teleconnections and feedback loops in the planetary machinery?" There are some dynamical patterns already recognized in the Earth System. An obvious example is ENSO, which creates a pattern of global climate that oscillates every three to seven years between a warm pool toward the eastern vs. western equatorial Pacific. On longer time scales, we see a pattern of glacial and interglacial cycles driven by orbital variations, and modulated by a number of interactions with the Earth System. Are there other dynamical patterns that are not yet recognized, but control important aspects of the Earth System? Are there ways in which human activities could alter such patterns in the future? These are critical research areas for the coming decades, and depend, in part, on identified feedback loop and telectonnections within the Earth System. Well-recognized feedback loops include ice albedo, for example, which leads to positive feedback, or cloud condensation nuclei from dimethyl sulfide (DMS) produced in the Surface Ocean, and production of DMS, leading to a negative feedback. How many critical feedback loops remain to be discovered that would impact the system's resilience to anthropogenic perturbations? Again, this is a critical area for research. Teleconnections are known throughout the system, including ENSO's effect on climate in North America and Europe, or the production of North Atlantic Bottom water in controlling the silica content of Pacific Ocean water.

A detailed explanation of the various Hilbertian Questions would be too long to include here. Each one of the Hilbertian Questions is rich in potential research activities, and each one will take decades to answer, if it can be answered at all at the level of detail needed to optimize human decision-making regarding management of the Earth System. Still, some progress is already being made, and in a few key areas, enough is already known to inform decision makers and drive what could someday become a coherent environmental policy process. An example of such a topic is anthropogenic emissions of greenhouse gases and global climate. We know with very little uncertainty the cause of twentieth and twenty-first-century climate change (which includes, but is not limited to, global warming). We also know to what extent this change will drive sea level to rise, thus encroaching on critical coastal ecosystems and densely inhabited coastlines. We even know (and have already observed) that storm intensity will increase as a result of warming sea surface temperatures. While quantitative details of atmospheric temperature projections, sea level rise, and storm tracks are only beginning to emerge, it is not necessary to wait for the "final answers" to begin preparing coastal communities, for instance, for the impacts associated with sea level rise and intensifying storms, such as a few that have recently struck New York (2012), the Philippines (2013), Puerto Rico, and Houston (2017).

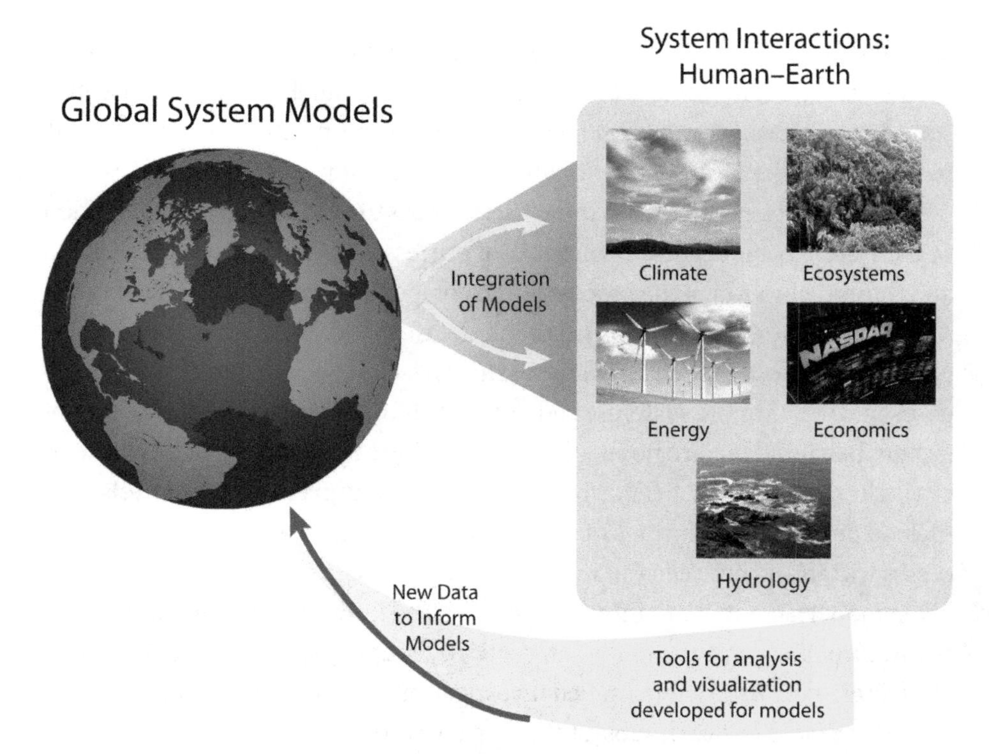

FIGURE 14.4 There is a spectrum of Earth System models ranging from simple "daisy-world" models, through various levels of intermediate complexity models (EMICs), to full-form models (e.g., GCMs). Each has their strengths and weaknesses.

Earth System Modeling

In order to test our understanding of the components of the Earth System and enhance our understanding of emergent behavior of the system as a whole, scientists have developed numerical Earth System models of a variety of types (Figure 14.4). Prognostic models are predictive models that show how reservoirs change in time due to inflow and outflow. Models may also be deterministic or stochastic. Deterministic models behave exactly the same way each time, whereas stochastic models involve an element of randomness in their behavior to account for statistical variability of various internal workings of the Earth System. Models may also be either equilibrium or transient, where equilibrium models are run to steady state equilibrium (where stocks and fluxes don't change anymore) for a given set of boundary conditions, whereas transient models involve boundary conditions that change through time, or a system that does not run long enough to reach steady state.

In addition, models can run forward and backward for different purposes. A forward model is based on a set of equations that characterize the scientist's understanding of the processes that control a system. It is run forward and creates output in the form of data that can be used for predictions or other purposes. Simulations are comprehensive versions of forward models. In contrast, inverse models use observational data and known processes to work backward and determine the values for specific model parameters, thus characterize other parts of the system that are not readily observable. Inverse models are used, for instance, to take observed spatial distributions of atmospheric gas concentrations and work backward to find their sources, on the basis of an understanding of climate systems, wind patterns, etc.

Models can be grouped into three varieties based on the detail of description of the system components, the number of processes that drive system behavior, and the extent to which the processes and components interact with each other. At the simplest extreme in process and detail (yet fully interacting) are the **conceptual models** such as **Daisyworld**. These models are limited to special cases and very simple systems, and are useful for illustrating key concepts of system behavior, but are not useful for exploring any sort of realistic behavior of the Earth System. At the other extreme are **comprehensive models** that include detailed descriptions of subsystem components, involve as many processes within and between the subsystems as possible, and to the extent possible given modern computers, integrate these to the limit of our understanding of the nature of each of the many interactions, couplings, and feedbacks. These models are very useful for detailed investigation of system behavior, especially at the interfaces between subsystems (e.g., ocean–atmosphere). However, they are so complex that they are difficult to set up for a specific "experiment," the results are sometimes difficult to interpret due to the complexity of the models themselves, and they require long computational running times for each model run, so relatively few experiments can be conducted. Between these two model extremes are the **Earth System models of intermediate complexity** (**EMICs**) that are being developed to incorporate most of the critical processes and interactions between subsystems, and at a minimum level of descriptive detail that still enables them to be applied to realistic situations pertinent to the actual Earth System, so that they can be run on simple computers such as laptops in very short run times (Weber, 2010). This enables many model experiments to be conducted, and also makes it possible to run for very long model times that can span hundreds of thousands or even millions of model years of the past and into the near future. It was EMICs that were able to determine the conditions that led to the collapse of the Saharan ecosystem 5,400 years ago. Evidence of this was subsequently found in Atlantic deep sea cores showing a dramatic increase in terrigenous dust after ecosystem collapse. EMICs also established the role of ice during glacial times in controlling thermohaline circulation and global climate changes. While all three types of models have their uses individually, the Earth System modeling community will need to use all types together to fully analyze and understand the component-by-component, as well as emergent, behavior of the Earth System.

What to Do? Mitigation and Adaptation to Global Change

Human behavior can modify and be modified by global change. **Mitigation** is the set of activities and actions that serve to reduce the changes caused by human activities. As such, mitigation involves stopping unwanted changes from occurring at all, or at least limiting them to tolerable levels. **Adaptation** focuses on accepting the fact that changes are occurring, and taking steps to reduce vulnerability to these changes—essentially preparing to live as comfortably as possible in a different environment from the one we grew up in, and indeed, in which humans evolved as a species (Le Kama et al., 2017; Doncaster et al., 2017). There are possible future environmental conditions that could certainly be created by human activities, but that would be considered unacceptable on physical, chemical, economic, and moral grounds. These are in the "Cassandra" realm (Fig. 10.9), in which there is a high degree of certainty that serious harm would arise to human health, welfare and sustainability of the societies upon which we depend. These are the futures to be avoided by mitigation measures, enabling humanity to adapt to future conditions that are deemed acceptable, even if they are markedly different from past or present conditions.

There are a number of mitigation measures that can reduce the rate of global changes, both with regard to climate and to ecosystem function. Stabilization of global climate depends on our ability to maintain surface temperatures at or near pre-industrial levels. This can be most directly done by curtailing fossil fuel burning, the primary cause of recent and expected climate change. This would involve both hard path and soft path solutions, in which the generation capacity of wind, solar, geothermal, hydro, and other sustainable energy sources would be greatly increased (hard path), and the demand for energy would be greatly reduced by buying, using, and wasting, less energy and products and transforming infrastructure to more efficient systems (soft path).

The social and political will to convert our energy productions systems from fossil fuels to sustainable energy sources (on any timescale to be effective) appears to be lacking, however. There are other ways to at least partially mitigate climate change. One is to capture carbon at the source of fossil fuel burning, and sequester it deep underground or in the deep ocean. Another is to find means to withdraw CO_2 from the atmosphere and mineralize it for permanent geologic storage. This may require large amounts of energy, which, in order to not defeat its own purpose, must be provided by non-fossil fuel sources. Alternatively, fertilization of the ocean for enhanced carbon uptake by marine phytoplankton has been considered, but initial experiments have been unsuccessful. An apparent attempt to reduce fossil fuel burning by converting corn into ethanol to add to gasoline has proven to be counterproductive—it uses more fossil fuel energy to produce and deliver than you get from the fuel, once added to gasoline. Other means of reducing fossil fuel consumption will have to be found.

Geoengineering: The intentional and controversial alteration of one part of the Earth System for mitigating a perceived problem in another part of the system, usually pertaining to climate change and global warming.

Another approach is to engage in **geoengineering** that serves to cool the planet while still increasing anthropogenic greenhouse gases (Jones et al., 2018; Pasztor, 2017). Such schemes could involve orbiting mirrors, trillions of orbiting reflectors, stratospheric aerosols, and other means for reducing the strength of sunlight reaching the earth's surface. The danger in these schemes is that even though they may reduce temperatures, the high CO_2 concentrations would still promote ocean acidification and associated ecosystem problems. Further, unanticipated run-away effects could lead to irreversible cooling and drive us into glaciation of the kind exaggerated in science fiction stories. Some have contemplated reducing the rate of sea level rise by impounding increasing volumes of water on the land surface behind dams such as on the Congo River or in natural depressions such as the Tarim Basin or Lake Chad. This would, however, only buy some time because once a reservoir fills up, it no longer reduces the rate of sea level rise. Further, there are serious environmental consequences to such continental flooding, and with our burgeoning human population, we need all the land surface we can preserve for human habitation and agriculture, in addition to any attempts to preserve ecosystems.

It is possible that the best that can be realistically done to mitigate climate change and its impacts is to slightly reduce the rate of emissions from fossil fuel burning by greatly enhancing the efficiency of heating and cooling systems in buildings, electricity generation, transportation systems, and industrial production processes. Efficiency can also be enhanced by **reducing consumption**. The production and distribution of products requires enormous amounts of energy, so the longer we can maintain and use the products we already have, the less fossil fuels will be burned and less irreplaceable resources will be wasted. In that sense, shopping can no longer be a recreational activity for affluent Americans, and learning to maintain and repair our goods and products will become essential. There is no "green consumer"—it is an oxymoron. One can only be a "green preserver." It is noteworthy that as a culture, the name we give ourselves is "consumers" rather than anything like "preservers" or "restorers." Further, we can reduce ecosystem loss created by conversion to agriculture by not wasting food. Americans waste more food per day than many in the developing world eat per day. The portion of American food that DOES get eaten is great enough to lead to an epidemic in obesity nationwide. In quality, because of energy loss between trophic levels, eating meat that is fed anything other than grass requires about ten times as much agricultural land than eating the corn or other vegetation we could eat that is instead fed to the livestock. (We cannot digest grass, but cows and sheep can.)

Reducing consumption: Just buy and waste less stuff.

Resiliance: The ability of a community or other system to recover from a disruption caused by an extreme event such as a storm, flood, drought, epidemic, etc.

Resistance: The ability of a community or other system to withstand an extreme event without causing disruption to operations, infrastructure or services. Resistance and resilience are often lumped together as "resilience" but the distinction is important. If a system is sufficiently resistant, it will not need to count on resilience. The Fukushima Daiichi nuclear plant was insufficiently resistant to a tsunami in 2011 and was disrupted to the point that it was completely shut down, so was also insufficiently resilient.

In the likely case that we can find the political will to only very partially mitigate global changes, means will need to be found to deal with these changes, amounting to adaptation to the inevitable alterations we can expect. Adaptation is also difficult, as reduction in vulnerability to environmental conditions can be costly, difficult, and politically unpalatable. Many cities are becoming concerned with resilience which is the ability to recover from a disruption in infrastructure and services caused by insufficient resistance to extreme events such a storms. As sea level rises, for example, coastal cities (there are many) are increasingly vulnerable to storms (e.g., New Orleans for Hurricanes Katrina and Rita in 2005, and Houston for Hurricane Harvey in 2017) and storm surge (e.g., New York for Hurricane Sandy in 2012). It is politically difficult to even discuss moving cities en masse out of harm's way, so they remain where they are, in the path of hurricanes, with rising sea level, and in some cases, on subsiding passive continental margins. However, there are some other ways to adapt to global changes. For example, farmers are fairly clever about altering their crops in response to changing precipitation expectations. While sometimes caught by droughts or floods, long-term climate changes can be at least partially adapted to in the agricultural sector by strategic cropping. More difficult is the fate of island nations that have nowhere to go as the rising ocean inundates the entirety of the low-lying land of certain islands. These, such as Kiribati and Tuvalu, are already considering plans for abandonment and relocation of their entire populations to other nations. On

WHAT CAN I DO THAT MATTERS FOR THE PLANET?

With all the knowledge and wisdom acquired through the study of environmental science, it may still seem that one's own personal actions are far too insignificant to make any difference at all to the fate of the planet. Nothing could be further from the truth. It is our own actions that define the cumulative impact of human activities. We can each reduce the size of our "ecological footprint," or the impact we each have on the Earth System. Specifically, reduction in energy consumption (soft path) is a most effective way to reduce the rate of climate change as well as local pollution. Applying political pressure to your elected officials to hasten the transition to more productive sustainable energy sources (hard path) can reduce or eliminate the degree of social and economic inconvenience the transition will entail. No longer consider yourself a "consumer" but rather a preserver, sustainer, and restorer. This means simply buying less stuff—keep what you have and make it last. When something breaks, fix it rather than replace the whole thing. (When is the last time you went to a cobbler for shoe repair, for instance?) If you really need to buy things in glass, paper, plastic, or aluminum packaging, recycle it. Eat smart—buy or fill your plate with only what you really need, and do not waste any food by throwing it away. Eat lower on the food web, recalling that 90% of the energy is lost between trophic levels (i.e., less meat). This way, less farmland will be needed and more ecosystem can be preserved. If you can, take the stairs rather than the elevator—bike or walk rather than drive a car—take a bus or train if available. Put on a sweater rather than turn up the heat. When the time comes someday, resist the urge to exacerbate overpopulation and associated resource stress by having more than two children over your lifetime. These all may seem like trivial things that individually would not affect the operation of the planet, but they do. In general, **do what would be best if everybody did it**.

a more local level, adaptation measures can be taken by individuals, who can prepare for more intense coastal storms, broader floodplains of rivers, and more power outages, by either moving homes to less vulnerable locations or enhancing resilience of their existing homes or places of work so that the expected events less severely impact their lives.

Agriculture has already altered the face of the earth, with land use being the most disruptive means of destroying ecosystems. Yet, agriculture has enabled us to avoid hunting (and gathering) to extinction many terrestrial animal (and plant) species. Given the realities of marine trophic levels, however, if we want to eat something other than phytoplankton and seaweed, we need to establish a maricultural system that is rooted in plants that use sunlight to create biomass. Such as system has been described by Michael Pollan, for example, in *The Omnivore's Dilemma,* in which terrestrial dairy and chicken farmers can really be no more than "grass farmers," allowing natural processes of prey-predator-scavenger relations, trophic levels, and biogeochemical cycling to provide a steady state of marketable food for the farmer. If this sort of thinking were to be applied to the ocean, fishermen would become **phytoplankton farmers**, directly supporting the higher trophic levels of interest to global food markets. We are not even close to establishing such a system, yet the functionality of the Earth System demands it in order to provide a sustainable marine source of food for humanity.

Phytoplankton farmers: A perspective on sustainable mariculture that uses farmers to enhance the production of primary producers that will, in turn, support production of higher trophic level organisms for human consumption. It is equivalent to dairy farmers and ranchers being considered "grass farmers," as this is the primary producer on land (albeit with less trophic levels in between the grass and what we eat).

The balance of mitigation vs. adaptation can be planned strategically, as there are some environmental changes that can be readily adapted to but difficult to mitigate, and others that can be mitigated but not adapted to in any reasonable manner. Adapting to temperature increase is easy. If it gets hotter, especially in winter, wear less warm clothing. Simple. Adapting to ocean acidification is another matter altogether, however. The chemical stress on shell-forming organisms, from phytoplankton to coral reefs, could, when added to other stresses such as temperature, pollution, turbidity, etc., lead to a collapse of many parts of the global marine ecosystem, while eliminating the CO_2 emissions discount provided by the marine biological pump. Adapting to this fundamental change in the global ecosystem would be much more difficult and remains a leading motivation for mitigation by reducing atmospheric CO_2 concentrations. In the end, we must decide, as a global society, the optimal balance between mitigation and adaptation (and thus answer that particular Hilbertian Question). In essence, we must "Avoid the unmanageable so that we can manage the unavoidable." For taking effective steps toward either mitigation or adaptation, however, it is necessary to first understand the functioning of the Earth System and how it responds to natural variability and anthropogenic perturbations at all levels. This is what Environmental Science is about, and as we gain a deeper understanding of the earth and its components, we gather the tools necessary for better management of the planet and of ourselves.

References

Anderson, B. T., Feldl, N., & Lintner, B. R. (2018). Emergent behavior of arctic precipitation in response to enhanced arctic warming. *Journal of Geophysical Research Atmospheres*, 123(5), 2704–2717.

Dagon, K., & Schrag, D. P. (2017). Regional climate variability under model simulations of solar geoengineering. *Journal of Geophysical Research Atmospheres*, 122(22), 12106–12121.

Doncaster, C., P., Tavoi, A., & Dyke, J. G. Using adaptation insurance to incentivize climate-change mitigation. Ecological Economics, 135, 246–258.

Hopcroft, P. O., Valdes, P. J., Harper, A. B., & Beerling, D. J. (2017). Multi vegetation model evaluation of the green Sahara climate regime. *Geophysical Research Letters*, 44(13), 6804–6813.

Jennerjahn, T. C., Ittekkot, V., Arz, H. W., Behling, H., Patzold, J., & Wefer, G. (2004). Asynchronous terrestrial and marine signals of climate change during Heinrich events. *Science*, 306(5705), 2236–2239.

Jones, A. C., Hawcroft, M. K., Haywood, J. M., Jones, A., Guo, X., & Moore, J. C. (2018). Regional climate impacts of stabilizing global warming at 1.5 K using solar geoengineering. *Earth's Future*, 6(2), 230–251.

Le Kama, A. A., & Pommeret, A. (2017). Supplementing domestic mitigation and adaptation with emissions reduction abroad to face climate change. *Environmental & Resource Economics*, 68(4), 875–891.

Pasztor, J. (2017). The need for governance of climate geoengineering. *Ethics & International Affairs*, 31(4), 419–430.

Pauly, D. (2018a). Fishing lessons artisanal fisheries and the future of our oceans. *Science*, 360(6385), 161.

Pauly, D. (2018b). A vision for marine fisheries in a global blue economy. *Marine Policy*, 87, 371–374.

Pollan, Michael. (2007). *The omnivore's dilemma: a natural history of four meals.* New York: Penguin

Postel, S. L., Daily, G. C., & Ehrlich, P. R. Human appropriation of renewable fresh water. *Science*, 271(5250), 785–788.

Rabassa, J., & Federico Ponce, J. (2013). The Heinrich and Dansgaard-Oeschger climatic events during marine isotopic stage 3: searching for appropriate times for human colonization of the Americas. *Quaternary International*, 299, 94–105.

Weber, S. L. (2010). The utility of Earth system Models of Intermediate Complexity (EMICs). Wiley Interdisciplinary Reviews—*Climate Change*, 1(2), 243–252.

Figure Credits

Fig. 14.1: Source: https://commons.wikimedia.org/wiki/File:Bloom_in_the_Barents_Sea.jpg

Fig. 14.3: Source: https://commons.wikimedia.org/wiki/File:Insolation.gif

Fig. 14.4a: Source: http://commons.wikimedia.org/wiki/File:Globe.svg

Fig. 14.4b: Copyright © 2014 Depositphotos Inc./Natashamam.

Fig. 14.4c: Copyright © 2010 Depositphotos Inc./WDGPhoto.

Fig. 14.4d: Copyright © 2007 by bfishadow, (CC by 2.0) at http://commons.wikimedia.org/wiki/File:NASDAQ_stock_market_display.jpg.

Fig. 14.4e: Copyright © 2010 by John Murphy, (CC by 2.0) at http://commons.wikimedia.org/wiki/File:California_State_Route_1_(10528045804)_(2).jpg.

Index

D

E

S

T

CPSIA information can be obtained
at www.ICGtesting.com
Printed in the USA
LVHW050111200123
737353LV00004B/30

9 781793 518514